PROGRESS IN

Molecular Biology and Translational Science

Volume 91

PROGRESS IN

Molecular Biology and Translational Science

Membrane Proteins as Drug Targets

edited by

Charles A. Lunn

Merck Research Laboratories
Kenilworth, New Jersey, USA

Volume 91

ELSEVIER

AMSTERDAM • BOSTON • HEIDELBERG • LONDON
NEW YORK • OXFORD • PARIS • SAN DIEGO
SAN FRANCISCO • SINGAPORE • SYDNEY • TOKYO
Academic Press is an imprint of Elsevier

Academic Press is an imprint of Elsevier
32 Jamestown Road, London, NW1 7BY, UK
Radarweg 29, PO Box 211, 1000 AE Amsterdam, The Netherlands
30 Corporate Drive, Suite 400, Burlington, MA 01803, USA
525 B Street, Suite 1900, San Diego, CA 92101-4495, USA

This book is printed on acid-free paper. ∞

Library of Congress Cataloging-in-Publication Data
A catalog record for this book is available from the Library of Congress

British Library Cataloguing in Publication Data
A catalogue record for this book is available from the British Library

ISBN: 978-0-12-381288-9
ISSN: 1877-1173

For information on all Academic Press publications
visit our website at elsevierdirect.com

Printed and Bound in the USA
10 11 12 13 10 9 8 7 6 5 4 3 2 1

Contents

Contributors

Numbers in parentheses indicate the pages on which the authors' contributions begin.

Vincent J. Aloyo, Drexel University College of Medicine, Philadelphia, Pennsylvania, USA (1)

Kelly A. Berg, University of Texas Health Science Center, San Antonio, Texas, USA (1)

Levi L. Blazer, Department of Pharmacology, University of Michigan, Ann Arbor, Michigan, USA (81)

Enric I. Canela, CIBERNED, Department of Biochemistry and Molecular Biology, Faculty of Biology, University of Barcelona, Barcelona, Spain (41)

Vicent Casadó, CIBERNED, Department of Biochemistry and Molecular Biology, Faculty of Biology, University of Barcelona, Barcelona, Spain (41)

Arthur Christopoulos, Drug Discovery Biology Laboratory, Monash Institute of Pharmaceutical Sciences; and Department of Pharmacology, Monash University, Parkville, Victoria, Australia (53)

William P. Clarke, University of Texas Health Science Center, San Antonio, Texas, USA (1)

Antoni Cortés, CIBERNED, Department of Biochemistry and Molecular Biology, Faculty of Biology, University of Barcelona, Barcelona, Spain (41)

Lakshmi A. Devi, Department of Pharmacology and Systems Therapeutics, Mount Sinai School of Medicine, New York, USA (121)

Edward A. Esposito, Blue Sky Biotech, Inc., Worcester, Massachusetts, USA (209)

Sergi Ferré, National Institute on Drug Abuse, IRP, NIH, DHHS, Baltimore, Maryland, USA (41)

Rafael Franco, CIBERNED, Department of Biochemistry and Molecular Biology, Faculty of Biology, University of Barcelona, Barcelona, Spain; and CIMA Neurociencias, Avda Pio XII 55, Pamplona, Spain (41)

Scott Gridley, Blue Sky Biotech, Inc., Worcester, Massachusetts, USA (209)

John A. Harvey, Drexel University College of Medicine, Philadelphia, Pennsylvania, USA (1)

Debbie L. Hay, School of Biological Sciences, University of Auckland, Symonds Street, Auckland, New Zealand (53)

Carme Lluís, CIBERNED, Department of Biochemistry and Molecular Biology, Faculty of Biology, University of Barcelona, Barcelona, Spain (41)

Josefa Mallol, CIBERNED, Department of Biochemistry and Molecular Biology, Faculty of Biology, University of Barcelona, Barcelona, Spain (41)

Gemma Navarro, CIBERNED, Department of Biochemistry and Molecular Biology, Faculty of Biology, University of Barcelona, Barcelona, Spain (41)

Richard R. Neubig, Department of Pharmacology; and Department of Internal Medicine (Cardiovascular Medicine), University of Michigan, Ann Arbor, Michigan, USA (81)

Kyle R. Palmer, Redpoint Bio Corporation, Ewing, New Jersey, USA (151)

Patrick M. Sexton, Drug Discovery Biology Laboratory, Monash Institute of Pharmaceutical Sciences; and Department of Pharmacology, Monash University, Parkville, Victoria, Australia (53)

Anthony L. Shrout, Blue Sky Biotech, Inc., Worcester, Massachusetts, USA (209)

John Simms, Drug Discovery Biology Laboratory, Monash Institute of Pharmaceutical Sciences; and Department of Pharmacology, Monash University, Parkville, Victoria, Australia (53)

Benita Sjögren, Department of Pharmacology, University of Michigan, Ann Arbor, Michigan, USA (81)

Dennis Sprous, Redpoint Bio Corporation, Ewing, New Jersey, USA (151)

Umberto Spampinato, INSERM U862, Neurocentre Magendie, Physiopathology of Addiction Group, University of Bordeaux, Bordeaux, France (1)

Dumaine Williams, Department of Pharmacology and Systems Therapeutics, Mount Sinai School of Medicine, New York, USA (121)

Denise L. Wootten, Drug Discovery Biology Laboratory, Monash Institute of Pharmaceutical Sciences; and Department of Pharmacology, Monash University, Parkville, Victoria, Australia (53)

Preface

The cell membrane is a complex, multifaceted subcellular organelle. A primary function of this organelle is as a barrier, segregating the cytoplasmic chemistry from the aqueous environment. This barrier function requires highly specific transduction mechanisms to transmit selected environmental keys from the environment to the cytoplasmic machinery—these sensing molecules (the seven transmembrane receptors and ion channels) provide an accessible path to modulate intracellular biology and have proved to be valid drug targets. The membrane is also a scaffold, organizing selected structural and biochemical components to modulate the efficiencies of cytoplasmic biochemistries. Affecting these structures represents another class of potential drug targets. Finally, the membrane is a source of biological substrates that can be enzymatically manipulated to generate therapeutically relevant signaling molecules—a third class of potential drug targets.

This volume seeks to investigate recent advances in our understanding of the cell membrane as a source for drug targets. We will start with progress in our understanding of seven transmembrane receptors, reviewing an increasingly important pharmacology for seven traditional transmembrane receptors (inverse agonists) and the increasing role of receptor heteromers as potentially unique drug targets. We will also consider the possible role of proteins seen to modulate receptor function (RAMPs, RGSs) as well as proteins that control receptor folding and trafficking (chaperones) as possible drug targets. We will next investigate strategies for targeting ion channels, a class of membrane-associated molecule generating increasing therapeutic interest outside the CNS. Finally, we will review the biology of enzymes associated with the cell membrane and the unique challenges of targeting these proteins.

Charles A. Lunn
New Jersey, USA

Inverse Agonism at Serotonin and Cannabinoid Receptors

Vincent J. Aloyo,* Kelly A. Berg,† William P. Clarke,† Umberto Spampinato,‡ and John A. Harvey*

*Drexel University College of Medicine, Philadelphia, Pennsylvania, USA

†University of Texas Health Science Center, San Antonio, Texas, USA

‡INSERM U862, Neurocentre Magendie, Physiopathology of Addiction Group, University of Bordeaux, Bordeaux, France

Contemporary receptor theory was developed to account for the existence of constitutive activity, as defined by the presence of receptor signaling in the absence of any ligand. In vitro studies with a variety of cell types have revealed the existence of constitutive activity and inverse agonism at a large number of receptors and also additional complexities of ligand–receptor interactions. Thus, ligands acting at a constitutively active receptor can act as agonists, antagonists, and/or inverse agonists, and these pharmacological characteristics

Progress in Molecular Biology and Translational Science, Vol. 91
DOI: 10.1016/S1877-1173(10)91001-6

1877-1173/10 $35.00

can differ for an individual ligand depending upon the receptor response measured and the physiological state of the system under study. Studies with a variety of cell types have established that the serotonin 5-HT_{2A} and 5-HT_{2C} receptors and the cannabinoid CB1 receptor demonstrate constitutive activity and inverse agonism *in vitro*. Serotonin and cannabinoid receptors are involved in a large number of physiological and behavioral functions. The possible existence of constitutive activity and inverse agonism at these receptors *in vivo* would provide new avenues for drug development. Recent studies have provided compelling evidence that both the serotonin 5-HT_{2A} and 5-HT_{2C} receptors and cannabinoid CB1 receptor demonstrate inverse agonism and constitutive activity also *in vivo*. This chapter describes our current knowledge of constitutive activity *in vitro* and then examines the evidence for constitutive activity *in vivo*.

I. Serotonin Receptors

Serotonin, one of the oldest neurotransmitters in evolution, is a monoamine neurotransmitter that is able to regulate a wide variety of behavioral and physiological functions including body temperature, cardiovascular regulation, sleep, pain, mood, body weight regulation, sexual behavior, motor function, and cognition.[1–4] This wide variety of functions is mediated by at least 13 members of the 7-transmembrane spanning (7-TMS) receptor superfamily of G-protein-coupled receptors (GPCRs) and one ligand gated channel. As shown in Table I, these 14 receptors have been assigned to seven distinct families based on their structural and functional characteristics.[4]

As can be seen from Table I, we are still in the early stages of understanding the function of these receptors and their possible role in therapeutics. At this time, some of these receptors have no known agonists other than 5-HT or suggested physiological function (5-ht_{1E}, 5-ht_{5A}), while the 5-ht_{5B} receptor has no known coupling, agonists, or function, raising the possibility that some of these receptor subtypes may represent vestigial proteins that no longer function. The absence of physiological function for these receptors is indicated by the use of a lower case (5-ht). 5-HT receptors exist both pre- and postsynaptically throughout the brain. Attempts to understand the functional characteristics of these GPCRs have become even more complicated by their extensive posttranscriptional mRNA editing and splice variants of some receptors, for example, 5-HT_{2C}, 5-HT_4, and 5-HT_7. An additional complication comes from recent *in vitro* studies of ligand–receptor interactions, which have led to revisions of traditional receptor theory, and this is described below.

TABLE I
SEROTONIN 5-HT RECEPTOR SUBTYPES

Receptor subtypes	Signaling	Agonists	Suggested functions
5-HT_{1A}	$G_{i/o}$ ⇓ cAMP: ⇑ K^+	8-OH-DPAT	Mood
5-HT_{1B}	$G_{i/o}$ ⇓ cAMP: ⇑ K^+	Sumatriptan, L-694247	Mood, cognition
5-HT_{1D}	$G_{i/o}$ ⇓ cAMP: ⇑ K^+	Sumatriptan, L694247 PNU109291	Body weight
5-ht_{1E}	$G_{i/o}$ ⇓ cAMP: ⇑ K^+	Not known	Not known
5-HT_{1F}	$G_{i/o}$ ⇓ cAMP: ⇑ K^+	LY334370	Possible involvement in migraine
5-HT_{2A}	$G_{q/11}$ ⇑ IP_3 & AA: ⇓ K^+	LSD, DOI	Hallucinations, psychosis, cognition, motor function
5-HT_{2B}	$G_{q/11}$ ⇑ IP_3 & AA: ⇓ K^+	BW723C86	Body weight
5-HT_{2C}	$G_{q/11}$ ⇑ IP_3 & AA: ⇓ K^+	RO600175	Mood, body weight, psychosis
5-HT_3	⇑ Na^+ & Ca^{2+} ⇓ K^+	SR57227	Vomiting reflex
5-HT_4	G_s ⇑ cAMP: ⇓ K^+	BIMU8, ML10302, RS67506	Cognition, mood, body weight
5-ht_{5A}	$G_{i/o}$ or Gs? ⇓ cAMP: ⇑ K^+	Not known	Not known
5-ht_{5B}	Not known	Not known	Not known
5-HT_6	G_s ⇑ cAMP: ⇓ K^+	LSD	Hallucinations, psychosis
5-HT_7	G_s ⇑ cAMP: ⇓ K^+	8-OH-DPAT	Cognition

Abbreviations: 8-OH-DPAT, 8-hydroxy-2-(di-*n*-propylamine)tetralin; L-694,247, 2-[5-[3-(4-methylsulfonylamino)benzyl-1,2,4-oxadiazol-5-yl]-1*H*-indole-3-yl]ethylamine; PNU109291, (*S*)-(−)-1-[2-[4-(4-methoxyphenyl)-1-piperazinyl]ethyl]-*N*-methyl-isochroman-6-carboxamide; LY33470, 4-fluoro-*N*-[3-(1-methyl-4-piperidinyl)-1*H*-indol-5-yl]benzamide hydrochloride; LSD, D-lysergic acid diethylamide; DOI, 2,5-dimethoxy-4-iodophenylisopropylamine; BW723C86, α-methyl-5-(2-thienylmethoxy)-1*H*-indole-3-ethanamine; RO600175, (*aS*)-6-chloro-5-fluoro-*a*-methyl-1*H*-indole-1-ethanamine fumarate; SR57227, 1-(6-chloro-2-pyridinyl)-4-piperidinamine hydrochloride; BIMU8, [endo-*N*-8-methyl-8-azabicyclo(3.2.1)oct-3-yl]-2,3-dihydro-3-isopropyl-2-oxo-1*H*-benzimidazol-1-carboxamide hydrochloride; ML10302, 4-amino-5-chloro-2-methoxybenzoic acid 2-(1-piperidinyl)ethyl ester hydrochloride; RS67506, 1-(4-amino-5-chloro-2-methoxyphenyl)-3-[1-2-methylsulfonylamino)ethyl-4-piperidinyl]-1-propanone hydrochloride; PLA_2, phospholipase A_2; PLC, phospholipase C; AA, arachidonic acid; cAMP, cyclic adenosine monophosphate; IP_3, inositol phosphate; ⇓ K^+, decreased potassium conductance, depolarization; ⇑ K^+, increased potassium conductance, hyperpolarization; ⇑ Na^+ and Ca^{2+}, increased conductance, depolarization.

II. Constitutive Receptor Activity and Inverse Agonism

Traditional receptor theory held that receptors in a population were quiescent unless activated by the binding of a ligand that possessed intrinsic efficacy (agonists). Intrinsic efficacy is the property of a drug that defines the strength of the receptor stimulus, which is ultimately converted into a response by the cellular signaling machinery. The magnitude of the total cellular receptor stimulus is proportional to the number of receptors occupied by the agonist and the nature of the conformational change in the receptor induced by the agonist. The magnitude of the response produced by an agonist (efficacy) is the consequence of both the size of the total receptor stimulus and the efficiency with which the cell converts the stimulus into a response. Receptor density and stimulus–response efficiency are cell/tissue-dependent properties, whereas intrinsic efficacy is a drug property unique for each drug–receptor pair.

In the traditional pharmacological world, there are two types of drugs. Drugs that bind (have affinity) and generate a receptor stimulus (have intrinsic efficacy) are called agonists. Agonists can differ in the magnitude of the receptor stimulus they produce (differ in intrinsic efficacy), which leads to their characterization as "full" or "partial" agonists. The second type of drug is one with affinity but zero intrinsic efficacy. These drugs, called antagonists, bind but do not induce a conformational change in the receptor (do not generate a receptor stimulus) and therefore do not produce a response. However, because receptor occupancy (at the orthosteric binding site) is mutually exclusive, the presence of an antagonist reduces occupancy of the receptor by an agonist to an extent commensurate with the antagonist concentration relative to its affinity for the receptor.

Twenty-five years ago, the traditional pharmacological world underwent a radical transformation. The pioneering work of Cerione et al.[5] with purified β-adrenergic receptors and Gαs proteins, followed by that of Costa and Herz[6] studying opioid receptors expressed natively in intact cells, led to the realization that receptors in a population were not necessarily quiescent but could spontaneously adopt an active conformation in the absence of an activating ligand. These spontaneously active receptors could generate a receptor stimulus and thus contribute to the basal regulation of cellular responses. The discovery that receptors could be constitutively active led to the development of the two-state model of receptor function (Fig. 1) in which receptors in a population exist in equilibrium between an inactive conformation (denoted as *R*) and an active conformation (denoted as *R**). The receptors in the active

$$\mathrm{R} \overset{L}{\rightleftharpoons} \mathrm{R}^* \rightarrow \textit{Stimulus} \underset{k_e}{\longrightarrow} \textit{Response}$$

Fig. 1. Two-state model for receptor function. Receptors in a population exist in one of two conformations. R represents the proportion of the receptor population that is in an inactive conformation, whereas R^* represents the active receptor population that generates a stimulus leading to the production of a cellular response. L is an allosteric transition constant that defines the fraction of receptors that are active versus inactive. K_e represents the efficiency by which the cell can convert the receptor stimulus into a response.

conformation generate a receptor stimulus in the absence of an activating ligand and thereby contribute to the production of a response. The fraction of a receptor population that is in an active conformation at any point in time is defined by an allosteric transition constant, usually denoted as L. The value of L is a unique constant for each receptor and reflects the propensity of the receptor to transition between the inactive and active conformations. In molecular terms, the value of L is related to the number and strength of intramolecular constraints, which in turn govern the conformational flexibility of the receptor protein.

Within the framework of the two-state model of receptor function, the presence of a ligand with affinity for the receptor will alter the relative distribution of the receptor population between R and R^* conformations. As shown in Fig. 2, a ligand will have affinity constants K_A and K_{A^*} for both the receptor conformations R and R^*, respectively. A ligand with equal affinity constants ($K_A = K_{A^*}$) will not influence the distribution of receptors between R and R^*. Such ligands will be antagonists, as they will not alter the receptor stimulus, but do occupy the receptor and therefore will compete for occupancy with other

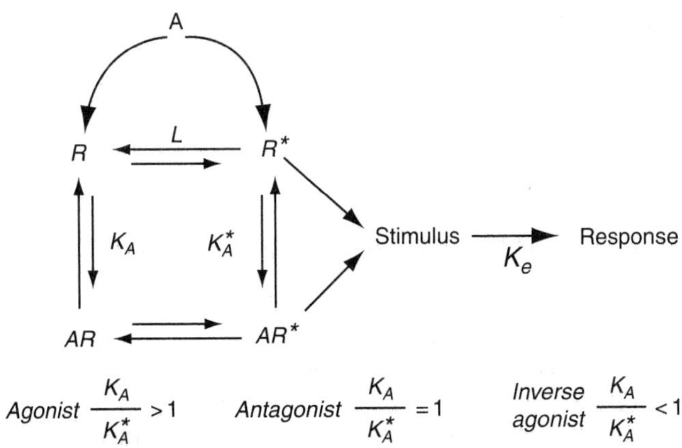

FIG. 2. The influence of a ligand in the two-state model of receptor function. Ligands (A) bind to R and R^*, with equilibrium dissociation constants of K_A or K_{A^*}, respectively. L is an allosteric transition constant (see Ref. 6a) that defines the probability of finding a receptor in the active conformation. Agonists (ligands with $K_A/K_{A^*} > 1$) enrich the R^* population, leading to increased receptor stimulus. The cellular signal transduction machinery converts the receptor stimulus into a response based upon the efficiency of stimulus–response coupling (K_e). Inverse agonists ($K_A/K_{A^*} < 1$) enrich the R population and thus deplete R^* and receptor stimulus and reduce basal response. Antagonists ($K_A = K_{A^*}$) do not alter the proportions of R versus R^* and thus do not change basal response, but because they occupy the receptor population, they can antagonize the action of both agonists and inverse agonists. Derived from the work of Leff.[6b]

ligands. However, if a ligand has higher affinity for R or R^*, then the equilibrium will be shifted such that the R conformation is enriched at the expense of R^* (which is depleted), or R^* will be enriched at the expense of R. Ligands with higher affinity for R^* ($K_{A^*} < K_A$) will be agonists, enriching the active receptor population, leading to generation of a larger receptor stimulus and larger response. Intrinsic efficacy can thus be defined as the ratio of K_A/K_{A^*}; the greater the value of this ratio, the greater will be the enrichment of R^* by the ligand and the greater will be the stimulus/response that results.

The two-state model also provides for a new class of ligand beyond agonists and antagonists. Ligands that have higher affinity for the inactive receptor ($K_A < K_{A^*}$) will deplete the receptor population of R^*, thereby decreasing the receptor stimulus and response. These ligands are called inverse agonists. As with agonists, the intrinsic efficacy of inverse agonists can be defined by the magnitude of the ratio of K_A/K_{A^*}. The smaller this ratio, the greater is the ability of the ligand to reduce the receptor stimulus/response.

Because these ligands are orthosteric, only one ligand can occupy a receptor at a time (binding is mutually exclusive). Consequently, if two ligands are present, each will compete for binding with the other. For example, an antagonist will hinder the occupancy of a receptor by an agonist or an inverse agonist. In the presence of an antagonist, concentration–response curves to both agonists and inverse agonists will be shifted to the right by an amount proportional to the antagonist concentration relative to its affinity for the receptor. Likewise, receptor occupancy of an inverse agonist can be reduced by an agonist, and vice versa.

A. Factors That Influence Constitutive Receptor Activity and Inverse Agonism

The maximal effect of inverse agonists (efficacy) is directly dependent upon the differential affinity of the ligand for the inactive receptor conformation (R) versus that for the active receptor conformation (R^*). However, inverse agonist efficacy is also dependent upon the magnitude of constitutive receptor activity. From the two-state model of receptor function (Fig. 2), it can be seen that there are three factors that influence the magnitude of constitutive receptor activity and thus the efficacy of inverse agonists.

The first of these factors is the allosteric transition constant, L. Receptors that have few intramolecular constraints are capable of a wider range of exploration of conformational space than are receptors with strong intramolecular interactions. Consequently, the proportion of receptors in a population that are in an active conformation in the absence of an agonist will be greater for those receptors that are more flexible. Since the magnitude of the receptor stimulus is proportional to the number of active conformation receptors, the response generated will be greater for flexible receptors. L is an inherent property of a receptor and thus will differ for different receptors.

The second factor that regulates the magnitude of constitutive receptor activity is the total receptor density, R_T. At any particular value of L, a percentage of the receptor population is in an active conformation. As receptor density increases, the total number of active (and inactive) receptors increases proportionately. Unlike L, which is a receptor property, R_T is a cell/tissue-dependent property. Consequently, measured constitutive activity (and inverse agonist efficacy) associated with a single receptor subtype will be different in different cells or tissues, depending upon receptor expression levels.

The third factor involved in regulating the magnitude of constitutive receptor activity is the efficiency with which the cell converts the receptor stimulus into a response, K_e. This in turn is dependent upon the number and type of signal transduction molecules (e.g., G proteins) and effectors. As such, K_e, like R_T, is also cell/tissue-dependent.

As a result of these cell/tissue-dependent factors, the effect of an inverse agonist to reduce a basal response associated with a single receptor subtype will in turn be different when tested in different cells or tissues in which there are differences in R_T or K_e. Differential agonist efficacy in different cells or tissues is well known (see Ref. 7). An important caveat associated with cell phenotype-dependent constitutive receptor activity is that the magnitude of receptor-dependent basal response or inverse agonist efficacy cannot be used as an indicator of the capacity of a receptor to isomerize into an active conformation. A receptor with a large number of intramolecular constraints, and thus with a low propensity to adopt an active conformation, may have a high degree of constitutive receptor activity if receptor density or stimulus–response coupling efficiency is high.

B. Response-Dependent Constitutive Receptor Activity

Receptors, especially GPCRs, can regulate the activity of multiple effector mechanisms in cells. Upon the discovery that different receptor agonists could differentially regulate each of the signaling pathways coupled to a single receptor subtype, a phenomenon known as "functional selectivity" (see Refs. 8,9), it became apparent that the two-state model of receptor function required modification. The two-state model could not allow for agonists to produce multiple receptor stimuli (to possess multiple, response-dependent intrinsic efficacies), which is necessary to accommodate functional selectivity of agonists. Thus, multistate receptor models were born.

The simplest multistate model is the three-state model (Fig. 3) in which receptors exist in equilibrium between an inactive (R) conformation and two active conformations R^* and R^{**}. Two allosteric transition constants (L and M) define the proportions of the receptor population that are in the R^* or R^{**} conformations at any single point in time. Ligands with different affinities for the three receptor conformations will bias the conformational proportions of the population, enriching or depleting R, R^*, or R^{**}. As a result, a single ligand can generate different stimuli (have different intrinsic efficacies), leading to differential regulation of each response coupled to the receptor (functional selectivity).

Because the values of L and M may differ, resulting in different resting (ligand-independent) quantities of R^* and R^{**} and because the stimulus–response functions (K_{e1} and K_{e2}) may also differ, constitutive receptor activity and inverse agonist efficacy may be response-dependent. Furthermore, ligands may have different K_A/K_{A^*} and $K_A/K_{A^{**}}$ ratios and thereby differentially enrich or deplete R^* and R^{**} resulting in increases or decreases in the cognate stimulus produced. In fact, it is possible for a ligand to enrich one active conformation at the expense of both the inactive and the other active conformation (Fig. 4). Such a ligand would be an agonist (for the enriched conformation) and an inverse agonist (for the depleted active conformation) at the same time. There are several examples of such ligands,[10–14] which are known as protean ligands.[15–18]

Although the three-state model is the simplest of multistate models, there is no reason not to expect that receptors may adopt many conformations as they explore their conformational space and that some of these conformations will be active (generate stimuli to regulate cellular effector mechanisms). Multistate models greatly expand the repertoire of pharmacological activities that ligands can have. Thus, constitutive receptor activity and inverse agonist

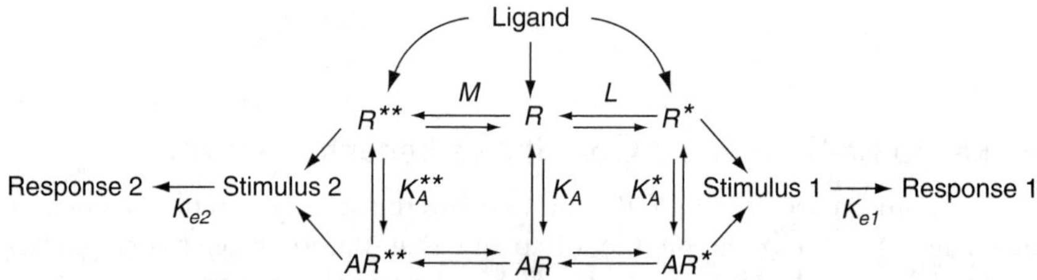

FIG. 3. Three-state model of receptor function. The receptor population consists of an inactive receptor conformation (R) in equilibrium with two active receptor conformations (R^* and R^{**}). In the absence of a ligand, different receptor stimuli (stimulus 1 and stimulus 2) result from different quantities of the active receptor conformations (R^* and R^{**}) due to different allosteric transition constants (L and M). Ligands can differentially enrich/deplete R versus R^* versus R^{**}, depending upon the value of the equilibrium dissociation constants K_A, K_{A^*}, and $K_{A^{**}}$. Adapted from Leff *et al.*[9a] and Berg *et al.*[9b]

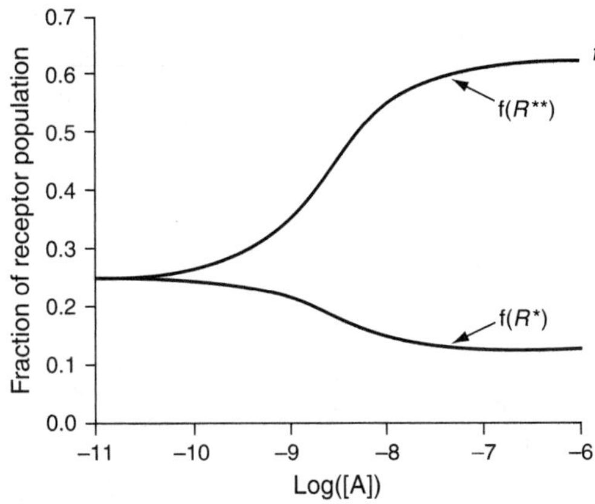

$$fR^{**} = \frac{\frac{1}{M} + \frac{[A]}{M \bullet K_A^{**}}}{\left(1 + \frac{1}{L} + \frac{1}{M}\right) + [A]\left(\frac{1}{K_A} + \frac{1}{L \bullet K_A^{*}} + \frac{1}{M \bullet K_A^{**}}\right)}$$

$$fR^{*} = \frac{\frac{1}{L} + \frac{[A]}{L \bullet K_A^{*}}}{\left(1 + \frac{1}{L} + \frac{1}{M}\right) + [A]\left(\frac{1}{K_A} + \frac{1}{L \bullet K_A^{*}} + \frac{1}{M \bullet K_A^{**}}\right)}$$

FIG. 4. Simulation of the distribution of receptors in active conformations in response to ligand A. Curves were drawn substituting values of 5×10^{-9}, 5×10^{-9}, 1×10^{-9}, 2, and 2 for K_A, K_{A^*}, $K_{A^{**}}$, L, and M, respectively. With these parameters, ligand A behaves as an agonist (enriching R^{**}) and an inverse agonist (depleting R^*) at the same time.

efficacy are not only dependent upon the receptor structure (allosteric isomerization constants, e.g., L, M, etc.) but also on the specific ligand pharmacological properties (relative conformational affinity values), the response measured, and the cell/tissue studied (R_T and K_{ex}). Therefore, it should not be surprising that the same ligand acting at the same receptor subtype may display different pharmacological properties when examined in different cells or tissues. Moreover, cell-dependent properties (R_T and K_{ex}) may change with the physiological state of the cell. For example, homologous and heterologous desensitization mechanisms can reduce receptor density, stimulus–response coupling, or both. The variable nature of R_T and K_{ex} will result in constitutive receptor activity and inverse agonist efficacy that is physiological-state-dependent. This diversity emphasizes the need to characterize the pharmacological properties of a ligand in the system to be studied, under the conditions of the study, because characterizations obtained in other systems or the same system under different experimental conditions may not generalize to all.

III. Ligands Identified *In Vitro* as Inverse Agonists at 5-HT Receptors

Studies in various cell lines have revealed an abundance of putative inverse agonists at most of the receptor subtypes (Table II). These putative inverse agonists appear to lack selectivity since many of them are identified at a number of different receptor subtypes.

TABLE II
PUTATIVE INVERSE AGONISTS AT 5-HT RECEPTORS *IN VITRO*

Receptor	Inverse agonists	References
5-HT_{1A}	Spiperone	19,20
	Spiperone	21
	p-MPPI, spiperone	22
	Methiothepin, spiperone	23
	Methiothepin, spiperone	24
	WAY 100635, spiperone	25
	Haloperidol, chlorpromazine, thioridazine, pimozide, sertindole	26
	Methiothepin, (+)butaclamol, spiperone	27
	Rec 27/0224, Rec 27/0074	28
5-HT_{1B}	Ethiothepin	29
	SB 224289	30
	SB 236057	31
	SB 236057-A	32
	SB 224289, methiothepin	33
	SB 224289	34
	SB 236057-A	35
5-$HT_{1B/1D}$	Methiothepin, ketanserin, ritanserin	36
	SB 224289, methiothepin	37
	Haloperidol, chlorpromazine, olanzapine, clozapine, risperidone, ocaperidone	38
5-HT_{2A}	Chlorpromazine, clozapine, haloperidol, loxapine, risperidone	39
	All antipsychotic drugs including ketanserin	40
	40 antipsychotic drugs and AC-90179	41
	ACP-103, haloperidol, risperidone	42
	ACP-103	43
	Pimavanserin	44
5-HT_{2C}	Mianserin, ketanserin	45
	Deramciclane	46
	SB 243213	47
	Sertindole, clozapine, olanzapine, ziprasidone, risperidone, zotepine, tiospirone, fluperapine, tenilapine	48
	SB 243213	49
	Chlorpromazine, mesoridazine, loxapine, fluphenazine, M100907, ketanserin, mianserin, ritanserin, amitriptyline	50
	Sertindole, clozapine	51
	Deramciclane	52
	SB 206553	10
	SB 206553, mirtazapine, mianserin	53

(*Continues*)

TABLE II (*Continued*)

Receptor	Inverse agonists	References
5-HT_4	GR125487	54
5-HT_6	All antipsychotic drugs	55
5-HT_7	Methiothepin, metergoline, mesulergine, clozapine, spiperone, ritanserin, methysergide, ketanserin	56
	SB 258719, SB 258741	57
	All typical/atypical antipsychotics and aripiprazole, SSR181507	58
	SB 691673, SB 269970, SB 258719, methiothepin, mesulergine, metergoline	59
	SB 269970, clozapine, mesulergine, methiothepin	60
	SB 269970	61

For example, ketanserin, which is usually considered to be a selective antagonist at the 5-HT_{2A} receptor, is identified as an inverse agonist at the 5-$HT_{1B/1D}$, 5-HT_{2A}, 5-HT_{2C}, and 5-HT_7 receptor. The apparent efficacy differences of these ligands may be due to the fact that these experiments were carried out in a variety of cell lines displaying varying degrees of constitutive activity and, as noted in the previous sections, such differences can result in differing actions of a ligand. At any rate, there appears to be no paucity of inverse agonists that need to be further characterized by additional studies both *in vitro* and *in vivo* to identify their possible role in physiology and therapeutics.

Some investigators have taken the next step of determining whether these putative inverse agonists identified by *in vitro* studies would also demonstrate inverse agonism *in vivo* (Table III). While these results are preliminary, in that experiments required for a more definitive demonstration of inverse agonism and constitutive activity (see Section V) were not carried out, they do represent an important start. Recently, evidence has been accumulated to allow for a reasonable conclusion that serotonin 5-HT_{2A} and 5-HT_{2C} receptors demonstrate constitutive activity both *in vitro* and *in vivo*. Consequently, we will now focus on those two receptors.

IV. Characteristics of Serotonin$_{2A}$ (5-HT_{2A}) and 5-HT_{2C} Receptors

The 5-HT_{2A} and 5-HT_{2C} receptors are widely expressed in the brain. They are located postsynaptically on pyramidal cells and GABAergic interneurons,[62,63] and presynaptically on the axon terminals of dopamine (DA), acetylcholine, glutamitergic, and noradrenergic neurons.[64,65] These diverse locations

TABLE III
LIGANDS TESTED *IN VIVO* THAT WERE SUGGESTED TO BE INVERSE AGONISTS

Receptor	Ligand	System employed	References
5-HT$_{1A}$	Rec 27/0224, 27/0074	Brain slices	28
5-HT$_{1B}$	SB 236057A	Microdialysis	32
	SB 224289	Learning	34
5-HT$_{2A}$	ACP-103	Reduced dyskinesias	43
	ACP-103	Animal model	42
5-HT$_{2C}$	Deramciclane	Anxiolytic, in clinical trials	52
	SB 243213	Anxiolytic, animal models	49
	SB 243213	Anxiolytic, animal models	47

of serotonin 5-HT$_{2A}$ and 5-HT$_{2C}$ receptors emphasize their ability to have a widespread influence on brain function. Both the 5-HT$_{2A}$ and 5-HT$_{2C}$ receptors have been implicated in the actions of hallucinogenic (psychotomimetic) drugs,[1,2,66,67] and many of the antipsychotic drugs are inverse agonists at these receptors.[41,48,68] In addition, the 5-HT$_{2A}$ receptor has been shown to regulate cognitive function,[62,69,70] while the 5-HT$_{2C}$ receptor is known to modulate fear and anxiety in both humans and animal models.[71–76]

The 5-HT$_{2A}$ and 5-HT$_{2C}$ receptor subtypes share a high degree of amino acid sequence homology ($\approx$ 80% in the transmembrane regions). As might therefore be expected, the pharmacological characteristics of these receptors are quite similar, with relatively few selective ligands available. In addition, some of the signaling cascades that are activated by these receptors are the same (e.g., phospholipase C (PLC) and phospholipase A$_2$ (PLA$_2$)). In spite of these similarities, there are also significant differences between the 5-HT$_{2A}$ and 5-HT$_{2C}$ receptor systems.[66,77–82] 5-HT$_{2A}$ and 5-HT$_{2C}$ receptors are widely distributed throughout the brain, notably present in corticolimbic areas such as the anygdala, hippocampus, frontal cortex, ventral tegmental area, nucleus accumbens, and hypothalamus. Consequently, it is not surprising that these receptors have been implicated in the etiology of various affective disorders, including depression and anxiety.[1,83–85]

A. 5-HT$_{2A}$ Receptor Constitutive Activity and Inverse Agonism *In Vitro*

Constitutive activity of the 5-HT$_{2A}$ receptor *in vitro* has been studied by several groups, frequently using the phospholipase C–inositol phosphate–Ca^{2+} (PLC–IP–Ca^{2+}) pathway as a readout response. However, all studies have resorted to either making mutations in the receptor[86–88] or increasing the expression of G proteins[41] to enhance constitutive activity. Constitutively active receptor mutants (CAMs) increase the observed constitutive activity (basal response) either by influencing the allosteric transition constant (e.g., L) as a result of structural changes in the receptor which alter the number or strength of intramolecular constraints or by altering the efficiency of the coupling of the receptor to its effector mechanisms (K_e). Although constitutive 5-HT_{2A} signaling to PLC appears to be rather weak in HEK cells, it has been reported that constitutive activity of the 5-HT_{2A} receptor is stronger for a reporter gene assay performed with NIH-3T3 cells.[41,88] It is not clear whether this difference in constitutive 5-HT_{2A} receptor activity is due to differences in stimulus–response coupling efficiency between the PLC and the reporter gene assay or differences in cell phenotype.

B. 5-HT_{2C} Receptor Constitutive Activity and Inverse Agonism *In Vitro*

Most of the work studying 5-HT_{2C} receptor constitutive activity and effects of 5-HT_{2C} inverse agonists *in vitro* comes from the study of the 5-HT_{2C}–PLC–IP pathway. We, and others, have shown that the native 5-HT_{2C} receptor displays a relatively high degree of constitutive activity toward PLC.[89–93] The 5-HT_{2C} receptor also activates PLA_2,[79] which results in the release of arachidonic acid (AA) from membrane phospholipids. In accord with the predicted response-dependent nature of inverse agonism (see above), 5-HT_{2C} inverse agonist efficacy is less pronounced for the PLA_2–AA response compared with the PLC–IP response.[91] Emphasizing the response-dependent nature of the pharmacological properties of ligands, SB 242084 is a protean 5-HT_{2C} receptor ligand (Fig. 5). Compared with a prototypical 5-HT_{2C} inverse agonist SB 206553, SB 242084 is a strong inverse agonist for reducing 5-HT_{2C} receptor-mediated activation of Gαi, measured using GTPγS binding. SB 242084 is a strong inverse agonist also for the PLA_2–AA response, but is an agonist for the PLC–IP response, even when both responses were measured simultaneously from the same cell population.[10]

Although not often considered to be a receptor-mediated "response" *per se*, constitutive receptor activity, like agonist-stimulated receptor activity, can activate a variety of cellular mechanisms that lead to desensitization (a time-dependent loss of receptor system responsiveness).[91,94] The cellular mechanisms that underlie desensitization are complex, but generally include uncoupling of the receptor from an effector mechanism, often resulting from receptor or signaling molecule phosphorylation, and/or internalization/downregulation of receptor protein from the cell surface.[95] Under nonstimulated conditions, the 5-HT_{2C} receptor, expressed at a density of 5 pmol/mg protein in NIH-3T3 cells, is phosphorylated; however, treatment with an inverse agonist, mianserin, did not reduce basal

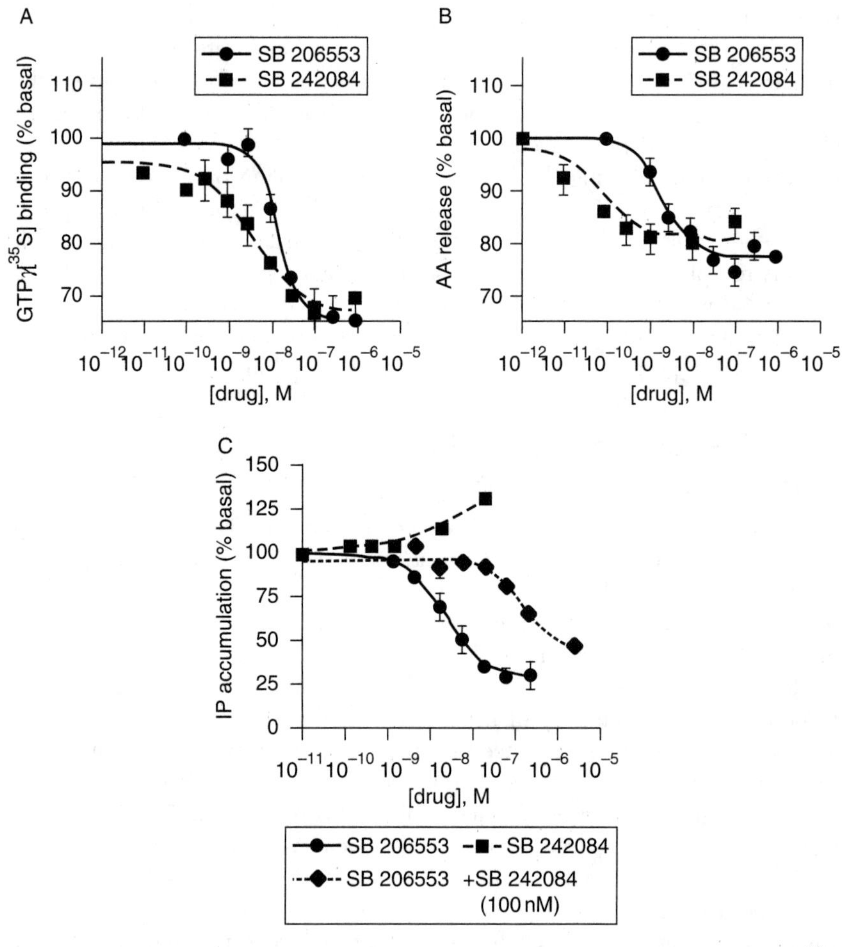

FIG. 5. SB 242084, the protean ligand. SB 242084 acts as an inverse agonist for GTPγ[^{35}S] binding to Gαi and PLA_2–AA, but is an agonist for PLC–IP. (A) Concentration–response curves to SB 206553 and SB 242084 for activation of Gαi measured with GTPγ[^{35}S] binding. Both drugs are strong inverse agonists. (B) Concentration–response curves to SB 206553 and SB 242084 for AA release. Both drugs are strong inverse agonists. (C) Concentration–response curves for SB 242084 and the prototypical inverse agonist, SB 206553 (alone and in the presence of 100 nM SB 242084) measuring IP accumulation. SB 206553 is a strong inverse agonist, whereas SB 242084 is a weak agonist for PLC–IP. As is characteristic of a low efficacy ligand, SB 242084 caused a rightward shift in the concentration–response curve to SB 206553. Reproduced with permission from Ref. 10.

phosphorylation.[96] In this study, however, mianserin was applied for only 15 min, and longer pretreatment periods may be needed to reverse the effect of prolonged constitutive receptor activity (see below).

Prolonged treatment with an inverse agonist can reduce constitutive activation of desensitization mechanisms, which can be visualized as a subsequent increase in responsiveness to both ligand-dependent and ligand-independent receptor activity toward a signaling pathway or by increased cell surface receptor levels, or both following washout of the inverse agonist. Reduction in constitutive 5-HT_{2C} receptor activity by prolonged treatment with inverse agonists in Chinese hamster ovary (CHO) cells heterologously and stably expressing 5-HT_{2C} receptors increases receptor-mediated PLC–IP, but not PLA_2–AA signaling,[91] suggesting that the 5-HT_{2C}–PLC effector pathway is selectively and constitutively desensitized in these cells. This constitutive desensitization is especially sensitive to constitutive receptor activity, occurring at low levels of 5-HT_{2C} receptor expression (250 fmol/mg protein). At this low level of expression, ligands that behave as simple competitive antagonists toward 5-HT_{2C} receptor-mediated activation of PLC and PLA_2 behave as strong inverse agonists for the desensitization response. The mechanism for the constitutive desensitization of PLC signaling in CHO cells involves decreased levels of Gαq/11; inverse agonist treatment results in increased Gαq/11 levels.[91] Interestingly, the increase in Gαq/11 levels in response to 5-HT_{2C} inverse agonist treatment promotes an increase in the responsiveness of a purinergic receptor, which is coupled to PLC and endogenously expressed in the cells (heterologous sensitization). This result suggests that constitutive activity (and inverse agonist action) of one receptor subtype can influence function of heterologous receptor systems.

The 5-HT_{2C} receptor is unique among the 7-transmembrane-spanning family of receptors in that it undergoes a posttranscriptional process known as mRNA editing. mRNA editing is the result of adenosine deaminase action on mRNA, which converts adenosine residues to inosine in specific portions of the mRNA molecule, thereby changing the genomic coding sequence for the receptor (for a recent excellent review, see Ref. 99). For the 5-HT_{2C} receptor, mRNA editing affects up to three amino acids in the putative second intracellular domain. The genomically coded, unedited sequence of the 5-HT_{2C} receptor contains the amino acids isoleucine (I), asparagine (N), and isoleucine (I) at positions 156, 158, and 160, respectively, in humans[98] and at positions 157, 159, and 161, respectively, in rats.[98] mRNA editing results in up to 32 permutations in coding sequence, generating up to 24 different isoforms (due to codon redundancy) of the 5-HT_{2C} receptor. Further, the distribution across brain and relative levels of each receptor isoform differ across species and with disease.[97]

In general, it appears that constitutive activity of RNA-edited 5-HT_{2C} receptor isoforms toward PLC is reduced compared with that of the nonedited receptor (Fig. 6). It has been suggested that the reason for the lack of constitutive activity toward PLC is the reduced capacity of the edited isoforms to couple to G proteins (K_e) as the second intracellular domain of 7-

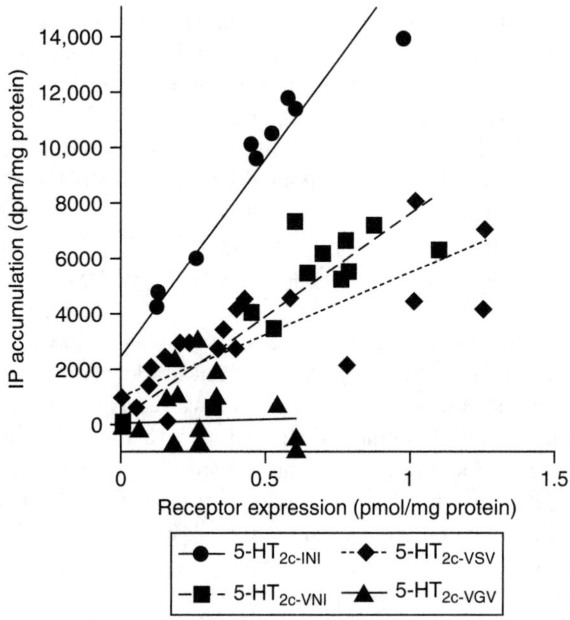

FIG. 6. Constitutive receptor activity of 5-HT_{2C} receptor isoforms to the PLC–IP pathway. HEK-293 cells were transfected with various amounts of nonedited ($5\text{-HT}_{2C\text{-INI}}$), partially edited ($5\text{-HT}_{2C\text{-VNI}}$), and fully edited ($5\text{-HT}_{2C\text{-VSV}}$ or $5\text{-HT}_{2C\text{-VGV}}$) receptor cDNA. The amount of IP accumulated for 30 min in the absence of ligand was measured. Receptor density was determined from radioligand binding using a saturating concentration of [^{3}H]-mesulergine (10 nM). The linear regression line between receptor density and IP accumulation is a measure of constitutive receptor activity. Reproduced with permission from Ref. 101.

transmembrane-spanning receptors is known to play a role in G protein coupling.[100–102] However, Niswender et al.[89] reported that mRNA-editing does not reduce the maximal PLC response to 5-HT, suggesting that G protein coupling efficiency is not altered, thus pointing to a reduced capacity of edited receptor isoforms to isomerize into an active conformation (L).

It was reported[99] that the slope of the regression line between receptor density and basal IP accumulation, which is a quantitative measure of constitutive receptor activity for fully edited 5-HT_{2C} receptor isoforms, was substantially reduced (about twofold) from that of the nonedited 5-HT_{2C} receptor, confirming previous reports of reduced constitutive receptor activity of edited isoforms.[89,90,98,103,104] The capacity of fully edited 5-HT_{2C} receptor isoforms to isomerize (measured as the susceptibility of the receptor to denature at elevated temperature)[105,106] was not different from that of the nonedited receptor. Moreover, using partial agonists (LSD and DOI) whose efficacy measure is not confounded by possible receptor reserve (as can be that of full agonists, such as

5-HT), it was found[99] that there was no difference in the ability of agonist-occupied edited receptor isoforms to activate PLC, suggesting no difference in G protein coupling efficiency. Since there is no difference in the G protein coupling efficiency or the capacity to isomerize between edited and nonedited 5-HT isoforms, it is possible that the reduced constitutive activity of the edited 5-HT_{2C} isoforms is caused by reduced efficiency of G protein coupling of the *unoccupied* receptor. Agonist occupancy could promote an active receptor conformation(s) that differs from that of the unoccupied receptor such that *agonist*-occupied edited and nonedited receptors have equal ability to couple to the PLC–IP signaling pathway.

V. Measurement of Constitutive Activity and Inverse Agonism *In Vivo*

Demonstration of inverse agonism and constitutive activity *in vivo* is a difficult but not impossible task. Based on *in vitro* studies in heterologous receptor systems, classification of a drug as an inverse agonist and demonstration of constitutive activity *in vivo* must meet the following criteria.

(1) In order to have relevance to actions in humans, the *in vivo* measure chosen should be shown to be similar in the animal model and in humans.
(2) The receptor in question must be shown to share the pharmacological profile obtained in humans.
(3) The *in vivo* measure being employed to measure drug effects must be shown to be mediated by that receptor.
(4) Ligands acting at the receptor must be shown to fall into three categories: agonists, antagonists, and inverse agonists.
(5) An inverse agonist must be shown to block the actions of an agonist.
(6) The effects of an agonist and inverse agonist must be blocked by an antagonist.
(7) Depletion of the native neurotransmitter and/or blockade of its release should eliminate the effects of an antagonist, but not affect the actions of agonists or inverse agonists.

This last demonstration is, of course, the most difficult since it is currently not possible to demonstrate total elimination of a transmitter *in vivo*. Indeed, initial *in vitro* demonstrations of inverse agonism and constitutive activity were criticized on the basis that there might be some transmitter still present. The inability to carry out complete transmitter depletion precludes any definitive statement concerning constitutive activity. However, if some amount of depletion is possible, one may be able to make an inference as to whether the *in vivo* function being employed is dependent on transmitter release and/or on constitutive activity.

Figure 7 illustrates three possible outcomes of transmitter depletion. Figure 7 (top panel) assumes that there is an absence of constitutive activity or that the constitutive activity is too low to have any significant influence on the function being measured, so that the function is primarily regulated by transmitter release. In such a case, an agonist would increase the function being measured, but obviously, antagonists and putative inverse agonists would both have the ability to block that function. Depending on the extent of transmitter depletion, the effects of antagonists and inverse agonists would be decreased or eliminated, while agonist action would be increased. Figure 7 (middle panel) assumes that an *in vivo* function being measured is dependent solely on constitutive activity and is not detectably affected by transmitter release. In such a case, one would be able to demonstrate agonism, antagonism, and inverse agonism, and the degree of transmitter depletion would have no effect on these actions. However, if constitutive activity is maximal, agonist action may not be seen either before or after transmitter depletion. Finally, Fig. 7 (bottom panel) assumes that the function under consideration is significantly affected by some combination of constitutive activity and transmitter release. In such a case, there would be effects of agonists, antagonists, and inverse agonists, but the effects of the inverse agonist would be greater than that of the antagonist. Depletion of the transmitter would still allow for agonist and inverse agonist actions, but agonist action would be increased while inverse agonist actions would be present but reduced in magnitude. Antagonist action would be reduced or eliminated depending on the amount of depletion. However, the inverse agonist would produce a greater reduction than the antagonist.

VI. Inverse Agonism and Constitutive Activity of the Serotonin 5-HT$_{2A}$ Receptor *In Vivo*

The following experiments were carried out to examine the possible existence of inverse agonism and constitutive activity at the 5-HT$_{2A}$ receptor as outlined by the seven criteria listed above:

Criterion 1. We had chosen to examine the role of the 5-HT$_{2A}$ receptor in learning as measured by acquisition of the rabbit's eyeblink response (extension of the nictitating membrane) during trace Pavlovian (classical)

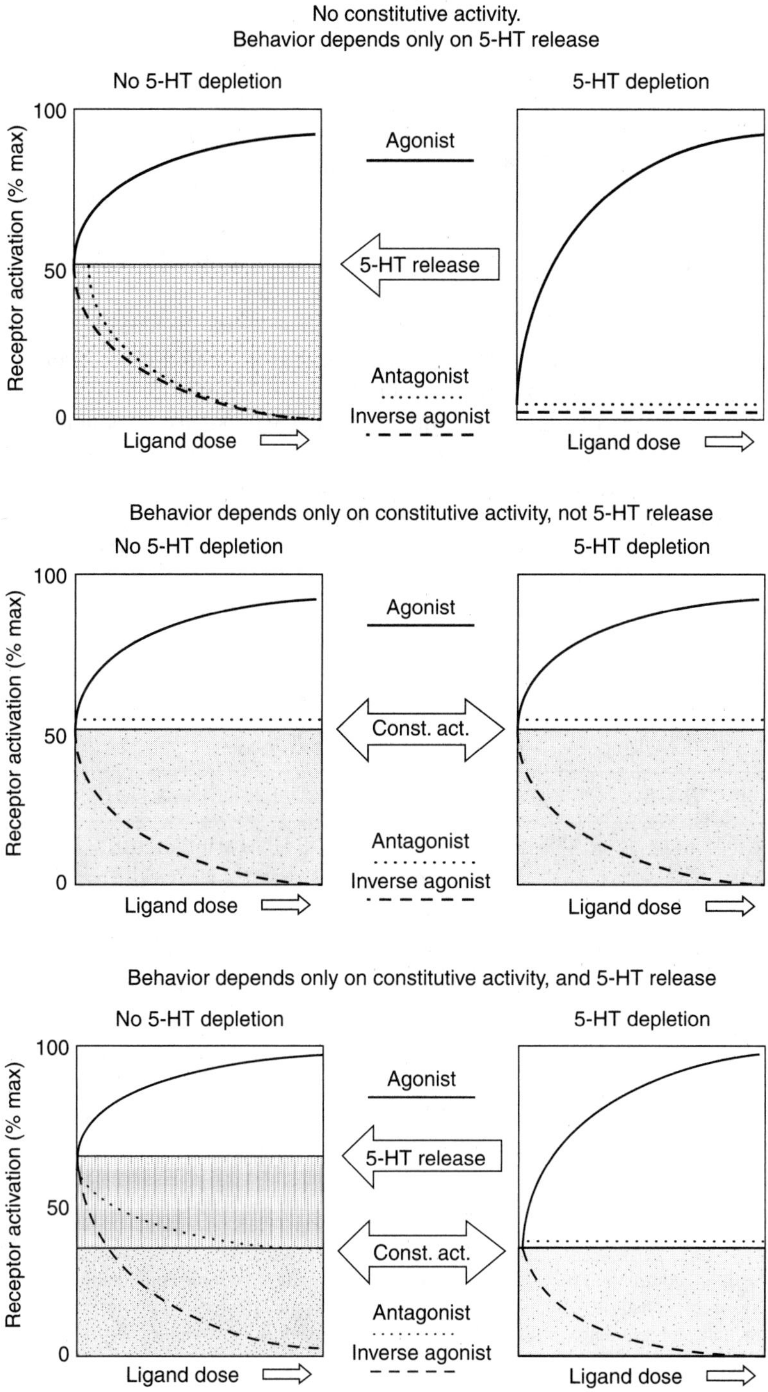

FIG. 7. Effects of neurotransmitter depletion in vivo on the actions of agonists, antagonists, and inverse agonists at the relevant receptor. The figure presents a receptor whose physiological or behavioral function is primarily dependent on transmitter release (top panel), a constitutively active

conditioning. This form of learning in the rabbit has been shown to exhibit all of the associative processes seen in human eyeblink conditioning and to be mediated by the same brain regions in rabbits and humans[107–113] and thus meets the requirements of Criterion 1.

Criterion 2. The 5-HT_{2A} receptor of the rabbit is pharmacologically similar to that of the human receptor as measured by a correlation of 0.91 between ligand affinity in the two species.[114]

Criterion 3. Hallucinogenic drugs that are putative 5-HT_{2A} agonists[2] produce cognitive effects in humans at comparable doses seen in the rabbit. For example, the threshold dose of LSD for producing hallucinations is 1.0 μg/kg in humans, and 0.8 μg/kg for producing a half maximal enhancement of eyeblink conditioning (see Ref. 3).

Criterion 4. Various putative 5-HT_{2A} ligands were found to fall into three categories: agonists that enhanced learning, antagonists that had no effect on learning, and inverse agonists that retarded learning (Table IV).

TABLE IV
EFFECT OF PUTATIVE 5-HT_{2A} LIGANDS ON LEARNING

Dose range (μmol/kg)		
Agonist[a]	Antagonist[b]	Inverse agonist[c]
LSD, 0.001–0.300, i.v.[115–118]	BOL, 0.003–5.8, i.v., s.c.[118,119]	M100907, 0.027–2.7, s.c.[d]
DOM, 0.3–3.0, i.v.[118]	LY53857, 0.067–6.7, s.c.[120]	MDL11939, 0.067–6.7, s.c.[120]
MDA, 1–10, s.c.[121]		Ritanserin, 0.067–6.7, s.c.[117,120]
MDMA 4.1–16.5, s.c.[122]		Mianserin, 0.1–10.0, s.c.[119,120]
		SR46349B, 0.26, s.c.[e]
		Ketanserin, 1.0–3.0, s.c.[e]
		Pizotifen, 8.4–33.8, s.c.[123]

[a]Agonists: drugs that significantly increased learning.
[b]Antagonists: drugs that had no significant effect on learning.
[c]Inverse agonists: drugs that significantly retarded learning.
[d]A. G. Romano, J. A. Harvey (unpublished data).
[e]L. S. Scarlota (unpublished data).

state with no requirement for transmitter release (middle panel), or some combination of transmitter release and constitutive activity (bottom panel). Each panel depicts the predicted effects of an agonist, antagonist, and inverse agonist in control animals (left side) and after transmitter depletion (right side). See text for further details.

Criterion 5. The enhancement of learning produced by LSD was blocked by both inverse agonist ritanserin[117] and MDL11939.[69]

Criterion 6. The enhancement of learning produced by LSD was blocked by the antagonist LY53857,[69] and the retardation of learning produced by the inverse agonist mianserin was blocked by d-2-bromo-lysergic acid diethylamide(BOL).[119]

Criterion 7. Reduction of 5-HT release by blockade of 5-HT raphe firing produced by the 5-HT_{1A} agonists 8-OH-DPAT and lisuride[124–126] had no effect on learning.[117] Similarly, the depletion of 5-HT by the neurotoxin 5,7-dihydroxytryptamine (5,7-DHT) by 71–72% in the frontal cortex and hippocampus (areas known to be critical for acquisition of eyeblink conditioning) had no effect on learning or on the ability of the inverse agonist MDL11939 to retard learning.[127] This result suggests that MDL11939 was producing its effects by enriching the inactive (R) form of the receptor and depleting the active (R^*) receptor rather than blocking the ability of 5-HT to combine with the receptor. Using a higher dose of 5,7-DHT, an even larger 5-HT depletion of 85–90% accompanied by a 96% decrease in the density of the 5-HT transporter also had no effect on learning.[127] Taken together, these results lead to the conclusion that the 5-HT_{2A} receptors involved in acquisition of the eyeblink response are constitutively active and fit the model of Fig. 7 (middle panel).

Just as the *in vitro* demonstration of constitutive activity and inverse agonism of a receptor is dependent on the particular cell type employed, so does the *in vivo* demonstration depend on the precise behavior being measured. For example, measurement not of learning but of motor performance of the rabbit eyeblink response[128] revealed a difference between the actions of ketanserin on learning and on the magnitude of the unconditioned eyeblink response (Table V). LSD, methylenedioxyamphetamine (MDA), and methylenedioxymethamphetamine

TABLE V

EFFECT OF PUTATIVE 5-HT_{2A} LIGANDS ON THE MAGNITUDE OF A CRANIAL REFLEX

Dose range (μmol/kg)		
Agonists[a]	Antagonists[b]	Inverse agonists[c]
LSD (0.001–0.3)[129]	BOL (0.06–5.8)[128]	MDL11939 (0.1–10.0)[120,128]
MDA (1–10)[121]	LY53857 (0.1–10)[120,128]	Ritanserin (0.1–10.0)[120,128]
MDMA (4.1–16.5)[122]	Ketanserin (0.1–10)[128]	Mianserin (0.1–10.0)[120,128]

[a]Agonists: drugs that significantly increased the unconditioned reflex magnitude.
[b]Antagonists: drugs that had no significant effect on reflex magnitude.
[c]Inverse agonists: drugs that significantly decreased reflex magnitude.

(MDMA) were the agonists that enhanced learning (Table IV) and they were the agonists that increased the magnitude of the unconditioned reflex (Table V). Similarly, MDL11939, ritanserin, and mianserin were inverse agonists for both learning and motor performance. However, while BOL and LY53857 were antagonists for learning and reflex magnitude, ketanserin proved to be an inverse agonist for learning (Table IV) but an antagonist for motor performance (Table V). This result emphasizes the complexity of ligand/receptor interactions and provides evidence for ligand-directed stimulus trafficking *in vivo*.

VII. Inverse Agonism and Constitutive Activity of the Serotonin 5-HT$_{2C}$ Receptor *In Vivo*

Evidence for a physiological role of 5-HT$_{2C}$ constitutive receptor activity in the living brain was first provided by intracerebral microdialysis studies assessing the role of central 5-HT$_{2C}$ receptors in the control of DA ascending pathways activity in the rat brain, thereby demonstrating that 5-HT$_{2C}$ constitutive receptor activity participates in the tonic inhibitory control of DA release *in vivo*.[10] Systemic administration of purported 5-HT$_{2C}$ receptor antagonists (SB 242084, SB 243213, SB 206553) significantly enhances basal DA release in DA-innervated areas of the rat brain, such as the frontal cortex, the nucleus accumbens, and the striatum.[65,92,130,131] The magnitude of this effect, however, differs with different antagonists, SB 206553 being the most efficacious in enhancing basal DA release (Fig. 8).

As discussed elsewhere,[10] the relative efficacy difference observed reflects distinct intrinsic pharmacological properties of these compounds. In fact, as shown by *in vitro* experiments in CHO cells expressing the 5-HT$_{2C}$ receptor, SB 206553, at variance with SB 242084 or SB 243213, behaves as a strong inverse agonist at the PLC pathway.[10,92] Consistent with these studies,[10,92] it has been shown that SB 206553 behaves *in vivo* as an inverse agonist at 5-HT$_{2C}$ receptor. Indeed, SB 206553-stimulated DA release is insensitive to the decrease in 5-HT terminal activity induced either by intraraphe injections of 5,7-dihydroxytryptamine neurotoxin or by peripheral administration of the 5-HT$_{1A}$ receptor agonist 8-OH-DPAT.[10] Furthermore, in agreement with the principle of mutual exclusivity, postulating that the effect of inverse agonist should be reduced by the occupancy of the receptor with ligands of weaker inverse efficacy (partial inverse agonist), antagonists, or even by agonists,[132] *in vivo* experiments have shown that the 5-HT$_{2C}$R antagonists SB 242084 and SB 243213 prevent the increase in striatal and accumbal DA release induced by SB 206553 and reverse the decrease in DA release produced by the 5-HT$_{2C}$R agonist Ro 60-0175 in both brain regions (Refs. 10,92 and Fig. 8). Taken

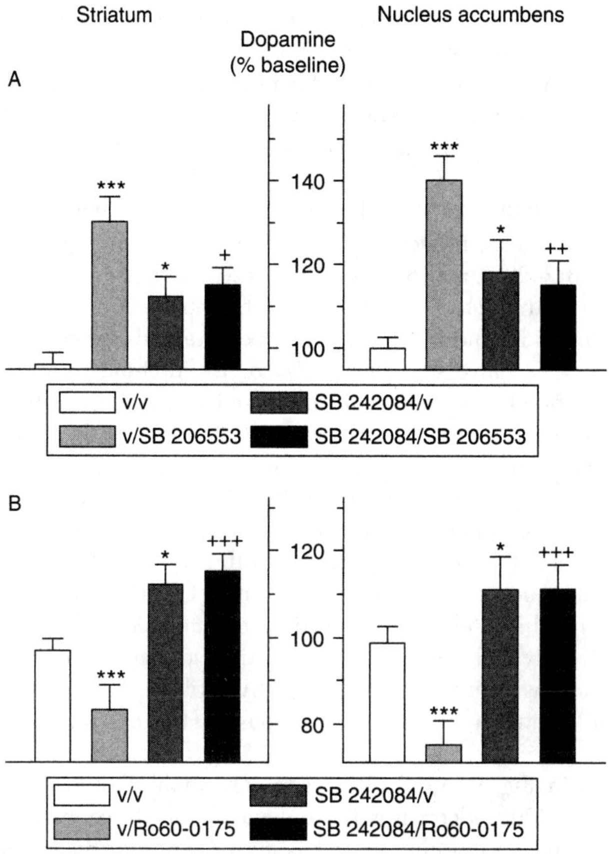

FIG. 8. Effect of central 5-HT$_{2C}$ receptor blockade on 5-HT$_{2C}$ receptor-mediated regulation of dopamine (DA) release *in vivo*. Reversal by SB 242084 (1 mg/kg, i.p.) of the DA effects elicited by the inverse agonist SB 206553 (5 mg/kg, i.p.) (A) and the agonist Ro 60-0175 (3 mg/kg, i.p.) (B) in the rat striatum and the nucleus accumbens. SB 242084 was administered 30 min before SB 206553 or Ro 60-0175. Data represent mean ± S.E.M. percentages of baseline averaged over 2 h monitoring ($n = 6$–9 animals/group). $^{*}p < 0.05$, $^{**}p < 0.01$, $^{***}p < 0.001$ versus the vehicle–vehicle (v/v) group and $^{+}p < 0.05$, $^{++}p < 0.01$, $^{+++}p < 0.001$ versus the v/SB 206553 or v/Ro60-0175 groups (Fisher's PLSD test). Data reproduced with permission from De Deurwaerdère et al.[10]

together, these findings demonstrate that the effect of SB 206553 on *in vivo* DA release is independent of the changes in extracellular levels of 5-HT and is likely related to its inverse agonist action at central 5-HT$_{2C}$ receptors, thus strengthening the identification of constitutive activity of 5-HT$_{2C}$ receptor as a physiological mechanism controlling the excitability of midbrain DA neuron *in vivo*.

Additional support for this conclusion is provided by subsequent studies[133] evaluating the influence of 5-HT_{2C} receptor inverse agonists and antagonists on the release of DA elicited by the antipsychotic drugs haloperidol and clozapine, a strong 5-HT_{2C} receptor inverse agonists.[92] Indeed, haloperidol-induced increase in accumbal and striatal DA release is potentiated by the 5-HT_{2C} receptor inverse agonist SB 206553 but unaltered by the 5-HT_{2C} receptor antagonists SB 242084 and SB 243213. Conversely, the increase in accumbal and striatal DA release induced by clozapine is unaffected by SB 206553 but blocked by SB 242084 and SB 243213. These findings indicate that 5-HT_{2C} receptor constitutive activity participates in the dopaminergic effects of clozapine and haloperidol and that clozapine modulates subcortical DA release by acting as a 5-HT_{2C} inverse agonist *in vivo*. In addition, they provide clear evidence that 5-HT_{2C} receptor inverse agonists can have different effects than 5-HT_{2C} receptor antagonists *in vivo*. This last point is strengthened by recent *in vivo* studies assessing the influence of 5-HT_{2C} receptors located within the medial prefrontal cortex (mPFC) in the control of DA release in the rat nucleus accumbens.[134] In this context, it is important to note that, at variance with 5-HT_{2C} receptors located within subcortical regions (ventral tegmental area, nucleus accumbens),[135,136] mPFC 5-HT_{2C} receptors have no influence on basal DA release but afford a facilitatory control on accumbal DA release under activated conditions.[137] Also, the fact that blockade of mPFC 5-HT_{2C}Rs *per se* has no influence on basal DA outflow (see Fig. 9) suggests the existence of a low, or absent, endogenous 5-HT tone at mPFC 5-HT_{2C} receptors.[134,137]

As shown in Fig. 9, intra-mPFC administration of the 5-HT_{2C} receptor antagonist SB 242084 decreases the enhancement of accumbal DA release induced by 10 mg/kg morphine, but has no influence on that induced by 2.5 mg/kg morphine, the DA receptor antagonist haloperidol, or the DA reuptake blocker GBR 12909. In contrast, intra-mPFC administration of the 5-HT_{2C} receptor inverse agonist SB 206553 inhibits the increase in accumbal DA release induced by both doses of morphine, as well as that elicited by haloperidol or GBR 12909 (Fig. 9). As discussed elsewhere,[134] the failure of the 5-HT_{2C} receptor antagonist SB 242084 to modulate the effect of 2.5 mg/kg morphine, haloperidol, or GBR 12909 is dependent on the fact that, at variance with high dose morphine, low dose morphine as well as haloperidol and GBR are not capable of enhancing 5-HT extracellular levels in the mPFC.[134] Indeed, the effect of an antagonist *in vivo* results from its ability to reduce receptor occupancy by an endogenous ligand, whereas that of an inverse agonist occurs as a result of its ability to reduce constitutive activity of the receptor in addition to blocking occupancy by the endogenous ligand.[138,139] In agreement with these considerations, intra-mPFC injection of the same dose of SB 242084[137]

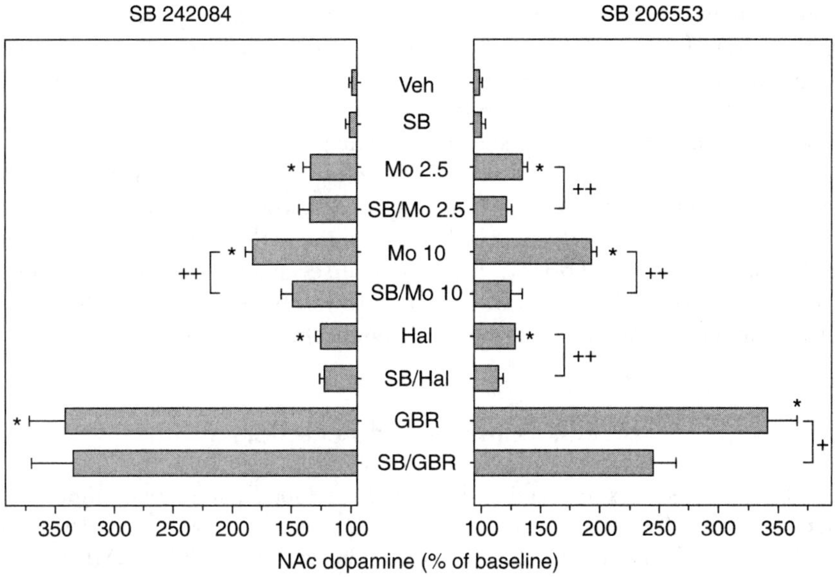

FIG. 9. Effects of intramedial prefrontal cortex (mPFC) injections of SB 242084 and SB 206553. Depicted are the effects of the 5-HT_{2C} receptor antagonist SB 242084 and the 5-HT_{2C} receptor inverse agonist SB 206553 on the increase in accumbal DA outflow induced by the peripheral administration of morphine (Mo, 2.5 or 10 mg/kg, s.c.), haloperidol (Hal, 0.01 mg/kg, s.c.), or GBR 12909 (GBR, 2.5 mg/kg, i.p.). SB 242084 (SB, left panel) or SB 206553 (SB, right panel) was injected into the mPFC at 0.5 μg/0.2 μl, 15 min before each drug administration. Each bar histogram represents the mean ± S.E.M. percentages of baseline averaged over 2 h monitoring ($n = 4$–7 animals/group). $^{*}p < 0.001$ versus the respective vehicle (Veh) group; $^{+}p < 0.01$, $^{++}p < 0.001$ versus the respective Mo 2.5, Mo 10, Hal, GBR groups (Fisher's PLSD test). NAc, nucleus accumbens. Data reproduced with permission from Leggio *et al.*[134]

or SB 206553 (unpublished results) inhibits the increase in accumbal DA outflow induced by cocaine, a drug known to enhance 5-HT outflow in the mPFC.[140]

Taken together, the results described above show that mPFC 5-HT_{2C} receptor constitutive activity participates in the control of activated accumbal DA release and provide clear evidence that 5-HT_{2C} receptor antagonist and inverse agonist behave differently *in vivo*, thereby emphasizing the therapeutic potential of inverse agonism for pathological conditions depending upon mesolimbic DA dysfunction, such as depression or schizophrenia.[132,141] Indeed, 5-HT_{2C} receptor inverse agonists, which act by suppressing 5-HT_{2C} receptor constitutive activity, provide an additional pharmacological tool for the fine-tuning of midbrain DA neuron excitability by the 5-HT_{2C} receptor.[142]

Interestingly, 5-HT_{2C} receptor constitutive activity has been proposed to be of broader significance to the induction and control of depressive states,[143] and some antidepressant drugs have been recently shown to behave as inverse agonists at 5-HT_{2C}Rs.[53] Furthermore, as mentioned above, 5-HT_{2C} receptor constitutive activity has been shown to participate in the DAergic effects of antipsychotic drugs *in vivo*,[133] and 5-HT_{2C} receptor inverse agonist activity of some antipsychotic drugs has been suggested to contribute to antipsychotic drugs atypicity.[48,50] However, at variance with the 5-HT_{2A} receptor,[144] clinical trials using well-defined 5-HT_{2C} receptor inverse agonists are not yet available to our knowledge, and further investigation is warranted to better evaluate the clinical and/or therapeutic relevance and applicability of inverse agonists.[141]

VIII. Cannabinoid Receptors

The endogenous cannabinoid agonists (endocannabinoids) mimic the actions of delta9-tetrahydrocannabinol and interact with two major classes of GPCRs named CB1 and CB2 whose activation typically inhibits cAMP formation. Endogenous and synthetic CB1 receptor agonists have been documented to modulate stimulation of [^{35}S] GTPγS incorporation into G-proteins and MAP kinase activity. Additionally, CB1 receptor agonists inhibit forskolin-stimulated cAMP production and voltage-dependent calcium currents. CB1 receptors are mainly located in the brain[145] where their activation is correlated with the production of analgesia, hypothermia, catalepsy, decreased locomotor activity, appetite, and memory disturbances. Therapeutic uses have been suggested for the treatment of anorexia, as an antiemetic during chemotherapy, as an analgesic in chronic pain, for the treatment of glaucoma, and in movement disorders including Parkinson's disease. CB2 receptors are mainly located in the periphery including testes, spleen, leukocytes, and lymphocytes, where they are able to suppress gonadal function and the immune system. Both antagonists and agonists that are specific for the two receptor subtypes have been synthesized. This chapter focuses on the CB1 receptor.

The endocannabinoids include the well-studied 2-arachidonoylglycerol (2AG) and arachidonoylethanolamine (anandamide) plus other less studied compounds.[146] The endocannabinoids are synthesized upon demand and rapidly degraded.[147] Endogenous and synthetic CB1 receptor agonists have been documented to modulate a variety of physiological processes. Prather and coworkers[148] demonstrated that, in rat brain tissue, CB1 receptors couple to multiple forms of Gαi and Gα subunits. Furthermore, the concentration of the CB1 agonist WIN 55212-2 required to activate the various G proteins differed by about 30 fold. These data suggest that different levels of endocannabinoids may differentially activate different G proteins. Similarly, Glass and Northup[149] demonstrated functional selectivity using membranes from *Spodoptera frugiperda* cells expressing CB1 receptors. They showed that HU210, WIN 55,212-2, and anandamide were full agonists for activating Gi, but only HU210 was a full agonist for activating Go.

A. Constitutive Activity and Inverse Agonism at the Cannabinoid CB1 Receptor

Considerable suggestive evidence has been published leading to the concept that CB1 receptors are constitutively active. This evidence includes both *in vitro* and *in vivo* experiments, but mainly relies on the properties of several drugs that produce effects opposite to those produced by agonists and are thus labeled inverse agonists. One of the most widely investigated CB1 antagonists/inverse agonists is SR141716A (rimonabant). Rimonabant was first determined to be a selective CB1 versus CB2 receptor antagonist,[150] but was soon recognized as a CB1 receptor inverse agonist.[151] For GPCRs, GTP or its analogues inhibit agonist binding. In contrast, Bouaboula and coworkers[151] demonstrated that rimonabant binding was enhanced by the GTP analogue GTPγS, providing indirect evidence that rimonabant is an inverse agonist. Furthermore, they demonstrated that rimonabant inhibited basal CB1-receptor-promoted mitogen activated protein kinase (MAPK) and adenylyl cyclase activity.[151] Likewise, in isolated rat superior cervical ganglion neurons expressing hCB1 receptors, rimonabant acts as an inverse agonist by enhancing voltage-dependent calcium currents, an effect opposite to that of cannabinoid agonists.[152] Since endogenous cannabinoids inhibit these calcium currents, it is possible that the enhancing effect of rimonabant is due to the inhibition of endogenous cannabinoids. In an attempt to eliminate this possibility, Pan and coworkers[152] performed the experiment under calcium-free conditions designed to prevent the calcium-dependent synthesis of the endocannabinoids, anandamide, and 2AG. They demonstrated that, even in calcium-free medium, rimonabant was able to enhance voltage-dependent calcium currents.[152] Although they established that rimonabant was not blocking the agonist anandamide or 2AG, their experiment does not rule out the possible presence of other endocannabinoids.

MK-0364 (taranabant, Ref. 155), another putative CB1 inverse agonist, exhibits approximately 1000-fold selectivity of CB1 compared to CB2 receptors. MK-0364 was classified as an inverse agonist at CB1R because it increases forskolin-induced cAMP in CB1R-CHO cells, an effect that is opposite to that observed with a CB1 receptor agonist.[153] A potential endogenous CB1 receptor inverse agonist, hemopressin, has been identified.[154] Hemopressin was classified as an inverse CB1 agonist since it inhibited basal GTPγS incorporation and stimulated adenylyl cyclase activity, effects that are opposite to those of the agonist HU-210.[154] *In vivo*, both hemopressin and another putative CB1 receptor inverse agonist, AM251, exhibited antihyperalgesic activity.[154]

In order to observe inverse agonist activity, the receptor under consideration must exhibit constitutive activity. In the absence of measurable constitutive activity, antagonists and inverse agonists will inhibit agonist-mediated effects equally. In some systems involving CB1 receptors, putative antagonists and inverse agonist do not exhibit differential effects.

For example, CB1 receptors are involved in the regulation of body temperature. Both systemic administration and direct injection of a CB1 receptor agonists into the hypothalamus result in hypothermia.[155,156] The CB1 receptor agonist (CP 55,940)-mediated reduction in core body temperature was attenuated by both the antagonist AM 4113 and the inverse agonist AM251. However, neither the antagonist (AM4113) nor several putative inverse agonists (AM251, rimonabant, or MK-0364) alter core body temperature by themselves.[150,153,156] These data suggest that either the CB1 receptor involved in the hypothermic response is not constitutively active, there is no endogenous agonist tone, or both. Although AM4113 binds with high affinity to rat brain CB1 receptors, it does not alter intracellular cAMP levels in HEK293 cell expressing hCB1 receptors, leading to its designation as an antagonist.[156,157] Similarly, 5-(4-chlorophenyl)-3-[(*E*)-2-cyclohexylethenyl]-1-(2,4-dichlorophenyl)-4-methyl-1*H*-pyrazole and 6″-Azidohen-2″-yne-cannabindiol have been classified as CB1 receptor antagonists (VCHSR; Refs. 158–160). O-2050 appears to be an antagonist in the guinea pig myenteric longitudinal muscle and the mouse vas deferens (for a review see Ref. 161). However, as discussed above, pharmacological properties of ligands are dependent upon the system studied and physiological state. We need to exercise caution when generalizing the actions of drugs based on tests done in one system.

Further complicating the classification of drugs as CB1 receptor antagonist or inverse agonist is the fact that endocannabinoids are rapidly synthesized in response to stimuli. Thus, the presence of endogenous agonist is difficult to rule out. *In vitro*, for example, exhaustive washing of cells or tissue slices may remove endogenous agonists but will not necessarily prevent resynthesis of endocannabinoids. *In vivo* ensuring the complete absence of endogenous cannabinoids is even more difficult. The presence of endogenous agonists will allow both antagonists and inverse agonists to inhibit what appears to be basal activity. In contrast, in a system with constitutive activity and in the absence of endogenous agonists, an antagonist will have no effect but an inverse agonist will inhibit. Furthermore, if the inverse agonist is acting on the CB1 receptor, an antagonist will block this effect, thus ensuring that the inverse agonist activity is indeed mediated by the CB1 receptor. Thus, the recent availability of CB1 receptor antagonists will greatly facilitate the classification of drugs as inverse agonist at CB1 receptors and the determination of which CB1 receptor systems demonstrate constitutive activity.

B. Therapeutic Potential of Cannabinoid CB1 Receptor Inverse Agonists

Cannabinoids derived from *Cannabis sativa* (marijuana) mediate a wide variety of physiological effects mediated by CB1 receptors. In addition to the well-known euphoria and alteration of sensory perception, these compounds also stimulate appetite and have anxiolytic and analgesic activity among other effects. Thus, it would be expected that CB1 receptor inverse agonists would have opposite effects, some of which could be of medicinal value. Since cannabinoids stimulate food consumption, CB1 receptor inverse agonist have been tested for their appetite suppressive value with the goal of weight reduction in overweight individuals. In food-restricted rats, rimonabant decreased the consumption of standard chow as well as highly palatable sucrose pellets.[162,163] Similarly, rimonabant reduced food intake in mice. Likewise, AM 251 reduced food intake in both lean and obese mice.[164] The anorectic effects of both MK-0364 and rimonabant were absent in CB1-receptor-deficient mice, demonstrating that these drug effects were mediated by CB1 receptors.[153,165] Furthermore, a series of 5,6-diaryl-2-amino-pyrazines were synthesized and examined for their antagonist-like properties at the CB1 receptor.[166] Four compounds of this series were found to have inverse agonist actions and were then tested *in vivo* along with rimonabant for their ability to affect food intake. All these four compounds were able to reduce food intake and body weight and lower levels of blood glucose, blood insulin, and plasma leptins.[166] All of these results suggest that CB1 receptor inverse agonists may be a therapeutic treatment for obesity.

Several clinical trials have examined the clinical utility of rimonabant for the treatment of obesity. Rimonabant (5 or 20 mg) or placebo was administered to obese individuals (body mass index 30 or greater). Following 1 year of treatment, the 20-mg group showed a significantly greater loss of weight, reduction of waist circumference and levels of triglycerides, as well as an increase in high-density lipoprotein.[167] Continued rimonabant use maintained these changes for a second year.[167] Similarly a European study demonstrated that, over 2 years, rimonabant administration resulted in significant weight loss and improvement in cardiometabolic risk factors.[168]

Rimonabant was approved for clinical use by the European Commission and the drug went on sale in July 2006, initially in the United Kingdom and subsequently in many other European countries. It was concluded that rimonabant was generally well tolerated, and the rates of adverse events, including depressed mood disorders and disturbances, were similar to placebo.[168] However, in January 2009, the European approval was suspended. Meta-analyses demonstrated that rimonabant resulted in an increased risk of mood disorders.[169,170] The withdrawal of rimonabant resulted in the cessation of clinical trials and further development of CB1 inverse agonists.[171]

It is unclear whether CB1 inverse agonists are required to reduce food intake. Recent basic studies have determined that CB1 antagonists can also reduce food intake. The antagonist O-2050 was able to decrease food intake and weight gain in nondeprived rats (similar to rimonabant[172]). Likewise, the CB1 receptor antagonist AM4113 reduced consumption of high-fat, high-carbohydrate, and standard food.[156,157] These results further question the therapeutic value of CB1 inverse agonists. However, peripherally restricted CB1 inverse agonists may have clinical utility (for a discussion, see Ref. 173).

IX. Summary

Recent studies carried out in various cell lines have revealed a surprising diversity of serotonin receptors and a complexity in ligand–receptor interactions. Serotonin as well as cannabinoid receptors display varying degrees of constitutive activity *in vitro*. Thus, ligands acting at a constitutively active receptor can act as agonists, antagonists, and/or inverse agonists and these pharmacological characteristics can differ for an individual ligand depending upon the receptor response being measured and the physiological state of the system being examined. This complexity of ligand actions also includes the existence of functional selectivity and protean ligands as well as the occurrence of extensive posttranscriptional mRNA editing and splice variants of some receptors. While these findings present an enormous challenge to pharmacologists, there is also a great opportunity to develop new tools for identifying ligands with more selective therapeutic drug action. The success of such a venture becomes more optimistic with recent identification of inverse agonism *in vivo* and the emerging evidence that constitutive activity is present in the intact organism.

References

1. Leysen JE. 5-HT$_2$ receptors. *Curr Drug Targets* 2004;**3**:11–26.
2. Nichols DE. Hallucinogens. *Pharmacol Ther* 2004;**101**:131–81.
3. Harvey JA. Role of the serotonin 5-HT(2A) receptor in learning. *Learn Mem* 2003;**10**:355–62.
4. Hannon J, Hoyer D. Molecular biology of 5-HT receptors. *Behav Brain Res* 2008;**195**:198–213.

5. Cerione RA, Codina J, Benovic JL, Lefkowitz RJ, Birnbaumer L, Caron MG. The mammalian beta 2-adrenergic receptor: reconstitution of functional interactions between pure receptor and pure stimulatory nucleotide binding protein of the adenylate cyclase system. *Biochemistry* 1984;**23**:4519–25.
6. Costa T, Herz A. Antagonists with negative intrinsic activity at delta opioid receptors coupled to GTP-binding proteins. *Proc Natl Acad Sci USA* 1989;**86**:7321–5(a)Christopoulos A, Kenakin T. G protein-coupled receptor allosterism and complexing. *Pharmacol Rev* 2002;**54**:323–74(b)Leff P. The two-state model of receptor activation. *Trends Pharmacol Sci* 1995;**16**:89–97.
7. Kenakin TP, Beek D. Is prenalterol (H133/80) really a selective beta 1 adrenoceptor agonist? Tissue selectivity resulting from differences in stimulus-response relationships. *J Pharmacol Exp Ther* 1980;**213**:406–13.
8. Urban JD, Clarke WP, von Zastrow M, Nichols DE, Kobilka B, Weinstein H, et al. Functional selectivity and classical concepts of quantitative pharmacology. *J Pharmacol Exp Ther* 2007;**320**:1–13.
9. Neve K. Functional selectivity of G protein-coupled receptor ligands: new opportunities for drug discovery. In: Neve K, editor. *The receptors*. New York: Humana Press; 2009(a)Leff P, Scaramellini C, Law C, McKechnie K. A three-state model of agonist action. *Trends Pharmacol Sci* 1997;**18**:355–62(b)Berg KA, Maayani S, Goldfarb J, Scaramellini C, Leff P, Clarke WP. Effector pathway-dependent relative efficacy at serotonin type 2A and 2C receptors: evidence for agonist-directed trafficking of receptor stimulus. *Mol Pharmacol* 1998;**54**:94–104.
10. De Deurwaerdére P, Navailles S, Berg KA, Clarke WP, Spampinato U. Constitutive activity of the serotonin2C receptor inhibits in vivo dopamine release in the rat striatum and nucleus accumbens. *J Neurosci* 2004;**24**:3235–41.
11. Ganguli SC, Park CG, Holtmann MH, Hadac EM, Kenakin TP, Miller LJ. Protean effects of a natural peptide agonist of the G protein-coupled secretin receptor demonstrated by receptor mutagenesis. *J Pharmacol Exp Ther* 1998;**286**:593–8.
12. Lane JR, Powney B, Wise A, Rees S, Milligan G. Protean agonism at the dopamine D2 receptor: (*S*)-3-(3-hydroxyphenyl)-*N*-propylpiperidine is an agonist for activation of Go1 but an antagonist/inverse agonist for Gi1, Gi2, and Gi3. *Mol Pharmacol* 2007;**71**:1349–59.
13. Newman-Tancredi A, Cussac D, Marini L, Touzard M, Millan MJ. h5-HT(1B) receptor-mediated constitutive Galphai3-protein activation in stably transfected Chinese hamster ovary cells: an antibody capture assay reveals protean efficacy of 5-HT. *Br J Pharmacol* 2003;**138**:1077–84.
14. Pauwels PJ, Rauly I, Wurch T, Colpaert FC. Evidence for protean agonism of RX 831003 at alpha 2A-adrenoceptors by co-expression with different G alpha protein subunits. *Neuropharmacology* 2002;**42**:855–63.
15. Kenakin T. Inverse, protean, and ligand-selective agonism: matters of receptor conformation. *FASEB J* 2001;**15**:598–611.
16. Neubig RR. Missing links: mechanisms of protean agonism. *Mol Pharmacol* 2007;**71**:1200–2.
17. Perez DM, Karnik SS. Multiple signaling states of G-protein-coupled receptors. *Pharmacol Rev* 2005;**57**:147–61.
18. Chidiac P. Considerations in the evaluation of inverse agonism and protean agonism at G protein-coupled receptors. *Methods Enzymol* 2002;**343**:3–16.
19. Newman-Tancredi A, Conte C, Chaput C, Spedding M, Millan MJ. Inhibition of the constitutive activity of human 5-HT1A receptors by the inverse agonist, spiperone but not the neutral antagonist WAY 100, 635. *Br J Pharmacol* 1997;**120**:737–9.
20. Newman-Tancredi A, Conte C, Chaput C, Verriele L, Millan MJ. Agonist and inverse agonist efficacy at human recombinant serotonin 5-HT$_{1A}$ receptors as a function of receptor:-G-protein stoichiometry. *Neuropharmacology* 1997;**36**:451–9.

21. Welsby PJ, Kellett E, Wilkinson G, Milligan G. Enhanced detection of receptor constitutive activity in the presence of regulators of G-protein signaling: applications to the detection and analysis of inverse agonists and low-efficacy partial agonists. *Mol Pharm* 2002;**61**:1211–21.
22. Barr AJ, Manning DR. Agonist-independent activation of Gz by the 5-hydroxytryptamine 1A receptor co-expressed in *Spodoptera frugiperda* cells. Distinguishing inverse agonists from neutral antagonists. *J Biol Chem* 1997;**272**:32979–87.
23. Newman-Tancredi A, Verriele L, Chaput C, Millan MJ. Labelling of recombinant human and native rat serotonin 5-HT_{1A} receptors by a novel, selective radioligand, [^{3}H]-S 15535: definition of its binding profile using agonists, antagonists and inverse agonists. *Naunyn-Schmiedebergs Arch Pharmacol* 1998;**357**:205–17.
24. McLoughlin DJ, Strange PG. Mechanisms of agonism and inverse agonism at serotonin 5-HT_{1A} receptors. *J Neurochem* 2000;**74**:347–57.
25. Cosi C, Koek W. The putative 5-HT(1A) receptor antagonist, WAY 100635, has inverse agonist properties at cloned human 5-HT(1A) receptors. *Eur J Pharmacol* 2000;**401**:9–15.
26. Cosi C, Koek W. Agonist, antgonist, and inverse agonist properties of antipsychotics at human recombinant 5-HT(1A) receptors expressed in HeLa cells. *Eur J Pharmacol* 2001;**433**:55–62.
27. Newman-Tancredi A, Verriele L, Millan J. Differential modulation by GTPgammasS of agonist and inverse agonist binding to h5-HT(1A) receptors revealed by [^{3}H]-WAY100, 635. *Br J Pharmacol* 2001;**132**:518–24.
28. Corradetti R, Mlinar B, Falsini C, Pugliese AM, Cilia A, Destefani C, et al. *J Pharmacol Exp Ther* 2005;**315**:109–17.
29. Beer MS, Heald MA, McAllister G, Stanton JA. Pharmacological characterisation of a cloned dog 5-HT1B receptor cell line. *Eur J Pharmacol* 1998;**360**:117–21.
30. Milan MJ, Gobert A, Audinot V, Dekeyne A, Newman-Tancredi A. Inverse agonists and serotonergic transmission: from recombinant, human serotonin $(5\text{-HT})_{1B}$ receptors to G-protein coupling and function in coticolimbic structures in vivo. *Neuropsychopharmaclogy* 1999;**21**:615–75.
31. Middlemiss DN, Gothert M, Schlicker E, Scott CM, Selkirk JV, Watson J, et al. SB-236057, a selective 5-HT_{1B} receptor inverse agonist, blocks the 5-HT human terminal autorecetor. *Eur J Pharmacol* 1999;**375**:359–65.
32. Roberts C, Hatcher P, Hagan JJ, Austin NE, Jeffrey P, Wyman P, et al. The effect of SB-236057A, a selective 5-HT_{1B} receptor inverse agonist, on in vivo extracellular 5-HT levels in the freely-moving guinea-pig. *Naunyn-Schmiedebergs Arch Pharmacol* 2000;**362**:177–83.
33. Newman-Tancredi A, Audinot V, Moreira C, Verriele L, Millan MJ. Inverse agonism and constitutive activity as functional correlates of serotonin h5-HT(1B) receptor/G-protein stoichiometry. *Mol Pharmacol* 2000;**58**:1042–9.
34. Meneses A. Could the 5-HT_{1B} receptor inverse agonism affect learning consolidation? *Neurosci Biobehav Rev* 2001;**25**:193–201.
35. Roberts C, Watson J, Price GW, Middlemiss DJ. SB-236057-A: a selective 5-HT_{1B} receptor inverse agonist. *CNS Drug Rev* 2001;**7**:433–44.
36. Pauwels PJ. 5-HT 1B/D receptor antagonists. *Gen Pharmacol* 1997;**29**:293–303.
37. Audinot V, Newman-Tancredi A, Millan MJ. Constitutive activity at serotonin(1D) receptors: detection by homologous GTPgammaS versus [(35)S]-GTPgammaS binding isotherms. *Neuropharmacol* 2001;**40**:57–64.
38. Audinot V, Newman-Tancredi A, Cussac D, Millan MJ. Inverse agonist properties of antipsychotic agents at cloned, human (h) serotonin (5-HT)(1B) and h5-HT(1D) receptors. *Neuropsychopharmacology* 2001;**25**:410–22.
39. Egan C, Herrick-Davis K, Teitler M. Creation of a constitutively activated state of the 5-hydroxytryptamine2A receptor by site-directed mutagenesis: inverse agonist activity of antipsychotic drugs. *J Pharmacol Exp Ther* 1998;**286**:85–90.

40. Egan C, Herrick-Davis K, Teitler M. Creation of a constitutively activated state of the 5-HT2A receptor by site-directed mutagenesis: revelation of inverse agonist activity of antagonists. *Ann NY Acad Sci* 1998;**861**:136–9.
41. Weiner DM, Burstein ES, Nash N, Croston GE, Currier EA, Vanover KE, et al. 5-hydroxytryptamine2A receptor inverse agonists as antipsychotics. *J Pharmacol Exp Ther* 2001;**299**:268–76.
42. Gardell LR, Vanover KE, Pounds L, Johnson RW, Barido R, Anderson GT, et al. *J Pharmacol Exp Ther* 2007;**322**:862–70.
43. Vanover KE, Betz AJ, Weber SM, Bibbiani F, Kielaite A, Weiner DM, et al. 5-HT_{2A} receptor inverse agonist, ACP-103, reduces tremor in a rat model and levodopa-induced dyskinesias in a monkey model. *Pharmacol Biochem Behav* 2008;**90**:540–4.
44. Abbas A, Roth BL. Pimavanserin tartrate: a 5-HT_{2A} inverse agonist with potential for treating various neuropsychiatric disorders. *Exp Opin Pharmacother* 2008;**9**:3251–9.
45. Hartman JL, Northup JK. Functional reconstitution in situ of 5-hydroxytryptamine2c (5-HT2c) receptors with alphaq and inverse agonism of 5-HT2c receptor antagonists. *J Biol Chem* 1996;**271**:22591–7.
46. Palvimaki EP, Majasuo H, Kuoppamaki M, Mannisto PT, Syvalahti E, Hietala J. Deramciclane, a putative anxiolytic drug, is a serotonin 5-HT_{2C} receptor inverse agonist but fails to induce 5-HT2C receptor down-regulation. *Psychopharmacology* 1998;**136**:99–104.
47. Bromidge SM, Dabbs S, Davies DT, Davies S, Duckworth DM, Forbes IT, et al. Biarylcarbamoylindolines are novel and selective 5-HT(2C) receptor inverse agonists: identification of 5-methyl-1-[[2-[(2-methyl-3-pyridil)oxy]-5-pyridil]carbamoyl]-6-trifluoromethylindoline (SB-243213) as a potential antidepressant/anxiolytic agent. *J Med Chem* 2000;**43**:1123–34.
48. Herrick-Davis K, Grinde E, Teitler M. Inverse agonist activity of atypical antipsychotic drugs at human 5-hydroxytryptamine2C receptors. *J Pharmacol Exp Ther* 2000;**295**:226–32.
49. Wood MD, Reavill C, Trail B, Wilson A, Stean T, Kennett GA, et al. SB-243213; a selective 5-HT_{2C} receptor inverse agonist with improved anxiolytic profile: lack of tolerance an withdrawal anxiety. *Neuropharmacology* 2001;**41**:186–99.
50. Rauser L, Savage JE, Meltzer HY, Roth BL. Inverse agonist actions of typical and atypical antipsychotic drugs at the human 5-hydroxytryptamine(2C) receptor. *J Pharmacol Exp Ther* 2001;**299**:83–9.
51. Hietala J, Kuonnamaki M, Palvimaki EP, Laakso A, Majasuo H, Syvalahti E. Sertindole is a serotonin 5-HT2c inverse agonist and decreases agonist but not antagonist binding to 5-HT2c receptors after chronic treatment. *Psychopharmacology* 2001;**157**:180187.
52. Koks S, Vasar E. Deramciclane (Egis). *Curr Opin Investig Drugs* 2002;**3**:289–94.
53. Chanrion B, Mannoury La Cour C, Gavarini S, Seimandi M, Vincent L, Pujol JF, et al. Inverse agonist and neutral antagonist actions of antidepressants at recombinant and native 5-HT_{2C} receptors: differential modulation of cell surface expression and signal transduction. *Mol Pharmacol* 2007;**73**:748–57.
54. Bockaert J, Claeysen S, Sebben M, Dumuis A. 5-HT4 receptors: gene, transduction and effects on olfactory memory. *Annals NY Acad Sci* 1998;**861**:1–15.
55. Purohit A, Smith C, Herrick-Davis K, Teitler M. Stable expression of constitutively activated mutant h5HT6 and h5HT7 serotonin receptors: inverse agonist activity of antipsychotic drugs. *Psychopharmacology* 2005;**179**:461–9.
56. Krobert KA, Levy FO. The human 5-HT7 serotonin receptor splice variants: constitutive activity and inverse agonist effects. *Br J Pharmacol* 2002;**135**:1563–71.
57. Mahe C, Loetscher E, Feuerbach D, Muller W, Seiler MP, Schoeffter P. Differential inverse agonist efficacies of SB-258719, SB-258741 and SB-269970 at human recombinant serotonin 5-HT7 receptors. *Eur J Pharmacol* 2004;**495**:97–102.

58. Rauly-Listienne I, Boutet-Robinet E, Ailhaud MC, Newman-Tancredi A, Cussac D. Differential profile of typical, atypical and third generation antipsychotics at human 5-HT7a receptors coupled to adenylyl cyclase: detection of agonist and inverse agonist properties. *Naunyn-Schmiedebergs Arch Pharmacol* 2007;**376**:93–105.
59. Romero G, Pujol M, Pauwels PJ. Reanalysis of constitutively active rat and human 5-HT7(a) receptors in HEK-293F cells demonstrates lack of silent properties for reported neutral antagonists. *Naunyn-Schmiedebergs Arch Pharmacol* 2006;**374**:31–9.
60. Krobert KA, Andressen KW, Levy FO. Heterologous desensitization is evoked by both agonist and antagonist stimulation of the human 5-HT(7) serotonin receptor. *Eur J Pharmacol* 2006;**532**:1–10.
61. Prause AS, Stoffel MH, Portier CJ, Mevissen M. Expression and function of 5-HT7 receptors in smooth muscle preparations from equine duodenum, ileum, and pelvic flexure. *Res Vet Sci* 2009;**87**:292–9.
62. Jakab RL, Goldman-Rakic PS. 5-Hydroxytryptamine$_{2A}$ serotonin receptors in the primate cerebral cortex: possible site of action of hallucinogenic and antipsychotic drugs in pyramidal cell apical dendrites. *Proc Natl Acad Sci USA* 1998;**95**:735–40.
63. Liu S, Bubar MJ, Lanfranco MF, Hillman GR, Cunningham KA. Serotonin$_{2C}$ receptor localization in GABA neurons of the rat medial prefrontal cortex: implications for understanding the neurobiology of addiction. *Neuroscience* 2007;**146**:1677–88.
64. Fink KB, Gothert M. 5-HT receptor regulation of neurotransmitter release. *Pharmacol Rev* 2007;**59**:360–417.
65. Alex KD, Pehek EA. Pharmacologic mechanisms of serotonergic regulation of dopamine neurotransmission. *Pharmacol Ther* 2007;**113**:296–320.
66. Roth BL, Willins DL, Kristiansen K, Kroeze WK. 5-Hydroxytryptamine2-family receptors (5-hydroxytryptamine2A, 5-hydroxytryptamine2B, 5-hydroxytryptamine2C): where structure meets function. *Pharmacol Ther* 1998;**79**:231–57.
67. Shapiro DA, Kristiansen K, Weiner DM, Kroeze WK, Roth BL. Evidence for a model of agonist-induced activation of 5-hydroxytryptamine 2A serotonin receptors that involves the disruption of a strong ionic interaction between helices 3 and 6. *J Biol Chem* 2002;**277**:11441–9.
68. Meltzer HY. The role of serotonin in antipsychotic drug action. *Neuropsychopharmacology* 1999;**21**(2 Suppl):106S–NaN.
69. Harvey JA. Serotonergic regulation of associative learning. *Behav Brain Res* 1996;**73**:47–50.
70. Harvey JA. Role of the serotnin 5-HT(2A) recetor in learning. *Learn Mem.* 2003;**10**:355–62.
71. Charney DS, Woods SW, Goodman WK, Heninger GR. Serotonin function in anxiety. II. Effects of the serotonin agonist MCPP in panic disorder patients and healthy subjects. *Psychopharmacology* 1987;**92**:14–24.
72. Klein E, Zohar J, Geraci MF, Murphy DL, Uhde TW. Anxiogenic effects of m-CPP in patients with panic disorder: comarison to caffeine's anxiogenic effects. *Biol Psychiatry* 1991;**30**:973–84.
73. Kennett GA, Bailey F, Piper DC, Blackburn TP. Effect of SB 200646A, a 5-HT$_{2C}$/5-HT$_{2B}$ receptor antagonist, in two conflict models of anxiety. *Psychopharmacology* 1995;**118**:178–82.
74. Martin JR, Bos M, Jenck F, Moreau J, Mutel V, Sleight AJ, et al. 5-HT$_{2C}$ receptor agonists: pharmacological characteristics and therapeutic potential. *J Pharmacol Exp Ther* 1998;**286**:913–24.
75. Bagdy G, Graf M, Anheuer ZE, Modos EA, Kantor S. Anxiety-like effects induced by acute fluoxetine, sertraline, or m-CPP treatment are reversed by pretreatment with the 5-HT$_{2C}$ receptor antagonist SB-242084 but not the 5-HT$_{1A}$ receptor antagonist WAY-100635. *Int J Neuropsychopharmacol* 2001;**4**:399–408.
76. Milan MJ. Serotonin 5-HT2C receptors as a target for the treatment of depressive and anxious states: focus on novel therapeutic strategies. *Therapie* 2006;**60**:441–60.

77. Grotewiel MS, Sanders-Bush E. Differences in agonist-independent activity of 5-Ht2A and 5-HT2c receptors revealed by heterologous expression. *Naunyn Schmiedebergs Arch Pharmacol* 1999;**359**:21–7.
78. Roth BL, Nakaki T, Chuang DM, Costa E. 5-Hydroxytryptamine2 receptors coupled to phospholipase C in rat aorta: modulation of phosphoinositide turnover by phorbol ester. *J Pharmacol Exp Ther* 1986;**238**:480–5.
79. Berg KA, Clarke WP, Sailstad C, Saltzman A, Maayani S. Signal transduction differences between 5-hydroxytryptamine type 2A and Type 2C receptor systems. *Mol Pharmacol* 1994;**46**:477–85.
80. Berg KA, Stout BD, Maayani S, Clarke WP. Differences in rapid desensitization of 5-hydroxytryptamine2A and 5-hydroxytryptamine2C receptor-mediated phospholipase C activation. *J Pharmacol Exp Ther* 2001;**299**:593–602.
81. Boulougouris V, Glennon JC, Robbins TW. Dissociable effects of selective 5-HT(2A) and 5-HT(2C) receptor antagonists on serial spatial reversal learning in rats. *Neuropsychopharmacology* 2008;**33**:2007–19.
82. Robinson ES, Dalley JW, Theobald DE, Glennon JC, Pezze MA, Murphy ER, et al. Opposing roles for 5-HT(2A) and 5-HT(2C) receptors in the nucleus accumbens on inhibitory response control in the 5-choice serial reaction time task. *Neuropsychopharmacology* 2008;**33**:2398–406.
83. Millan MJ. Serotonin 5-HT2C receptors as a target for the treatment of depressive and anxious states: focus on novel therapeutic strategies. *Therapie* 2005;**60**:441–60.
84. Millan MJ. Multi-target strategies for the improved treatment of depressive states: conceptual foundations and neuronal substrates, drug discovery and therapeutic application. *Pharmacol Ther* 2006;**110**:135–370.
85. Wood MD. Therapeutic potential of 5-HT2C receptor antagonists in the treatment of anxiety disorders. *Curr Drug Targets CNS Neurol Disord* 2003;**2**:383–7.
86. Aly SA, Hossain M, Bhuiyan MA, Nakamura T, Nagatomo T. Assessment of binding affinity to 5-hydroxytryptamine 2A (5-HT2A) receptor and inverse agonist activity of naftidrofuryl: comparison with those of sarpogrelate. *J Pharmacol Sci* 2009;**110**:445–50.
87. Egan C, Herrick-Davis K, Teitler M. Creation of a constitutively activated state of the 5-HT2A receptor by site-directed mutagenesis: revelation of inverse agonist activity of antagonists. *Ann N Y Acad Sci* 1998;**861**:136–9.
88. Teitler M, Herrick-Davis K, Purohit A. Constitutive activity of G-protein coupled receptors: emphasis on serotonin receptors. *Curr Top Med Chem* 2002;**2**:529–38(a)Shapiro DA, Kristiansen DM, Weiner DM, Kroeze WK, Roth BL. Evidence for a model of agonist-induced activation of 5-hydroxytryptamine 2A serotonin receptors that involves the disruption of a strong ionic interaction between helices 3 and 6. *J Biol Chem* 2002;**277**:11441–9.
89. Niswender CM, Copeland SC, Herrick-Davis K, Emeson RB, Sanders-Bush E. RNA editing of the human serotonin 5-hydroxytryptamine 2C receptor silences constitutive activity. *J Biol Chem* 1999;**274**:9472–8.
90. Herrick-Davis K, Grinde E, Niswender CM. Serotonin 5-HT2C receptor RNA editing alters receptor basal activity: implications for serotonergic signal transduction. *J Neurochem* 1999;**73**:1711–7.
91. Berg KA, Stout BD, Cropper JD, Maayani S, Clarke WP. Novel actions of inverse agonists on 5-HT2C receptor systems. *Mol Pharmacol* 1999;**55**:863–72.
92. Berg KA, Navailles S, Sanchez TA, Silva YM, Wood MD, Spampinato U, et al. Differential effects of 5-methyl-1-[[2-[(2-methyl-3-pyridyl)oxyl]-5-pyridyl]carbamoyl]-6-trifluoro methylindone (SB 243213) on 5-hydroxytryptamine(2C) receptor-mediated responses. *J Pharmacol Exp Ther* 2006;**319**:260–8.

93. Barker EL, Westphal RS, Schmidt D, Sanders-Bush E. Constitutively active 5-hydroxytryptamine2C receptors reveal novel inverse agonist activity of receptor ligands. *J Biol Chem* 1994;**269**:11687–90.
94. Wilbanks AM, Laporte SA, Bohn LM, Barak LS, Caron MG. Apparent loss-of-function mutant GPCRs revealed as constitutively desensitized receptors. *Biochemistry* 2002;**41**:11981–9.
95. Gainetdinov RR, Premont RT, Bohn LM, Lefkowitz RJ, Caron MG. Desensitization of G protein-coupled receptors and neuronal functions. *Annu Rev Neurosci* 2004;**27**:107–44.
96. Westphal RS, Backstrom JR, Sanders-Bush E. Increased basal phosphorylation of the constitutively active serotonin 2C receptor accompanies agonist-mediated desensitization. *Mol Pharmacol* 1995;**48**:200–5.
97. Werry TD, Loiacono R, Sexton PM, Christopoulos A. RNA editing of the serotonin 5HT2C receptor and its effects on cell signalling, pharmacology and brain function. *Pharmacol Ther* 2008;**119**:7–23.
98. Fitzgerald LW, Iyer G, Conklin DS, Krause CM, Marshall A, Patterson JP, et al. Messenger RNA editing of the human serotonin 5-HT2C receptor. *Neuropsychopharmacology* 1999;**21**:82S–90S. (a)Burns CM, Chu H, Rueter SM, Hutchinson LK, Canton H, Sanders-Bush E, Emeson E. Regulation of serotonin-2C receptor G-protein coupling by RNA editing. *Nature* 1997;**387**:303–8.
99. Berg KA, Dunlop J, Sanchez T, Silva M, Clarke WP. A conservative, single-amino acid substitution in the second cytoplasmic domain of the human Serotonin2C receptor alters both ligand-dependent and -independent receptor signaling. *J Pharmacol Exp Ther* 2008;**324**:1084–92.
100. Bourne HR. How receptors talk to trimeric G proteins. *Curr Opin Cell Biol* 1997;**9**:134–42.
101. Gudermann T, Schoneberg T, Schultz G. Functional and structural complexity of signal transduction via G-protein-coupled receptors. *Annu Rev Neurosci* 1997;**20**:399–427.
102. Wess J. Molecular basis of receptor/G-protein-coupling selectivity. *Pharmacol Ther* 1998;**80**:231–64.
103. Wang Q, O'Brien PJ, Chen CX, Cho DS, Murray JM, Nishikura K. Altered G protein-coupling functions of RNA editing isoform and splicing variant serotonin2C receptors. *J Neurochem* 2000;**74**:1290–300.
104. Berg KA, Cropper JD, Niswender CM, Sanders-Bush E, Emeson RB, Clarke WP. RNA-editing of the 5-HT(2C) receptor alters agonist-receptor-effector coupling specificity. *Br J Pharmacol* 2001;**134**:386–92.
105. Gether U, Ballesteros JA, Seifert R, Sanders-Bush E, Weinstein H, Kobilka BK. Structural instability of a constitutively active G protein-coupled receptor. Agonist-independent activation due to conformational flexibility. *J Biol Chem* 1997;**272**:2587–90.
106. Samama P, Bond RA, Rockman HA, Milano CA, Lefkowitz RJ. Ligand-induced overexpression of a constitutively active beta2-adrenergic receptor: pharmacological creation of a phenotype in transgenic mice. *Proc Natl Acad Sci USA* 1997;**94**:137–41.
107. Solomon PR, Vander Schaaf ER, Norbe AC, Weisz DJ, Thompson RF. Hippocampus and trace conditioning of the rabbit's nictitating response. *Behav Neurosci* 1986;**100**:729–44.
108. Gibbs CM, Powell DA. Single-unit activity in the dorsomedial prefrontal cortex during expression of discriminative bradycardia in rabbits. *Behav Brain Res* 1991;**43**:79–92.
109. McEchron MD, Disterhoft JF. Sequence of single neuron changes in CA1 hippocampus of rabbits during acquisition of trace eyeblink conditioned responses. *J Neurophysiol* 1997;**78**:1030–44.
110. Kronforst-Collins MA, Disterhoft JF. Lesions of the caudal area of rabbit medial prefrontal cortex impair trace eyeblink conditioning. *Neurobiol Learn Mem* 1998;**69**:147–62.
111. Clarke RE, Squire LR. Classical conditioning and brain systems: the role of awareness. *Science* 1998;**280**:77–81.

112. LaBar KS, Disterhoft JF. Conditioning, awareness, and the hippocampus. *Hippocampus* 1998;**8**:620–6.
113. Romano AG. Variations in CS associability and multiple unit hippocampal activity in the rabbit. *Behav Brain Res* 1999;**103**:163–73.
114. Aloyo VJ, Harvey JA. Antagonist binding at 5-HT2_A and 5-HT2_C receptors in the rabbit: high correlation with the profile for the human receptors. *Eur J Pharmacol* 2000;**406**:163–9.
115. Gimpl MP, Gormezano I, Harvey JA. Effects of LSD on learning as measured by classical conditioning of the rabbit nictitating membrane response. *J Pharmacol Exp Ther* 1978;**208**:330–4.
116. Siegel S, Freedman DX. Effects of LSD-25 on classical trace conditioning. *Pharmacol Biochem Behav* 1988;**30**:427–31.
117. Welsh SE, Kachelries WJ, Romano AG, Simansky KJ, Harvey JA. Effects of LSD, ritanserin, 8-OH-DPAT and lisuride on classical conditioning in the rabbit. *Pharmacol Biochem Behav* 1998;**59**:469–75.
118. Harvey JA, Gormezano I, Cool Valerie A. Effects of d-lysergic acid diethylamide, d-2-bromolysergic acid diethylamide, d, 1-2, 5-dimethoxy-4-methylamphetamine and d-amphetamine on classical conditioning of the rabbit nictitating membrane response. *J Pharmacol Exp Ther* 1982;**221**:289–94.
119. Romano AG, Hood H, Harvey JA. Dissociable effects of the 5-HT$_2$ antagonist mianserin on associative learning and performance in the rabbit. *Pharmacol Biochem Behav* 2000;**67**:103–10.
120. Welsh SE, Romano AG, Harvey JA. Effects of serotonin 5-HT$_{2A/2C}$ antagonists on associative learning in the rabbit. *Psychopharmacology* 1998;**137**:157–63.
121. Romano AG, Bormann NM, Harvey JA. A unique enhancement of associative learning produced by methylenedioxyamphetamine. *Behav Pharmacol* 1991;**2**:225–31.
122. Romano AG, Harvey JA. MDMA enhances associative and nonassociative learning in the rabbit. *Pharmacol Biochem Behav* 1994;**47**:289–93.
123. Ginn SR, Powell DA. Pizotifen attenuates classical eyeblink and heart rate conditioning in rabbits. *Physiol Psychol* 1986;**14**:36–41.
124. Newman-Tancredi A, Cussac D, Quentric Y, Tauzard M, Verriele L, Carpentier N, et al. Differential actions of antiparkinson agents at multiple classes of monoaminergic receptor: III. Agonist and antagonist properties at serotonin, 5-HT(1) and 5-HT(2), receptor subtypes. *J Pharmacol Exp Ther* 2002;**303**:815–22.
125. Sharp T, Bramwell SR, Grahame-Smith DG. 5-HT$_1$ agonists reduce 5-hydroxytryptamine release in rat hippocampus *in vivo* as determined by brain microdialysis. *Br J Pharmacol* 1989;**98**:989–97.
126. Sharp T, Bramwell SR, Hjorth S, Grahame-Smith DG. Pharmacological characterization of 8-OH-DPAT-induced inhibition of rat hippocampal 5-HT release *in vivo* as measured by microdialysis. *Br J Pharmacol* 1989;**96**:283–90.
127. Romano AG, Quinn JL, Liu R, Dave KD, Schwab D, Alexander G, et al. Effect of serotonin depletion on 5-HT$_{2A}$ mediated learning: evidence for constitutive activity of the 5-HT$_{2A}$ receptor *in vivo*. *Psychopharmacology* 2005;**184**:173–81.
128. Harvey JA, Welsh SE, Hood H, Romano AG. Effects of 5-HT$_2$ receptor antagonists on a cranial nerve reflex in the rabbit: evidence for inverse agonism. *Psychopharmacology* 1999;**141**:162–8.
129. Harvey JA, Gormezano I, Cool-Hauser VA, Schindler CW. Effects of LSD on classical conditioning as a function of CS-UCS interval: relationship to reflex facilitation. *Pharmacol Biochem Behav* 1988;**30**:433–41.
130. Gobert A, Rivet JM, Lejeune F, Newman-Tancredi A, Adhumeau-Auclair A, Nicolas JP, et al. Serotonin(2C) receptors tonically suppress the activity of mesocortical dopaminergic and adrenergic, but not serotonergic, pathways: a combined dialysis and electrophysiological analysis in the rat. *Synapse* 2000;**36**:205–21.

131. De Deurwaerdère P, Spampinato U. The nigrostriatal dopamine system: a neglected target for 5-HT_{2C} receptors. *Trends Pharmacol Sci* 2001;**22**:502–4.
132. Aloyo VJ, Berg KA, Spampinato U, Clarke WP, Harvey JA. Current status of inverse agonism at serotonin(2A) (5-HT(2A)) and 5-HT(2C) receptors. *Pharmacol Ther* 2009;**121**:160–73.
133. Navailles S, De Deurwaerdère P, Spampinato U. Clozapine and haloperidol differentially alter the constitutive activity of central serotonin2C receptors *in vivo*. *Biol Psychiatry* 2006;**59**:568–75.
134. Leggio GM, Cathala A, Neny M, Rouge-Pont F, Drago F, Piazza PV, et al. *In vivo* evidence that constitutive activity of $serotonin_{2C}$ receptors in the medial prefrontal cortex participates in the control of dopamine release in the rat nucleus accumbens: differential effects of inverse agonist *versus* antagonist. *J Neurochem* 2009;**111**:614–23.
135. Navailles S, Moison D, Ryczko D, Spampinato U. Region-dependent regulation of mesoaccumbens dopamine neurons in vivo by the constitutive activity of central serotonin2C receptors. J Neurochem. 2006;**99**:1311–319.
136. Navailles S, Moison D, Cunningham KA, Spampinato U. Differential regulation of the mesoaccumbens dopamine circuit by serotonin2C receptors in the ventral tegmental area and the nucleus accumbens: an *in vivo* microdialysis study with cocaine. *Neuropsychopharmacology* 2008;**33**:237–46.
137. Leggio GM, Cathala A, Moison D, Cunningham KA, Piazza PV, Spampinato U. Serotonin(2C) receptors in the medial prefrontal cortex facilitate cocaine-induced dopamine release in the rat nucleus accumbens. *Neuropharmacology* 2009;**56**:507–13.
138. Berg KA, Harvey JA, Spampinato U, Clarke WP. Physiological relevance of constitutive activity of 5-HT_{2A} and 5-HT_{2C} receptors. *Trends Pharmacol Sci* 2005;**26**:625–30.
139. Costa T, Cotecchia S. Historical review: negative efficacy and the constitutive activity of G-protein-coupled receptors. *Trends Pharmacol Sci* 2005;**26**:618–24.
140. Müller CP, Carey RJ, Huston JP, De Souza Silva MA. Serotonin and psychostimulant addiction: focus on 5-HT1A-receptors. *Prog Neurobiol* 2007;**81**:133–78.
141. Para S, Bond RA. Inverse agonism: from curiosity to accepted dogma, but is it clinically relevant? *Curr Opin Pharmacol* 2007;**7**:146–50.
142. Berg KA, Clarke WP, Cunningham KA, Spampinato U. Fine-tuning $serotonin_{2C}$ receptor function in the brain: molecular and functional implications. *Neuropharmacology* 2008;**55**:969–76.
143. Berg KA, Harvey JA, Spampinato U, Clarke WP. Physiological and therapeutic relevance of constitutive activity of 5-HT2A and 5-HT2C receptors for the treatment of depression. *Prog Brain Res* 2008;**172**:287–305.
144. Meltzer HY, Mills R, Revell S, Williams H, Johnson A, Bahr D, et al. Pimavanserin, a serotonin(2A) receptor inverse agonist, for the treatment of parkinson's disease psychosis. *Neuropsychopharmacology* 2010;**35**:881–92.
145. Butler H, Korbonits M. Cannabinoids for clinicians: the rise and fall of the cannabinoid antagonists. *Eur J Endocrinol* 2009;**161**:655–62.
146. Kogan NM, Mechoulan R. The chemistry of endocannabinoids. *J Endocrinol Invest* 2006;**29**:2–14.
147. Vandevoorde S, Lambert DM. Focus on the three key enzymes hydrolyzing endocannabinoids as new drug targets. *Curr Pharm Des* 2005;**11**:2647–68.
148. Prather PL, Martin NA, Breivogel CS, Childers SR. Activation of cannabinoid receptors in rat brain by WIN 55212-2 produces coupling to multiple G protein α-subunits with different potencies. *Mol Pharmacol* 2000;**57**:1000–10.
149. Glass M, Northup JK. Agonist selective regulation of G proteins by cannabinoid CB_1 and CB_2 receptors. *Mol Pharmacol* 1999;**56**:1362–9.

150. Rinaldi-Carmona M, Barth F, Héaulme M, Shire D, Calandra B, Congy C, et al. SR141716A, a potent and selective antagonist of the brain cannabinoid receptor. *FEBS Lett* 1994;**350**:240–4.
151. Bouaboula M, Perrachon S, Milligan L, Canat X, Rinaldi-Carmona M, Portier M, et al. A selective inverse agonist for central cannabinoid receptor inhibits mitogen-activated protein kinase activation stimulated by insulin or insulin-like growth factor 1: evidence for a new model of receptor/ligand interactions. *J Biol Chem* 1997;**272**:22330–9.
152. Pan X, Ikeda SR, Lewis DL. SR 141716A acts as an inverse agonist to increase neuronal voltage-dependent Ca^{2+} currents by reversal of tonic CB1 cannabinoid receptor activity. *Mol Pharmacol* 1998;**54**:1064–72.
153. Fong TM, Guan X-M, Marsh DJ, Shen C-P, Stribling DS, Rosko KM, et al. Antiobesity efficacy of a novel cannabinoid-1 receptor inverse agonist, *N*-[(1*S*,2*S*)-3-(4-chlorophenyl)-2-(3-cyanophenyl)-1-methylpropyl]-2-methyl-2-{[5-(trifluoromethyl)pyridin-2-yl]oxy}propanamide (MK-0364), in rodents. *J Pharmacol Exp Ther* 2007;**321**:1013–22.
154. Heimann AS, Gomes I, Dale CS, Pagano RL, Gupta A, de Souza LL, et al. Hemopressin is an inverse agonist of CB1 cannabinoid receptors. *Proc Natl Acad Sci USA* 2007;**194**:20588–93.
155. Fitton AG, Pertwee RG. Changes in body temperature and oxygen consumption rate of conscious mice produced by intrahypothalamic and intracerebroventricular injections of delta 9-tetrahydrocannabinol. *Br J Pharmacol* 1982;**75**:409–14.
156. Chambers AP, Vemuri VK, Peng Y, Wood JT, Olszewska T, Pittman QJ, et al. A neutral CB_1 receptor antagonist reduces weight gain in rat. *Am J Physiol Regul Integr Comp Physiol* 2007;**293**:R2185–93.
157. Sink KS, McLaughlin PJ, Wood JA, Brown C, Fan P, Vemuri VK, Peng Y, Olszewska T, Thakur A, Makriyannis A, Parker LA, Salamone JD. The novel cannabinoid CB1 receptor neutral antagonist AM4113 suppresses food intake and food-reinforced behavior but does not induce signs of nausea in rats. *Neuropsychopharmacology* 2008;**33**:946–55.
158. Hurst DP, Lynch DL, Barnett-Norris J, Hyatt SM, Seltzman HH, Zhong M, et al. *N*-(piperidin-1-yl)-5-(4-chlorophenyl)-1-(2,4-dichlorophenyl)-4-methyl-1*H*-pyrazole-3-carboxamide (SR141716A) interaction with LYS 3.28(192) is crucial for its inverse agonism at the cannabinoid CB1 receptor. *Mol Pharmacol* 2002;**62**:1274–87.
159. Hurst D, Umejiego U, Lynch D, Seltzman H, Hyatt S, Roche M, et al. Biarylpyrazole inverse agonists at the cannabinoid CB1 receptor: importance of the C-3 carboxamide oxygen/lysine3.28(192) interaction. *J Med Chem* 2006;**49**:5969–87.
160. Thomas A, Ross RA, Saha B, Mahadevan A, Razdan RK, Pertwee RG. 6"-Azidohex-2"-yne-cannabidiol: a potential neutral, competitive cannabinoid CB1 receptor antagonist. *Eur J Pharmacol* 2004;**487**:213–21.
161. Pertwee RG. Inverse agonism and neutral antagonism at cannabinoid CB1 receptors. *Life Sci* 2005;**76**:1307–24.
162. Arnone M, Maruani J, Chaperon F, Thiebot M-H, Poncelet M, Soubrie P, et al. Selective inhibition of sucrose and ethanol intake by SR 141716, an antagonist of central cannabinoid (CB1) receptors. *Psychopharmacology* 1997;**132**:104–6.
163. Colombo G, Agabio R, Diaz G, Lobina C, Reali R, Gessa GL. Appetite suppression and weight loss after the cannabinoid antagonist SR 141716. *Life Sci* 1998;**63**:PL113–7.
164. Chen RZ, Huang R-R C, Shen C-P, MacNeil DJ, Fong TM. Synergistic effects or cannabinoid inverse agonist AM251 and opioid antagonist nalmefene on food intake in mice. *Brain Res* 2004;**999**:227–30.
165. Wiley JL, Burston JJ, Leggett DC, Alekseeva OO, Razdan RK, Mahadevan A, et al. CB_1 cannabinoid receptor-mediated modulation of food intake in mice. *Br J Pharmacol* 2005;**145**:293–300.
166. Wustrow DJ, Maynard GD, Yuan J, Zhao H, Mao J, Guo Q, et al. Aminopyrazine CB1 receptor inverse agonists. *Bioorg Med Chem Lett* 2008;**18**:3376–81.

167. Pi-Sunyer FX, Aronne LJ, Heshmati HM, Devin J, Rosenstock J. Effect of rimonabant, a cannabinoid-1 receptor blocker, on weight and cardiometabolic risk factors in overweight or obese patients: RIO-North America: a randomized controlled trial. *JAMA* 2006;**295**:761–75.
168. Van Gaal LF, Scheen AJ, Rissanen AM, Rossner S, Hanotin C, Ziegler O. Long-term effect of CB1 blockade with rimonabant on cardiometabolic risk factors: two year results from the RIO-Europe Study. *European Heart J* 2008;**29**:1761–71.
169. Rucker D, Padwal R, Li SK, Curioni C, Lau DCW. Long term pharmacotherapy for obesity and overweight: updated meta-analysis. *BMJ* 2007;**335**:1194–9.
170. Christensen R, Kristensen PK, Bartels EM, Bliddal H, Astrup A. Efficacy and safety of the weight-loss drug rimonabant: a meta-analysis of randomised trials. *Lancet* 2007;**370**:1706–13.
171. Lefoll B, Gorelick DA, Goldberg SR. The future of endocannabinoid-oriented clinical research after CB_1 antagonists. *Psychopharmacology* 2009;**205**:171–4.
172. Gardner A, Mallet PE. Suppression of feeding, drinking, and locomotion by a putative cannabinoid receptor 'silent antagonist'. *Eur J Pharmacol* 2006;**530**:103–6.
173. Janero DR, Makriyannis A. Cannabinoid receptor antagonists: pharmacological opporunities, clinical experience, and translational prognosis. *Expert Opin Emerg Drugs* 2009;**14**:43–65.

G Protein-Coupled Receptor Heteromers as New Targets for Drug Development

SERGI FERRÉ,* GEMMA NAVARRO,[†] VICENT CASADÓ,[†] ANTONI CORTÉS,[†] JOSEFA MALLOL,[†] ENRIC I. CANELA,[†] CARME LLUÍS,[†] AND RAFAEL FRANCO[†,‡]

*National Institute on Drug Abuse, IRP, NIH, DHHS, Baltimore, Maryland, USA

[†]CIBERNED, Department of Biochemistry and Molecular Biology, Faculty of Biology, University of Barcelona, Barcelona, Spain

[‡]CIMA Neurociencias, Avda Pio XII 55, Pamplona, Spain

We now have a significant amount of experimental evidence that indicates that G protein-coupled receptor (GPCR) oligomerization, including homo- and heteromerization, is a general phenomenon. Receptor heteromers possess unique biochemical characteristics that are demonstrably different from those of its individual units. These properties include allosteric modulation(s) between units, changes in ligand recognition, G protein-coupling and trafficking. The discovery of GPCR oligomers have been related to the parallel discovery and application of a variety of resonance energy transfer (RET) techniques, such as bioluminescence, fluorescence and sequential RET (BRET, FRET and SRET, respectively), time-resolved FRET (T-FRET) and fluorescence recovery after photobleaching (FRAP) microscopy. However, RET techniques are difficult to implement in native tissues. For receptor heteromers, indirect approaches, such as the determination of a unique biochemical characteristic ("biochemical fingerprint"), permit their identification in native tissues and their use as targets for drug development. Dopamine and opioid receptor heteromers are the focus of intense research which is related to the possible multiple applications of their putative ligands in pathological conditions, which include basal ganglia disorders, schizophrenia and drug addiction.

Progress in Molecular Biology
and Translational Science, Vol. 91
DOI: 10.1016/S1877-1173(10)91002-8

I. Unique Biochemical Characteristics of GPCR Heteromers

G protein-coupled receptors (GPCRs) are classically believed to be functional monomeric entities. In fact, recent studies have shown that monomers of class A GPCRs (adrenergic β_2, rhodopsin, and opioid μ receptors), when reconstituted in lipid vesicles, couple and activate their respective G proteins upon agonist binding.[1–3] Also monomeric rhodopsin in solution can activate its G protein transducin.[4] Nevertheless, we now have a significant amount of experimental evidence that indicates that GPCR oligomerization, including homo- and heteromerization, is a general phenomenon (see below) and, in fact, it still needs to be determined if GPCR monomers are functionally present in the cellular plasma membrane.

There is some confusion about the terms used to describe GPCR oligomers. One of the main reasons has been the lack of knowledge about what constitutes a receptor unit. We should first consider the current definition of "receptor," which is, "a signal transducing unit, a cellular macromolecule or an assembly of macromolecules that is concerned directly and specifically with chemical signaling between and within cells."[5] It is therefore redundant to talk about a "functional receptor." Recommendations for definitions and criteria for identification of receptor heteromers have recently been proposed.[5] First we should make a distinction between "heteromeric receptor" or "homomeric receptor" and "receptor heteromer" or "receptor homomer." A heteromeric receptor is a "dimeric or oligomeric receptor for which the minimal functional unit is composed of two or more different subunits that are not functional on their own." This definition applies to ligand-gated ion channels (ionotropic receptors) such as glutamate *N*-methyl-D-aspartate (NMDA) receptors and most nicotinic acetylcholine receptors.[6,7] The term also applies to some tyrosine kinase receptors, such as receptors for the glial cell line-derived neurotrophic factor (GDNF) family of ligands, in which one subunit is responsible for the association with the ligand and the other subunit for the catalytic response.[8] Furthermore, there are also heteromeric GPCRs, such as the $GABA_B$ receptor, which is composed of at least two seven-transmembrane (7TM) protein units, $GABA_{B1}$ and $GABA_{B2}$, which constitute a ("functional") receptor. Thus, the $GABA_{B1}$ subunit is responsible for ligand binding and the $GABA_{B2}$ subunit is responsible for G protein activation and signaling.[9] In this case we have complete asymmetry in the functioning of both 7TM units in the heteromeric receptor. Some taste receptors, for which genetic deletion of one of the subunits leads to suppression of the receptor function,[9] should also be called heteromeric GPCRs. If the receptor subunits are identical, they constitute a "homomeric receptor." This is the case for some ionotropic receptors, such as the α7 nicotinic acetylcholine receptor.[7]

On the other hand, a receptor heteromer is "a macromolecular complex composed of at least two ("functional") receptor units with biochemical properties that are demonstrably different from those of its individual components."[5] Thus, a receptor homomer is the same as receptor heteromer, but combining two or more identical (functional) receptor units. In a recent study, using a functional complementation assay, D_2 receptor homodimers with a single G protein were suggested to be a minimal signaling unit (homodimeric D_2 receptors?), which is maximally activated by agonist binding to one of the protomer, whereas additional agonist or inverse agonist binding to the second promoter blunts or enhances signaling, respectively.[10] This allosteric modulation of signaling results from a direct interaction of the receptor homodimer (or homodimeric receptor?) with the G protein, rather than from a downstream effect.[10] A similar situation of two (functional?) protomers and one G protein is most probably found in a receptor heterodimer, but in this case, two different ligands interact in the heteromer, and allosteric interactions in the receptor heteromer have also been described.[5] It is currently believed that an allosteric interaction in a receptor heteromer involves an intermolecular interaction in which binding of a ligand to one of the protomers changes the binding properties of another promoter.[5,11,12] The same is believed to occur in receptor homomers, which is translated into either a positive or a negative cooperativity in ligand–receptor binding.[13,14]

In addition to allosteric modulations, GPCR oligomerization involves changes in ligand recognition, G protein-coupling and trafficking. Opioid receptor heteromers constitute an example of such changes in ligand recognition at the receptor heteromer. The opioid receptor subfamily comprises μ, δ, and κ receptors, but at least two pharmacological "subtypes" with different affinities for different ligands have been identified for each of the three cloned receptors.[15] The δ receptor can form heteromers with the μ[16,17] and also with the κ receptors.[18] It seems that the two pharmacological subtypes of δ receptors, δ_1 and δ_2 receptors, correspond to a δ–μ receptor heteromer and a δ receptor homomer, respectively.[19] With regard to changes in G protein-coupling, taking into account the fact that recent models support the binding of only one G protein to two receptor units,[10,20] this means that a receptor heteromer will at least have to "decide" which G protein it should bind to as these receptors are usually coupled to different G proteins. Dopamine receptors are classified as D_1-like, with the D_1 and D_5 receptor subtypes, which usually couple to $G_{s/olf}$ proteins; and D_2-like, with the D_2 and D_4 receptor subtypes, which couple to $G_{i/o}$ proteins.[21] Histamine H_3 receptors also couple to the $G_{i/o}$ protein when not forming heteromers.[21] Recent studies have shown that the D_1–D_3 receptor heteromer couples to the $G_{s/olf}$ protein and the D_1–H_3 receptor heteromer couples to the $G_{i/o}$ protein.[22,23] In some instances, the receptor heteromer can "choose" a completely new G protein. Thus, the D_1–D_2 and D_2–D_5 heteromers couple preferentially to G_q proteins.[24,25]

There is experimental evidence that indicates that most GPCR oligomers form in the endoplasmic reticulum (ER) and that their synthesis is a ligand-independent process. In fact, it has been shown that in some cases dimerization/oligomerization allows the 7TM protein or GPCR (depending on whether we are dealing with a potential subunit of a homomeric or heteromeric receptor or a unit of a receptor homomer or heteromer) to escape from its conditional retention in the ER. When expressed alone, the $GABA_{B1}$ subunit is retained in the ER because of the presence of an ER-retention sequence in the intracellular C-terminal tail. Coexpression of the $GABA_{B2}$ subunit results in an intermolecular interaction between both subunits that masks the retention motif and allows surface delivery of the heteromeric $GABA_B$ receptor.[9] The same situation occurs with some adrenergic receptors, such as the adrenergic α_{1D} receptor, which needs to heteromerize with α_{1B} in order to be transported to the cell membrane.[26] ER-retention motifs can, in fact, be used to demonstrate heteromerization. For instance, the cell surface delivery of the chemokine CXC_1 receptor was impaired when it was fused to an ER-retention motif of the adrenergic α_{2C} receptor, but ER retention was prevented by cotransfection with the close-related chemokine CXC_2 receptor.[27] Once on the cell surface, many GPCRs internalize spontaneously or upon agonist binding. Indeed, the capacity of a selective ligand of one GPCR to cause cointernalization of a second coexpressed GPCR that does not bind the same ligand is used, experimentally, as evidence for heteromerization. For instance, in cells coexpressing chemokine CXC_4 and CC_5 receptors, their respective ligands (CXCL12 and CCL5, respectively) induced cointernalization of both receptors.[28]

II. Energy Transfer-Based Techniques to Study GPCR Oligomerization

The use of the energy transfer-based techniques, bioluminescence resonance energy transfer (BRET) and fluorescence resonance energy transfer (FRET) has been fundamental in taking the issue of GPCR oligomerization to the forefront of GPCR research, providing evidence of the presence of an increasing number of GPCR receptor heteromers in living cells.[29–31] RET involves nonradioactive transfer of energy from a chromophore in an excited state, the "donor," to a fluorescent "acceptor" molecule. In FRET both molecules are fluorescent, whereas in BRET the donor molecule is an enzyme which becomes bioluminescent upon catalyzation of its substrate. Importantly, the efficacy of energy transfer is inversely proportional to the sixth power of the distance between the donor and acceptor chromophores. This implies that range of energy transfer from the donor to the acceptor is below 10 nm.[29–31]

When using this technology to study GPCR oligomerization, the donor and acceptor molecules are usually fused to the C-termini of GPCRs and cotransfected in mammalian cell lines. Since the diameter of the seven TM helical core is estimated to be about 5 nm, a positive energy transfer strongly suggests that oligomerization has occurred.[29–31]

It is important to point out that BRET and FRET do not demonstrate the existence of physical contact between the fused proteins, but only show that there exists a very close proximity, which could of course depend on oligomerization. Nevertheless, when using BRET, physical contact can be practically shown by using several different experimental approaches. Those possibilities include BRET saturation assays, BRET competition assays and experiments that show ligand-promoted changes in BRET.[32] In BRET saturation experiments, a constant amount of the donor fusion protein (usually a receptor fused to *Renilla Luciferase* (RLuc)) is coexpressed with increasing amounts of the acceptor fusion protein (usually a receptor fused to yellow fluorescent protein (YFP)). If there is oligomerization, saturation is reached when all receptor-Rluc molecules are specifically associated with their receptor-YFP counterparts. By contrast, if the BRET signal results from random collision promoted by high receptor density, a quasi-linear curve is obtained. In BRET competition assays, an untagged receptor A or B is coexpressed with the receptor A-Rluc and receptor B-YFP. Theoretically, if there is receptor A-receptor B oligomerization, then the BRET signal is specific and decreases as a consequence of the untagged receptor interacting with one of the fusion proteins and competing for the complementary BRET fusion protein. Finally, in some cases, when receptor A-Rluc and receptor B-YFP are coexpressed, changes in BRET signal due to ligand binding to the heteromer can be detected, which can be indicative of ligand-induced conformational changes in the heteromer. This has been, for instance, observed in experiments with melatonin (MT) receptor heteromers.[33] FRET-based methods have also been introduced to monitor signal transfer within a receptor heteromer upon ligand binding.[34] These techniques detect movements (conformational changes) within a GPCR homo- or heteromer by inserting fluorophores, for instance, cyan fluorescent protein (CFP) and YFP, in intracellular loops of both or just one of the protomers (intramolecular FRET and intermolecular FRET). A nice example is the recently reported application of this technique to the study of adrenergic α_{2A}-opioid μ receptor heteromer. In this case, when the double-inserted α_{2A} receptor was coexpressed with the μ receptor, the α_{2A} receptor decreases its response to noradrenaline in the presence of morphine.[34] This inhibitory effect was extremely fast (half-life of less than 500 ms), strongly suggesting that it was due to direct protein–protein interaction rather than competition for G-protein subunits.[34]

Sequential-BRET-FRET (SRET) has been introduced to identify oligomers formed by three different proteins.[35] In SRET, the oxidation of an RLuc substrate by an RLuc-fusion protein triggers the excitation of the acceptor GFP^2 by $BRET^2$ and subsequent energy transfer to the acceptor YFP by FRET.[35] $BRET^2$ differs from $BRET^1$ in the acceptor, GFP^2 instead of YFP, and the substrate for Rluc which is DeepBlue C instead of coelenterazine. This technique has been used to demonstrate the heteromultimerization between adenosine A_{2A}, dopamine D_2 and cannabinoid CB_1 receptors[35] and between A_{2A}, D_2 and glutamate $mGlu_5$ receptors.[36] The presence of A_{2A}–CB_1–D_2 and A_{2A}–D_2–$mGlu_5$ heteromers was also demonstrated by a combination of bimolecular fluorescence complementation (BiFC) and BRET techniques.[36,37] BiFC is based on the principle that a fluorescent complex from nonfluorescent constituents can be produced if a protein–protein interaction occurs. In this technique, two receptors are fused at their C termini with either the N-terminal or the C-terminal fragments of YFP, respectively; and receptor heterodimerization leads to YFP reconstitution. Then, if there is heterotrimerization, BRET can be obtained when the cells also coexpress the third receptor, which is fused to Rluc.[36,37]

The use of acceptor and donor molecules that are genetically fused to GPCRs in the classical FRET and BRET approaches can alter the functionality of the receptor heteromer under study. Furthermore, the fusion proteins can also be expressed in intracellular compartments, making it difficult to demonstrate that the resonance energy transfer resulted from a direct interaction of proteins at the cell surface. Finally, an additional limitation of the classical FRET is the low signal-to-noise ratio resulting from the intrinsic fluorescence of the cells and the overlap between the emission spectra of FRET donors and acceptors. Time-resolved FRET (TR-FRET) is a FRET variant designed to circumvent these problems.[38] TR-FRET is based on the prolonged fluorescence characteristics of certain rare earth tracers such as the lanthanide compounds, which enable the processes of excitation and detection to be separated temporally. Upon excitation, lanthanide compounds such as europium cryptate show a long-lived emission fluorescence, which can induce a long-lived emission from the acceptor molecule (e.g., Alexa Fluor 647, DY-647, or d2). As lanthanide compounds have very low fluorescence emission at the emission wavelength of the acceptor, the signal-to-noise ratio increases enormously. In TR-FRET, the acceptor and the donor fluorophore molecules are usually conjugated to antibodies against N-terminal-epitope-tagged sequences and membrane-impermeant antibodies are used to exclusively assess cell surface GPCR oligomerization. However, the bivalent nature of antibodies could potentially stabilize large complexes and the large size, and multiple labeling of the antibodies can easily increase FRET due to random collision. A combination of TR-FRET with snap-tag technology has been recently introduced.[39]

A snap tag, which is two-thirds the size of YFP, is introduced at the N-terminal end of a GPCR. The snap tag is derived from O^6-guanine alkyltransferase, which covalently reacts with benzyl guanine, and the nonpermeant benzyl guanine derivatives which are covalently labeled with a fluorophore compatible with TR-FRET measurements are used. This technology has been used to demonstrate the existence of homomers of the heteromeric $GABA_B$ receptor and to confirm the predominant arrangement of glutamate mGlu receptors as homodimers.[39]

Finally, another recently introduced new RET methodology to study GPCR oligomerization is called fluorescence recovery after photobleaching (FRAP) microscopy.[40] In this technique, one receptor carries YFP at its extracellular N-terminus and another receptor carries CFP at its intracellular C-terminus. An area of the cell membrane is bleached and the exponential time-course of fluorescence recovery is monitored. The binding of an antibody to the YFP-tag markedly impairs fluorescence recovery of receptor-YFP, but does not modify recovery of receptor-CFP, unless they are in close proximity. This technique can provide information about the dynamics of the interactions between protomers in GPCR oligomers.[40] Thus, a moderate delay or a lack of delay in fluorescence recovery upon immobilization of a partner receptor indicates either a transient or a stable interaction, respectively. For instance, the results of Dorsch et al. indicate that the adrenergic β_1 and β_2 receptors form transient and robust homomers, respectively.[40]

III. Receptor Heteromers as Pharmacological Targets

Receptor heteromers, in particular, open up many opportunities in pharmacology, since they constitute new targets for drug development. An important step in this field has been the identification of receptor heteromers in native tissues. RET techniques can be applied only when studying receptor heteromerization in artificial cell systems. Direct identification of heteromerization can be achieved by taking advantage of selective probes (e.g., specific antibodies or labeled selective ligands) that can discriminate between the receptor heteromer and other configurations of the individual components. Bivalent ligands, with selective binding for each of the two protomers connected by a spacer of variable length, can become a good strategy, as was recently reported in the study of the A_{2A}–D_2 receptor heteromer.[41] No specific receptor heteromer antibodies have yet been reported. An indirect but valid strategy is to determine a "biochemical fingerprint" of the receptor heteromer, that is a biochemical characteristic of a receptor heteromer, which can be used for its identification in native tissue.[5] A strong suggestion that a biochemical fingerprint is specific for a receptor heteromer can be obtained by showing that it is abolished or altered when the heteromerization is disrupted, or alternatively when the quaternary structure of the heteromer is significantly modified without disrupting heteromerization. This, of course, could be shown with RET in artificial cell systems but requires identification of the domains or epitopes (of at least one of the receptors) that form the interaction surface in the heteromer. This is a new area of research which uses computation techniques and bioinformatics.[42] These studies can guide experiments using mutated/chimeric receptors or enable design of peptides that can selectively occupy and disrupt the receptor heteromer interface.[5]

Opioid and dopamine receptor heteromers are presently the focus of intense research, which is related to the possible multiple applications of their putative ligands in pathological conditions, which include basal ganglia disorders, schizophrenia, drug addiction and pain. In some cases, the allosteric interactions in the receptor heteromer can be used as a strategy to modify the effect of already existing drugs. The best example is the addition of an A_{2A} receptor antagonist as a means to potentiate the antiparkinsonian effects of L-dopa.[43] A_{2A}–D_2 receptor heteromerization has been demonstrated in mammalian transfected cells with coimmunoprecipitation and FRET and BRET (reviewed in Ref. 44). By using computerized modeling, pull-down techniques and mass spectrometric analysis, it was shown that A_{2A}–D_2 receptor heteromerization depends on an electrostatic interaction between an Arg-rich epitope located in the amino-terminal portion of the third intracellular portion of the D_2 receptor and a single phosphate group from a casein kinase phosphorylatable Ser localized in the distal portion of the carboxy-terminus of the A_{2A} receptor.[45] *In vitro* studies with peptides corresponding to both epitopes demonstrated that the Arg-phosphate interaction possesses a "covalent-like" stability. Hence, these bonds could withstand fragmentation by mass spectrometric collision-induced dissociation at energies similar to those that fragment covalent bonds.[46] The Arg-phosphate electrostatic interaction between epitopes located in intracellular domains is obviously not the only interaction responsible for A_{2A}–D_2 receptor heteromerization. Thus, a significant but not complete reduction of BRET was observed when transfected cells express mutated D_2 receptors that lack the key amino acids involved in the Arg-phosphate interaction,[45] indicating that other receptor domains are also involved. Most probably, transmembrane domains also play a role in A_{2A}–D_2 receptor heteromerization, as has been demonstrated for other GPCR homomers and heteromers.[47,48] Nevertheless, the significant modification of BRET with mutated receptors indicates that the Arg-phosphate interaction is necessary to provide the final quaternary structure of the heteromer, which in fact determines its function. Patch-clamp experiments in identified GABAergic enkephalinergic neurons demonstrated that a disruption of the Arg-phosphate interaction in A_{2A}–D_2 receptor heteromers (by intracellular addition of small peptides with the same sequence as the receptor epitopes involved in the Arg-phosphate interaction) completely eliminates the ability of the A_{2A} receptor to antagonistically modulate the D_2 receptor-mediated inhibition of neuronal excitability.[49]

The above-mentioned antagonistic interaction between A_{2A} and D_2 receptors is most probably related to the existence of an allosteric modulation of the A_{2A}–D_2 receptor heteromer. In other cases, targeting both protomers in the heteromer can induce some trafficking-related unwanted effects. For instance, the opioid δ–μ receptor heteromer seems to be a better target than either the μ or the δ receptors alone, since blockade of the δ receptor decreases tolerance to the analgesic effects of the most used μ receptor agonist, morphine. In keeping with this, bivalent ligands (with a μ receptor agonist and a δ receptor antagonist attached by a spacer with variable length) are being developed.[50] It is important to note that the existence of the δ–μ receptor heteromer is currently the focus of a debate, since a recent study has questioned the colocalization of the δ and μ receptors,[51] this being an obvious prerequisite for receptor heteromerization. Thus, using a recently generated δ receptor-eGFP knock-in mouse strain,[52] δ and μ receptor cellular colocalization in dorsal root ganglia was demonstrated using anti-GFP and anti-μ receptor antibodies; and was reported to be $\sim$5%.[51] But given the increased level of δ receptor expression in these knock-in mice[52] and the high avidity of the anti-GFP antibody as compared to the anti-μ receptor antibody, it is likely that the level of μ receptor coexpression with the δ receptor was underestimated in this study. Furthermore, previous studies have found that the GFP tag at the C-terminus affects the maturation of the δ receptors[53] and that the levels of δ receptor attenuate the maturation of μ receptors.[54] Finally, an important point to remember is that GPCR expression is altered during development and in pathology. Under these conditions, the level of receptor coexpression is likely to be significantly altered, making some particular GPCR heteromers attractive drug targets. The dopamine D_1–D_3 receptor heteromer is a putative example of a pathology-involved receptor heteromer. It is thus possible that an increase in the expression of D_1–D_3 receptor heteromers is directly involved in L-dopa-induced dyskinesia in patients with Parkinson's disease.[22,55]

Acknowledgments

This work was supported by the intramural funds of the National Institute on Drug Abuse, NIH, DHHS and by Grants from Spanish "Ministerio de Ciencia y Tecnología" (SAF2008-00146 and SAF2008-03229-E/for ERA-NET Coordination of Research Activities) and grant 060110 from "Fundació La Marató de TV3."

References

1. Whorton MR, Bokoch MP, Rasmussen SG, Huang B, Zare RN, Kobilka B, et al. A monomeric G protein-coupled receptor isolated in a high-density lipoprotein particle efficiently activates its G protein. *Proc Natl Acad Sci USA* 2007;**104**:7682–7.
2. Whorton MR, Jastrzebska B, Park PS, Fotiadis D, Engel A, Palczewski K, et al. Efficient coupling of transducin to monomeric rhodopsin in a phospholipid bilayer. *J Biol Chem* 2008;**283**:4387–94.
3. Kuszak AJ, Pitchiaya S, Anand JP, Mosberg HI, Walter NG, Sunahara RK. Purification and functional reconstitution of monomeric mu-opioid receptors: allosteric modulation of agonist binding by Gi2. *J Biol Chem* 2009;**284**:6732–26741.
4. Ernst OP, Gramse V, Kolbe M, Hofmann KP, Heck M. Monomeric G protein-coupled receptor rhodopsin in solution activates its G protein transducin at the diffusion limit. *Proc Natl Acad Sci USA* 2007;**104**:10859–64.
5. Ferré S, Baler R, Bouvier M, Caron MG, Devi LA, Durroux T, et al. Building a new conceptual framework for receptor heteromers. *Nat Chem Biol* 2009;**5**:131–4.
6. Dingledine R, Borges K, Bowie D, Traynelis SF. The glutamate receptor ion channels. *Pharmacol Rev* 1999;**51**:7–61.
7. Gotti C, Zoli M, Clementi F. Brain nicotinic acetylcholine receptors: native subtypes and their relevance. *Trends Pharmacol Sci* 2006;**27**:482–91.
8. Airaksinen MS, Saarma M. The GDNF family: signalling, biological functions and therapeutic value. *Nat Rev Neurosci* 2002;**3**:383–94.
9. Pin JP, Neubig R, Bouvier M, Devi L, Filizola M, Javitch JA, et al. International Union of Basic and Clinical Pharmacology. LXVII. Recommendations for the recognition and nomenclature of G protein-coupled receptor heteromultimers. *Pharmacol Rev* 2007;**59**:5–13.
10. Han Y, Moreira IS, Urizar E, Weinstein H, Javitch JA. Allosteric communication between protomers of dopamine class A GPCR dimers modulates activation. *Nat Chem Biol* 2009;**5**:688–95.
11. Ferré S, Ciruela F, Woods AS, Lluis C, Franco R. Functional relevance of neurotransmitter receptor heteromers in the central nervous system. *Trends Neurosci* 2007;**30**:440–6.
12. Milligan G, Smith NJ. Allosteric modulation of heterodimeric G-protein-coupled receptors. *Trends Pharmacol Sci* 2007;**28**:615–20.
13. Casadó V, Cortés A, Ciruela F, Mallol J, Ferré S, Lluis C, et al. Old and new ways to calculate the affinity of agonists and antagonists interacting with G-protein-coupled monomeric and dimeric receptors: the receptor-dimer cooperativity index. *Pharmacol Ther* 2007;**116**:343–54.
14. Casadó V, Cortés A, Mallol J, Pérez-Capote K, Ferré S, Lluis C, et al. GPCR homomers and heteromers: a better choice as targets for drug development than GPCR monomers? *Pharmacol Ther* 2009;**124**:248–57.
15. Waldhoer M, Bartlett SE, Whistler JL. Opioid receptors. *Annu Rev Biochem* 2004;**73**:953–90.
16. George SR, Fan T, Xie Z, Tse R, Tam V, Varghese G, et al. Oligomerization of mu- and delta-opioid receptors. Generation of novel functional properties. *J Biol Chem* 2000;**275**:26128–35.
17. Gomes I, Gupta A, Filipovska J, Szeto HH, Pintar JE, Devi LA. A role for heterodimerization of mu and delta opiate receptors in enhancing morphine analgesia. *Proc Natl Acad Sci USA* 2004;**101**:5135–9.
18. Jordan BA, Devi LA. G-protein-coupled receptor heterodimerization modulates receptor function. *Nature* 1999;**399**:697–700.
19. van Rijn RM, Whistler JL. The delta(1) opioid receptor is a heterodimer that opposes the actions of the delta(2) receptor on alcohol intake. *Biol Psychiatry* 2009;**66**:777–84.

20. Herrick-Davis K, Grinde E, Harrigan TJ, Mazurkiewicz JE. Inhibition of serotonin 5-hydroxytryptamine 2C receptor function through heterodimerization: receptor dimers bind two molecules of ligand and one G-protein. *J Biol Chem* 2005;**280**:40144–51.
21. Leurs R, Bakker RA, Timmerman H, de Esch IJ. The histamine H3 receptor: from gene cloning to H3 receptor drugs. *Nat Rev Drug Discov* 2005;**4**:107–20.
22. Fiorentini C, Busi C, Gorruso E, Gotti C, Spano P, Missale C. Reciprocal regulation of dopamine D1 and D3 receptor function and trafficking by heterodimerization. *Mol Pharmacol* 2008;**74**:59–69.
23. Ferrada C, Moreno E, Casadó V, Bongers G, Cortés A, Mallol J, et al. Marked changes in signal transduction upon heteromerization of dopamine D1 and histamine H3 receptors. *Br J Pharmacol* 2009;**157**:64–75.
24. Rashid AJ, So CH, Kong MM, Furtak T, El-Ghundi M, Cheng R, et al. D1-D2 dopamine receptor heterooligomers with unique pharmacology are coupled to rapid activation of Gq/11 in the striatum. *Proc Natl Acad Sci USA* 2007;**9**:654–9.
25. So CH, Verma V, Alijaniaram M, Cheng R, Rashid AJ, O'Dowd BF, et al. Calcium signaling by dopamine D5 receptor and D5-D2 receptor hetero-oligomers occurs by a mechanism distinct from that for dopamine D1-D2 receptor hetero-oligomers. *Mol Pharmacol* 2009;**75**:843–54.
26. Minneman KP. Heterodimerization and surface localization of G protein coupled receptors. *Biochem Pharmacol* 2007;**73**:1043–50.
27. Wilson S, Wilkinson G, Milligan G. The CXCR1 and CXCR2 receptors form constitutive homo- and heterodimers selectively and with equal apparent affinities. *J Biol Chem* 2005;**280**:28663–74.
28. Contento RL, Molon B, Boularan C, Pozzan T, Manes S, Marullo S, et al. CXCR4-CCR5: a couple modulating T cell functions. *Proc Natl Acad Sci USA* 2008;**195**:10101–6.
29. Milligan G. Applications of bioluminescence- and fluorescence resonance energy transfer to drug discovery at G protein-coupled receptors. *Eur J Pharm Sci* 2004;**21**:397–405.
30. Pfleger KD, Eidne KA. Monitoring the formation of the dynamic G-protein-coupled receptor-protein complexes in living cells. *Biochem J* 2005;**385**:625–37.
31. Marullo S, Bouvier M. Resonance energy transfer approaches in molecular pharmacology and beyond. *Trends Pharmacol Sci* 2007;**28**:362–5.
32. Ayoub MA, Pfleger KDG. Recent advances in bioluminescence resonance energy transfer technologies to study GPCR heteromerization. *Curr Opin Pharmacol* 2010;**10**:44–52.
33. Ayoub MA, Levoye A, Delagrange P, Jockers R. Preferential formation of MT1/MT2 melatonin receptor heterodimers with distinct ligand interaction properties compared with MT2 homodimers. *Mol Pharmacol* 2004;**66**:312–21.
34. Vilardaga JP, Nikolaev VO, Lorenz K, Ferrandon S, Zhuang Z, Lohse MJ. Conformational cross-talk between α2A-adrenergic and μ-opioid receptors controls cell signaling. *Nat Chem Biol* 2008;**4**:126–31.
35. Carriba P, Navarro G, Ciruela F, Ferré S, Casadó V, Agnati LF, et al. Detection of heteromerization of more than two proteins by sequential BRET-FRET. *Nat Methods* 2008;**5**:727–33.
36. Cabello N, Gandía J, Bertarelli DC, Watanabe M, Lluís C, Franco R, et al. Metabotropic glutamate type 5, dopamine D2 and adenosine A2A receptors form higher-order oligomers in living cells. *J Neurochem* 2009;**109**:1497–507.
37. Navarro G, Carriba P, Gandía J, Ciruela F, Casadó V, Cortés A, et al. Detection of heteromers formed by cannabinoid CB1, dopamine D2, and adenosine A2A G-protein-coupled receptors by combining bimolecular fluorescence complementation and bioluminescence energy transfer. *ScientificWorldJournal* 2008;**8**:1088–97.
38. Gandía J, Lluís C, Ferré S, Franco R, Ciruela F. Light resonance energy transfer-based methods in the study of G protein-coupled receptor oligomerization. *Bioessays* 2008;**30**:82–9.

39. Maurel D, Comps-Agrar L, Brock C, Rives ML, Bourrier E, Ayoub MA, et al. Cell-surface protein-protein interaction analysis with time-resolved FRET and snap-tag technologies: application to GPCR oligomerization. *Nat Methods* 2008;**5**:561–7.
40. Dorsch S, Klotz K-N, Engelhardt S, Lohse MJ, Bunemann M. Analysis of receptor oligomerization by FRAP microscopy. *Nat Methods* 2009;**6**:225–30.
41. Soriano A, Ventura R, Molero A, Hoen R, Casadó V, Cortés A, et al. Adenosine A2A receptor-antagonist/dopamine D2 receptor-agonist bivalent ligands as pharmacological tools to detect A2A-D2 receptor heteromers. *J Med Chem* 2009;**52**:5590–602.
42. Simpson LM, Taddese B, Wall ID, Reynolds CA. Bioinformatics and molecular modeling approaches to GPCR oligomerization. *Curr Opin Pharmacol* 2010;**10**:30–37.
43. Müller CE, Ferré S. Blocking striatal adenosine A2A receptors: a new strategy for basal ganglia disorders. *Recent Pat CNS Drug Discov* 2007;**2**:1–21.
44. Ferré S, Quiroz C, Woods AS, Cunha R, Popoli P, Ciruela F, et al. An update on adenosine A2A-dopamine D2 receptor interactions: implications for the function of G protein-coupled receptors. *Curr Pharm Des* 2008;**14**:1468–74.
45. Ciruela F, Burgueno J, Casado V, Canals M, Marcelino D, Goldberg SR, et al. Combining mass spectrometry and pull-down techniques for the study of receptor heteromerization. Direct epitope-epitope electrostatic interactions between adenosine A2A and dopamine D_2 receptors. *Anal Chem* 2004;**76**:5354–63.
46. Woods AS, Ferré S. The amazing stability of the arginine-phosphate electrostatic interaction. *J Proteome Res* 2005;**4**:1397–402.
47. Guo W, Urizar E, Kralikova M, Mobarec JC, Shi L, Filizola M, et al. Dopamine D2 receptors form higher order oligomers at physiological expression levels. *EMBO J* 2008;**27**:2293–304.
48. González-Maeso J, Ang RL, Yuen T, Chan P, Weisstaub NV, López-Giménez JF, et al. Identification of a serotonin/glutamate receptor complex implicated in psychosis. *Nature* 2008;**452**:93–7.
49. Azdad K, Gall D, Woods AS, Ledent C, Ferré S, Schiffmann SN. Dopamine D2 and adenosine A2A receptors regulate NMDA-mediated excitation in accumbens neurons through A2A-D2 receptor heteromerization. *Neuropsychopharmacology* 2009;**34**:972–86.
50. Daniels DJ, Lenard NR, Etienne CL, Law PY, Roerig SC, Portoghese PS. Opioid-induced tolerance and dependence in mice is modulated by the distance between pharmacophores in a bivalent ligand series. *Proc Natl Acad Sci USA* 2005;**102**:19208–13.
51. Scherrer G, Imamachi N, Cao YQ, Contet C, Mennicken F, O'Donnell D, et al. Dissociation of the opioid receptor mechanisms that control mechanical and heat pain. *Cell* 2009;**137**:1148–59.
52. Scherrer G, Tryoen-Tóth P, Filliol D, Matifas A, Laustriat D, Cao YQ, et al. Knockin mice expressing fluorescent delta-opioid receptors uncover G protein-coupled receptor dynamics in vivo. *Proc Natl Acad Sci USA* 2006;**103**:9691–6.
53. Wang HB, Guan JS, Bao L, Zhang X. Distinct subcellular distribution of delta-opioid receptor fused with various tags in PC12 cells. *Neurochem Res* 2008;**33**:2028–34.
54. Décaillot FM, Rozenfeld R, Gupta A, Devi LA. Cell surface targeting of mu-delta opioid receptor heterodimers by RTP4. *Proc Natl Acad Sci USA* 2008;**105**:16045–50.
55. Marcellino D, Ferré S, Casadó V, Cortés A, Le Foll B, Mazzola C, et al. Identification of dopamine D1-D3 receptor heteromers. Indications for a role of synergistic D1-D3 receptor interactions in the striatum. *J Biol Chem* 2008;**283**:26016–25.

Receptor Activity Modifying Proteins and Their Potential as Drug Targets

Denise L. Wootten,[*,†]
John Simms,[*,†] Debbie L. Hay,[‡]
Arthur Christopoulos,[*,†]
and Patrick M. Sexton[*,†]

[*]*Drug Discovery Biology Laboratory, Monash Institute of Pharmaceutical Sciences, Monash University, Parkville, Victoria, Australia*

[†]*Department of Pharmacology, Monash University, Parkville, Victoria, Australia*

[‡]*School of Biological Sciences, University of Auckland, Symonds Street, Auckland, New Zealand*

The receptor activity modifying proteins (RAMPs) are a family of membrane proteins that interact with some G protein-coupled receptors (GPCRs) to regulate their function. RAMPs can alter GPCR pharmacology and can regulate their signaling as well as trafficking to and from the cell surface. GPCRs have been successfully exploited as drug targets for many years.

DOI: 10.1016/S1877-1173(10)91003-X

1877-1173/10 $35.00

RAMPs therefore provide a new avenue for drug development, offering opportunities for regulating the function of therapeutically relevant RAMP-interacting GPCRs. RAMPs could be directly targeted themselves or advantage could be taken of the unique RAMP/GPCR interfaces for generating more selective drugs.

I. Introduction

Receptor activity modifying proteins (RAMPs) are single transmembrane spanning proteins that can associate with membrane proteins, specifically G protein-coupled receptors (GPCRs), potentially altering GPCR trafficking, pharmacology and/or signaling capabilities.[1] GPCRs constitute one of the largest protein superfamilies in mammals, are estimated to comprise 2–3% of the human genome, and regulate many physiological processes.[2] As a result, nearly 40% of all existing medications regulate GPCR function. Thus, these receptors are of enormous interest to the pharmaceutical industry.[3] Classifications based on sequence homology and ligand structure have identified several GPCR subfamilies.[4] To date, Family A is the largest family and includes receptors for neurotransmitters and hormones; Family B contains hormone receptors for large peptides, such as calcitonin (CT), glucagon and secretin; and Family C incorporates receptors for small molecules such as glutamate, GABA and calcium. Increasing evidence suggests that GPCRs can exist as homo- and/or heterodimers, or as a part of larger oligomeric complexes that can influence many aspects of GPCR activity and consequently cell function.[5] In addition, they can interact with a diverse range of other proteins such as RAMPs, that can also regulate GPCR function. The association of GPCRs with RAMPs creates novel prospects for the development of highly selective ligands by providing an opportunity to achieve a higher degree of selectivity than that of drugs that target the GPCR by itself.

II. RAMPs and Their Discovery

The receptors for calcitonin gene-related peptide (CGRP), amylin (AMY) and adrenomedullin (AM) were believed to be individual GPCRs for many years; however, despite numerous attempts, efforts to clone these receptors proved difficult. In the early 1990s, an orphan GPCR was cloned from rat and human species that contained approximately 50% sequence identity to the calcitonin receptor (CTR) and was therefore named the calcitonin receptor-like receptor[6,7]

(CRLR; this is now abbreviated to CLR[8]). This receptor was thought to be the receptor for CGRP, but when expressed in most cells, except for one clone of HEK 293 cells, it did not respond to CGRP. It was therefore hypothesized that the HEK 293 cells expressed a cofactor that allowed CLR to bind CGRP. The identity of the CGRP receptor was resolved in 1998 with the discovery of the RAMP family of accessory proteins. RAMP1 was first identified when it was found to be responsible for the translocation of CLR from the intracellular compartments of the cell to the plasma membrane which yields a functional CGRP receptor.[9] Subsequently, two additional RAMPs (RAMP2 and RAMP3) were identified, which when expressed with CLR, formed functional AM receptors.[9] The ability of a group of accessory proteins to change the ligand specificity of a GPCR altered the understanding of the molecular basis of GPCR phenotypes, highlighting a novel mechanism for producing greater diversity in receptor repertoire than was previously appreciated.

III. RAMPs and Their Interaction with GPCRs

RAMPs have been demonstrated to interact with a number of GPCRs and their interactions with the receptors for the calcitonin (CT) family of peptides are as yet the best characterized. In addition to interacting with the CLR, RAMPs 1, 2 and 3 were also shown to interact with the CT receptor (CTR) to yield functional AMY receptors, with each CT/RAMP complex displaying a distinct phenotype.

RAMPs are generally poorly expressed at the cell surface in the absence of an interacting partner protein. In screen of Family B GPCRs, this feature was exploited by using confocal microscopy to identify other partners, in addition to the CTR-family, which could traffic HA-tagged RAMPs to the cell surface.[10] Like CLR and CTR, the vasoactive intestinal peptide (VIP) 1 receptor (VPAC1R) interacted with all three RAMPs; however, the glucagon and parathyroid hormone 1 (PTH1) receptors specifically interacted with RAMP2, while the PTH2 receptor interacted specifically with RAMP3. The VPAC2R, growth hormone releasing hormone (GHRH), glucagon-like peptide 1 (GLP1) and GLP2 receptors did not appear to be affected by the presence of RAMPs in translocation assays, although this does not exclude the potential for weak interactions or nontrafficking interactions.[10] A recent study using bioluminescence resonance energy transfer and bimolecular fluorescence complementation demonstrated that the secretin receptor (SCTR), another Family B GPCR, can also specifically associate with RAMP3, but not with RAMP1 or RAMP2.[11] A chaperone role for RAMPs 1 and 3, but not RAMP2, in the trafficking of a Family C GPCR, the calcium sensing receptor (CaSR), has also been reported,[12,13] highlighting the potential for RAMP modulation of receptors outside of Family B GPCRs. There is also preliminary evidence suggesting that RAMP expression may alter β-adrenoceptor pharmacology, although in that study there was no direct evidence for a receptor/RAMP interaction.[14] RAMPs may therefore play a more generalized role in regulating the activity of GPCRs than is currently known.

RAMPs are widely distributed throughout the body, in many cases being present in tissues where no interacting GPCR partner has been identified. This raises the possibility that there may be other receptor partners and roles for RAMPs besides those associated with GPCRs, such as, an interaction with tubulin.[15] This broad distribution of RAMPs suggests that considerable work is still required to fully understand their role in physiology and disease; however, their potential to interact with many GPCRs and modulate their function provides novel avenues for drug discovery.

IV. RAMP Modulation of GPCR Pharmacology

The most well-studied consequence of RAMP interaction with GPCRs is its ability to alter the affinity and efficacy of ligands acting at the CLR and CTR, thus modulating receptor specificity. CLR by itself is not able to bind any known ligand; however, the CLR/RAMP1 complex forms a high-affinity CGRP receptor, whereas the presence of RAMP2 or RAMP3 in association with CLR gives rise to two distinct subtypes of the AM receptor, AM_1 (CLR/RAMP2), and AM_2 (CLR/RAMP3).[8] The AM_1 receptor displays a low affinity for CGRP (approx 100-fold lower than AM), whereas CGRP is recognized with moderate affinity at the AM_2 receptor depending on the species and the form of the CGRP studied.[16] Similarly, CGRP receptors also bind AM with reasonably high affinity.[8] Recently, two additional endogenous peptides of the CT family have been discovered. Intermedin (AM_2) displays high affinity for the CLR complexed with either RAMP1, 2, or 3 and ProCT interacts with the CGRP receptor (CLR/RAMP1).[17,18]

Unlike the CLR, the CTR will traffick efficiently to the cell surface and also bind to CT in the absence of RAMPs.[8] The CTR alone has low affinity for AMY, but in the presence of RAMPs 1, 2, or 3, three high-affinity AMY receptor subtypes are formed, which exhibit a lower affinity for human CT. Several CTR splice variants have also been identified and their interactions with RAMPs generate a large number of possible CTR/AMY receptor subtypes.[1] This is complicated by the knowledge that, in some systems, CTRs when coexpressed with RAMP1, also have relatively high affinity for CGRP. In addition to these peptides, three new members of the CT family have been identified, the CTR stimulating peptides (CRSPs), which bind to and stimulate CTR but not CLR or CTR/RAMP complexes.[19] Thus, in the examples discussed above, RAMPs introduce enormous complexity wherein a single GPCR, which can either be associated with RAMPs or not, can produce many pharmacologically distinct phenotypes, engendering extensive pharmacological diversity.

To date, the demonstrated ability of RAMPs to alter the pharmacological phenotype of GPCRs is limited to the CT family. However, the identification of novel receptor/RAMP pairings provides potential for this effect to be more widespread than is currently appreciated. There are two mechanisms by which RAMPs have been proposed to alter the pharmacological phenotype of their interacting GPCRs. The first is that they act allosterically, wherein their association with the receptor alters its conformational bias and exposes different binding epitopes. The second mechanism proposes that RAMPs contribute directly to the binding of the various peptides, providing epitopes that interact with the ligands in combination with the receptor. Although it is difficult to distinguish between these two models and, indeed, they are not necessarily mutually exclusive, there is strong experimental support for an interaction between the N-terminus of the RAMP and the N-terminus of the CTR/CLR which determines the specific receptor pharmacology of each complex.[20–23] Each RAMP/receptor interface may therefore create unique pockets which can provide the selectivity required for the individual ligands.

V. RAMPs as Receptor Chaperones

As previously mentioned, the association of RAMP with the CLR is required for its efficient transport from the endoplasmic reticulum (ER) to the cell surface. This retention in the cell is also true for RAMPs (particularly RAMP1) when they are not associated with a receptor partner.[9] In the absence of a receptor partner, RAMP1 is retained in the ER and exists as a homodimer. Interaction with a partner protein decreases the quantity of RAMP in the homodimeric form and the hetero-oligomeric form of the RAMP–GPCR complex enables the RAMP to transit from the ER, through the Golgi, to the cell surface.[24] C-terminal deletion studies have indicated a short sequence from Ser^{141} to Thr^{144} (Fig. 1) is critical for the retention of RAMP1 in the ER in the absence of receptor coexpression;[25] however, the role played by this sequence may be cell specific.[26] In contrast, studies using epitope-tagged RAMPs 2 and 3 revealed significant cell-surface expression of these proteins in the absence of receptor coexpression. This capacity for independent cell-surface expression of RAMPs 2 and 3 is attributed to their ability to be glycosylated. Substitution of these glycosylation sites eliminated receptor-independent but not receptor-dependent RAMP cell-surface expression.[27] Conversely, the introduction of equivalent glycosylation sites into RAMP1 allowed receptor-independent cell-surface expression. The extent to which

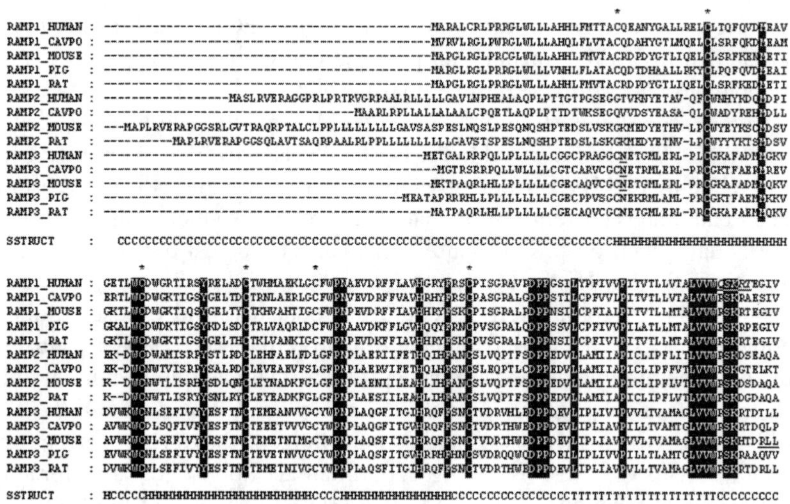

FIG. 1. Amino acid sequence alignment of RAMP1, RAMP2, and RAMP3 from various species. Residues with black backgrounds and white text are conserved throughout the family. Annotations bold/underlined represent glycosylation sites, italic/underlined show the ER retention sequence in RAMP1, underlined/normal text indicate the PDZ domain in the C-terminal tail of RAMP3, asterisk (*) represents cysteine residues that form disulphide bonds required for correct folding of the N-terminal domain of which four are conserved among all RAMPs. Secondary structure is illustrated using a three-letter nomenclature; C indicates coil region, H is helical, and T represents the predicted transmembrane domain.

RAMPs 2 and 3 reach the cell surface is also likely to be influenced by the type of epitope tag and/or leader sequence utilized in the study.[10,11] CLR is less dependent on RAMP association for cell-surface expression, as in some cases, appreciable levels of CLR can accumulate in the absence of RAMP.[27] However, the majority of reports in the literature suggest that very little CLR will be delivered to the cell surface in the absence of an interacting RAMP. Additionally, even if CLR does reach the cell surface without a RAMP, it is functionally inactive.[27,28]

A chaperone role for RAMPs in the efficient translocation of the CaSR to the cell surface has also been identified.[12,13] In Cos7 cells, which do not express RAMPs, the CaSR is retained within the ER. The cotransfection of RAMP1 and RAMP3, but not RAMP2, allowed translocation of the receptor to the Golgi, and subsequent glycosylation resulted in its transport to the cell surface. In addition, siRNA-mediated knockdown of endogenous RAMP1 in HEK 293 cells prevented the expression of transfected CaSR at the cell surface. This pattern of behavior parallels the effect of RAMP on CLR transport.[9,12] The effect of RAMP on the expression of other receptors has not been extensively explored; however, these findings raise the possibility that the chaperone function of RAMPs may extend to other receptors other than those that have already been studied.

Interestingly, a recent study revealed that the interaction of RAMP3 with the SCTR resulted in an increase in the expression of RAMP3 at the cell surface; however, no change in the pharmacological profile of the receptor or secretin-induced signaling to ERK, cAMP or calcium mobilization was observed.[11] This study revealed a novel role for receptor/RAMP interactions in the modulation of receptor pharmacology, as an increase in the expression of SCTR resulted in a decrease in the appearance of the AM phenotype when coexpressed with RAMP3 and CLR. Therefore, it was proposed that due to the ubiquitous distribution of RAMPs across many cell and tissue types, an additional role of RAMP–receptor interactions may be that these interactions act as a "sink" for RAMP proteins so as to modulate the expression level of the AM/CGRP/AMY receptors and other as yet unidentified receptor phenotypes. This may provide a novel avenue for drug development as it potentially provides a highly selective target for drug action.

VI. RAMPs and Receptor Internalization

Significant evidence exists for a role of RAMPs in modulating receptor trafficking following internalization. To date, these studies have only explored the role of RAMPs interacting with the CLR, and the extent to which other identified receptor/RAMP pairings affect internalization and trafficking of the receptor is unknown. Initial evidence suggested that RAMP/CLR complexes were internalized together following agonist stimulation and that most likely, they remain complexed together throughout the lifecycle of the proteins. Further investigations with the CLR/RAMP1 complex showed that this internalization was likely to be β-arrestin and dynamin-dependent.[28,29] Subsequent studies into the trafficking properties of RAMP2- and RAMP3-containing CLR complexes revealed that RAMP2-containing AM_1 receptors showed increased internalization compared with RAMP3-containing AM_2 receptors when expressed in HEK 293 cells, although the mechanism of this difference is unclear.[30] Additional research highlighted the importance of a PDZ-domain in the C-terminus of RAMP3, (Fig. 1) which appears to be responsible for the ability of the AM_2 receptors to be readily recycled in a cell-dependent manner. Interaction of *N*-ethylmaleimide-sensitive factor (NSF) with the PDZ domain of RAMP3 selectively allowed recycling of the CLR/RAMP3 complex back to the cell surface.[29,31] In contrast, in this study, AM_1 and CGRP receptors that do not contain a PDZ domain in their RAMP components were subsequently targeted to a degradative pathway.[31] However, in other cell systems, CGRP receptors are only degraded following sustained CGRP treatment, but recycled to the cell surface after transient stimulation.[32] In addition, following internalization, the CLR/RAMP3 is also recycled by another protein, NHERF-1, which binds to the PDZ domain on RAMP3 and blocks internalization of only the AM_2 receptors (but not that of CGRP or AM_1 receptors).[33] These differential properties may explain the existence of two AM receptors, which in all other respects, appear to behave in a similar manner. In addition, RAMP3 contains a dileucine motif that may be involved in additional protein–protein recognition. Furthermore, the C-terminal tail of all three RAMPs contain combinations of serine and threonine residues that are potential sites for phosphorylation. These in turn may play a role in desensitization and trafficking of these RAMP/receptor complexes. A study of CGRP receptor phosphorylation showed that while CLR was phosphorylated, RAMP1 was not.[34] However, to date, the role of potential phosphorylation sites has not been extensively explored.

VII. RAMPs and Receptor Signaling

In addition to the role of RAMPs in modifying receptor behavior, there is also increasing evidence that RAMPs may alter the signaling phenotype of the respective GPCR. The C-terminal tail of the RAMPs has been shown to play a significant role in modulating the G protein-coupling specificity of AMY receptors. Removal of the C-terminus of RAMP1 and its exchange for the C-terminus of RAMP2 reduced the potency of CGRP in stimulating cAMP formation at the AMY1 receptor.[26] In addition, AMY binding was decreased upon RAMP C-terminal truncation, which could be partially restored by the overexpression of Gαs. These data suggest that the C-terminus of the RAMPs plays a direct role in signaling, specifically in G protein coupling. More recently, studies on the relative coupling of AMY receptors to cAMP generation, intracellular Ca^{2+} and ERK1/2 activation in both Cos-7 cells and HEK 293 cells has revealed selective modulation of these signaling pathways, depending on the particular RAMP in complex with the CTR.[35] This study revealed a marked induction of AMY potency at AMY_1 and AMY_3 receptors for formation of cAMP, but only weak induction of AMY potency was seen for Ca^{2+} and ERK signaling compared to when CTR was expressed alone. This suggested that the CTR/RAMP-based AMY receptors are relatively less well coupled to Gq/Gi than the isolated CTR. In addition, the overexpression of Gα subunits displayed differential modulation in the induction of high-affinity AMY binding and Gαs overexpression increased binding at CTR/RAMP2 and CTR/RAMP3 complexes, whereas Gαq overexpression selectively increased binding only at the CTR/RAMP3 complexes.[35] These data demonstrate that RAMP-complexed CTRs have different signaling profiles compared to CTRs expressed in the absence of RAMPs and that these differences are likely to be due to the direct effects of the individual RAMPs on the G protein-coupling efficiencies.

It is also apparent that the generation of high-affinity AMY binding at CTR/RAMP2 complexes is very sensitive to experimental conditions. Early studies on CTR/RAMP complexes identified that the ability of RAMP2 to create a high-affinity AMY receptor was dependent on the cell background and receptor isoform being studied. In both Cos7 and rabbit aortic endothelial cells, RAMP2 induces only a weak AMY phenotype when expressed with the CTR_a isoform, but a strong AMY interaction is observed when expressed with the CTR_b isoform.[36–38] In contrast, a strong AMY phenotype was observed following coexpression of either isoform in CHO-P cells.[37] These discrepancies, most likely, reflect differences in their coupling efficiency to G proteins between the different cell types, even though other components within the cellular background may also be important in modulating RAMP/receptor interactions.

In contrast to AMY receptors, deletion of the C terminal tail of RAMPs had little effect on any property of the CGRP or AM receptors.[26,30] This may be linked to the presence of an additional protein, receptor component protein (RCP), which has been shown to enhance the coupling of CGRP and AM receptors to Gαs.[39] It is unclear if there is any direct association of RCP with RAMPs, but RCP has been shown to interact with the second intracellular loop of CLR and is crucial for cAMP signaling mediated by CGRP and AM receptors. RCP may also influence the stability of the CLR/RAMP complex.[40]

Modulation of receptor signaling by RAMPs has also been observed for the VPAC1R. This receptor interacts with all three RAMPs; however, its agonist binding properties were unaltered by these interactions. Nonetheless, while measurements of cAMP remained unchanged following overexpression of the VPAC1R with RAMP2, phosphoinositide turnover was enhanced in response to VIP.[10] It has been suggested that this may reflect a change in the compartmentalization of the receptor signaling complex rather than a direct effect on G protein-coupling. A change in the signaling profile was not observed upon interaction of the VPAC1R with RAMPs 1 or 3.[10] Furthermore, a recent study revealed that although RAMP3 interacted with the SCTR, no effect on cAMP signaling, intracellular calcium mobilization, or ERK1/2 signaling was observed.[11] In these cases, the outcome of receptor oligomerization with RAMPs may involve modification of different receptor parameters or alternative signaling pathways that have not yet been identified. Although RAMPs interact with several other receptor partners, the effect on the signaling profiles of these receptors has not yet been investigated. Collectively, these studies indicate that there is great potential for selective modulation of receptor signaling by targeting the receptor/RAMP complex.

VIII. Posttranslational Modifications of RAMPs—Glycosylation

As previously mentioned, RAMP1 lacks glycosylation sites and is unable to reach the cell surface in the absence of an interacting partner. Both RAMP2 and RAMP3 are glycosylated and *in vitro* assays have suggested that these RAMPs can reach the plasma membrane to some extent without an accompanying receptor.[10,27,34] As these types of experiments rely on tagged proteins, it is unclear to what extent RAMPs 2 and 3 traffic independently *in vivo*. However, it is known that hetero-oligomerization with a receptor facilitates RAMP delivery to the cell surface.[10] RAMP2 and RAMP3 have multiple glycosylation sites. The majority of the known RAMP3s contain four potential glycosylation sites (Asn^{29}, Asn^{58}, Asn^{71}, and Asn^{104} in humans) although only the fourth site (Asn^{104}) is absolutely conserved (Fig. 1). The equivalent of this fourth glycosylation site is also present in all known RAMP2s.[1] Studies using a mutant RAMP3 construct lacking the glycosylation sites demonstrated that this modification was not required for the association of RAMP3 with CLR, however, the cell-surface expression of the mutant complex was reduced by 50%.[41] Its affinity for AM binding was also reduced, implying that glycosylation of the RAMP is either directly or indirectly important for AM binding.

In the early literature, there was speculation that receptor glycosylation was also responsible for the differences in pharmacological profiles shown by the CGRP and AM_1 receptors.[9] This was due to observations that CLR was more heavily glycosylated when associated with RAMP1 than when associated with RAMP2. It is now accepted that this differential increase in glycosylation was an artifact and is an unlikely explanation for the changes observed in their pharmacology as RAMP2 and RAMP3, which in addition to RAMP1, increase the rate of terminal glycosylation of CLR. There is no evidence to suggest that differential glycosylation is relevant to the CGRP/AM phenotypes as only the CLR/RAMP hetero-oligomers that contain the fully glycosylated CLR are present at the cell surface for high-affinity CGRP or AM binding.[34]

IX. Correlation of RAMPs with Receptors *In Vivo*

While profiling interaction of RAMPs and their associated receptors in heterologous *in vitro* expression systems is informative, it is desirable to understand the contribution of the relevant RAMP/receptor pairings *in vivo*. This is inherently difficult, given the complexity that may arise when several RAMP and receptor partners are expressed in the same tissue. Resolving the presence of RAMPs and receptors at the cellular level is crucial to determining the important physiological interactions. Due to the lack of good quality antibodies to RAMPs and GPCRs, these types of studies mainly rely on the examination of mRNA expression levels. While useful, caution must be applied when interpreting this information, as mRNA expression does not always correlate to protein levels.[42]

X. Regulation of RAMPs in Disease

Given that RAMPs form a key part of the CGRP, AM, and AMY receptors, it is predictable that changes in RAMP expression would influence the expression of these receptors and hence the sensitivity of cells and tissues to CGRP, AM, and AMY, in addition to any drugs that may be targeted to a RAMP/receptor complex. Changes in receptor activity have indeed been reported as a consequence of altering RAMP expression levels. The relatively recent development of knockout mouse models for each of the individual RAMPs has provided new insight into the function of these proteins. RAMP1 knockout mice have elevated blood pressure with CGRP displaying a markedly reduced vasodilator activity in these animals and also a reduced ability to suppress proinflammatory cytokine production.[43] By contrast, RAMP1-overexpressing mice showed enhanced plasma extravasation when CGRP was injected into the whisker pad.[44] This is particularly interesting as CGRP action on the trigeminal system causes many of the symptoms of migraine and this study demonstrated that overexpression of RAMP1 in the trigeminal system enhanced CGRP sensitivity. The RAMP2 and RAMP3 knockout mice each displayed distinct phenotypes, revealing divergent roles for these two proteins despite their association with CLR to yield pharmacologically similar AM receptors.[45] Deletion of the RAMP2 gene resulted in a lethal phenotype, of which the likely causes were abnormalities in the blood and lymphatic systems. In contrast, RAMP3 knockout mice displayed no obvious phenotype until they reached old age, where they were lighter than their wild-type littermates.[45–47] A transgenic mouse model with overexpression of RAMP2 in the smooth muscle showed enhanced sensitivity to the vasodilatory actions of AM and displayed greater decreases in blood pressure upon administration of AM. Changes in RAMP expression levels on receptor function have many implications for the onset and progression of disease and also in the treatment of these diseases.

In addition to RAMP knockout mouse models, the regulation of RAMP expression levels has been investigated extensively in animal models of disease where it is possible that RAMP mRNA levels may be differentially regulated. From these studies it is clear that RAMP expression is differentially regulated under conditions of disease, drug treatment, pregnancy, and hypoxia. There are many examples where RAMP regulation has been shown to be involved in cardiovascular disease. In a rat model of heart failure induced by aortic stenosis, RAMP1 and RAMP3 mRNA expression was elevated in the atria and ventricles 6 months postsurgery, whereas RAMP2 expression levels were unchanged.[48] In two separate studies, RAMP2 mRNA (in addition to CLR and an increased AM phenotype) was shown to be upregulated in the atria and ventricles of rats with congestive heart failure.[49,50] This increase in RAMP2 could be reversed by a mixed endothelin A/B receptor antagonist.[50] An isoproterenol-induced rat model of myocardial hypertrophy and ischemia also displayed increased RAMP2 mRNA levels in the heart and aorta and greatly increased numbers of AM binding sites compared with vehicle-treated rats.[51] Furthermore, a selective increase in RAMP2 mRNA in cardiomyocytes isolated from myocardial tissue of rats 7 days after myocardial infarction was observed compared to control rats.[52] In the same study, RAMP3 expression levels were increased in both cardiomyocytes and noncardiomyocytes. RAMP regulation has also been examined in primary rat cardiomyocytes where RAMP1 and RAMP3 mRNA levels were increased in the presence of increasing concentrations of angiotensin II.[53]

Animal models have been used to investigate RAMP regulation in hypertension.[52,54–59] These studies have demonstrated that RAMP levels change under conditions of hypertension, however the precise nature of the change seems to be model- and tissue-dependent. Hormones can also regulate RAMP expression levels. Aldosterone treatment of mesenteric arteries from spontaneously hypertensive rats increased both CGRP sensitivity and RAMP1 expression but not CLR expression.[60] In other vascular beds, parallel changes in CLR and RAMP1 expression have been reported.

RAMP mRNA levels are also regulated in animal models of diabetes. RAMP2 levels were elevated in the kidney whereas RAMP3 levels were not altered compared to nondiabetic rats.[61] In addition, RAMP levels are subject to regulation during pregnancy with upregulation of RAMP1, 2, and 3 levels in the rat uterus, an effect that could be reversed with a progesterone antagonist.[62] In the rat uterine artery and the rat placenta, RAMP1 was also elevated. There is some evidence to suggest that RAMP1 and CLR decrease during preeclampsia in the fetoplacental vessels, resulting in a decrease in CGRP binding sites.[63]

RAMPs are also altered in renal failure, where they may play a protective role. A rat model of obstructive nephropathy revealed an upregulation of RAMP1 and RAMP2 mRNA, but RAMP3 levels were unchanged.[59] In addition, this study showed an increase in CLR expression; however, no change in the AM phenotype was seen. Similar observations were found in a rat model of renal fibrosis but with a more pronounced increase in CLR expression.[59] RAMP1, RAMP2, and CLR mRNA expression levels were reduced in mouse models of sepsis, but RAMP3 mRNA levels were increased.[64] In neuroblastoma cell lines treated with hypoxia mimetic agents, RAMP2 mRNA expression was reduced; however, increased expression of the AM gene was detected.[65]

In rat lung models of chronic hypoxia, RAMP1 and RAMP3, but not RAMP2, were upregulated and CLR levels remained unchanged.[66] Although caution must be applied when extrapolating to protein expression using mRNA levels alone, these studies provide clear evidence for the regulation of RAMP expression physiologically, which may, in turn, affect drug action. In this review, the impact of RAMP expression levels has been discussed mainly in the context of the CT family of peptides as these are as yet the best characterized. However, the potentially broader role of RAMPs also need to be considered as theoretically, targeting RAMPs may aid in the development of viable therapeutics that may help manage and control some of these disease states.

XI. RAMPs as Drug Targets

In principle, RAMPs could be utilized as potential drug targets to control and manage diseases such as those mentioned above and potentially others for which the role of RAMPs has not yet been understood. RAMPs could be targeted either directly themselves or by targeting the complex formed between RAMP and its interacting receptor. At this stage, due to limited structural data, structure-based drug design for RAMPs may be difficult; however, with the growing body of experimental evidence and future findings, structural information may prove useful in attempts to target RAMPs and their respective receptor complexes.

A. RAMP Structure

RAMP1, RAMP2, and RAMP3 share a common topology consisting of a large extracellular amino-terminus, a single transmembrane (TM) segment and a small intracellular C-terminus. RAMP1 and RAMP3 are both approximately the same length, and RAMP2 contains an additional 26 residues (Fig. 1).[24] There are currently no structures of the full length protein of any RAMP; however, recently the crystal structure of the extracellular domain of RAMP1 was solved.[67] This structure validated a previously predicted *ab initio* model of RAMP1, and confirmed that the N-terminus consists of three α helical domains containing three disulphide bonds between the conserved cysteine residues[67,68] (the positions of these cysteine residues are highlighted in Fig. 1). The specific disulphide bonding pattern has been shown to be important for folding of the N-terminus with interruption of these disulphide bonds by mutation of the cysteine residues to alanine resulting in impaired formation of the RAMP/CLR complexes, and consequently, disrupted CGRP or AM signaling.[68–70]

B. The Molecular Basis of the GPCR/RAMP Complex

The precise nature of RAMP interaction with their receptor partners is of interest when considering them as potential drug targets because RAMPs appear to produce their biological activities primarily by forming these interactions. Although it is now known that RAMPs can interact with a number of GPCRs, most structural information comes from their interaction with CLR. At present the stoichiometry of the RAMP/receptor complex is unclear. Early studies using cross-linking of bivalent ligands suggested that the complex consisted of one RAMP1 and one CLR.[34] Cross-linking experiments with [^{125}I]-CGRP also support a 1:1 complex, although the details of the labeling patterns depended on the cell line used.[28,71] However, a recent study that combined coimmunoprecipitation, bioluminescence resonance energy transfer with bimolecular fluorescence complementation in live cells has proposed a new complementation for the CGRP receptor composing of 2 CLR:1 RAMP1.[72] The data was consistent with the presence of at least two CLR molecules forming a homodimer where the presence of RAMP1 induces a conformational rearrangement of the oligomer rather than competing for its formation. In addition it was shown that although both CLR and RAMP1 form homo-oligmers in the ER, only the RAMP1 monomer can associate with the CLR homo-oligomer to form an asymmetric CGRP receptor that could reach the cell surface.[72] Nonetheless, the data do not preclude the possibility of the CLR homodimer interacting with two monomeric RAMPs.

The potential asymmetry of the RAMP/receptor complex raises important questions concerning the arrangement of the complex and the contribution of each of these subunits to ligand binding and G protein interaction. Given that [^{125}I]-CGRP can interact with both RAMP and CLR, it is proposed that domains from both the proteins participate or are in close contact to the binding pocket.[28,71] Initial work using RAMP1 and RAMP2 chimeras with the CLR indicated that the N-terminus is the principal component of the RAMPs that is involved in determining receptor pharmacology.[20] A recent study that utilized chimeras between RAMP1 and RAMP3 also supports this theory.[73] In addition to modulating phenotype, evidence suggests that the N-terminal domain of a RAMP is sufficient to maintain a functional interaction between the RAMP and the receptor. A chimeric protein in which the TM helix of RAMP1 was replaced with that of the platelet-derived growth factor receptor produced functional CGRP receptors; however, the affinity for CGRP was reduced 10-fold and no change in efficacy for cAMP signaling was observed.[21] In addition, the N-terminus of RAMP1 alone also allowed formation of receptors that were weakly responsive to CGRP and this was sufficient to promote terminal glycosylation of the CLR, thus supporting the proposition that the N terminus segment is sufficient to produce functional CGRP receptors.[21,25]

Despite this, there are inconsistencies with earlier work where partial deletion of the TM domain of RAMP1 resulted in a reduction in CGRP potency due to a reduction in stability of the RAMP/CLR complex.[25] Furthermore, a study using RAMP1/3 chimeras indicated that replacement of the TM domain and C-terminal tail of RAMP1 with RAMP3 resulted in a reduction in CGRP potency, whereas substitution of the equivalent region in RAMP3 by RAMP1 did not alter AM potency.[73] These data indicate that while the N-terminus may play the predominant role in determining receptor phenotype, the TM domain also confers some selectivity. Additionally, a role for the TM domain in stabilizing the interaction between RAMP and its partner GPCR is also evident from studies with the CTR. Chimeras of RAMP1 and RAMP2 demonstrated a level of AMY binding that was indeed strongly influenced by the presence of the TM domain.[74]

Subsequent work on the N-terminus has helped delineate some of the regions within the N-terminal domain that have specific roles in determining the formation of functional receptors and the specificity of the phenotype. Removal of residues 51–55 or 83–90 impaired HA-tagged RAMP1 translocation to the cell surface but did not greatly impair the ability of the CGRP to stimulate cAMP. Thus, the epitopes defined by these residues, while having some role in the formation of a functional receptor, are not required for conferring the specific phenotype. Deletion of residues between 41–50, 59–71, and 91–103 impaired both translocation of HA-tagged RAMP1 to the cell surface and its CGRP responsiveness. Thus, epitopes within these regions are of prime importance for the association of RAMP1 with CLR and perhaps the formation of a CGRP binding site. Deletion of residues 28–33, 78–80, and 88–90 gave receptors that responded to CGRP but which could not be activated by AM.[75] Chimeric protein studies have also proposed that residues 89–92 of human RAMP2 and the equivalent residues in human RAMP3[59–65] are required for AM binding,[76,77] whereas in RAMP1, residues 78–90 appear to be involved in AM binding.[75] This section is proximal to the domain identified in RAMPs 2 and 3. Interestingly, it was observed that the epitopes that cause a major loss of both CGRP-responsiveness and RAMP1 cell-surface expression are predominantly clustered at the opposite end of the N-terminus from those that are involved in AM binding, when these clusters were mapped onto the crystal structure of the RAMP1 extracellular domain.[67]

Attempts to identify individual amino acids responsible for CGRP receptor function have shown that helices 2 and 3 of RAMP1 are most important for CLR recognition.[73,78] An alanine scan of residues 91–103 in helix 3 identified a stretch of aromatic amino acids within RAMP1 that are likely to be important for its interaction with the CLR. Mutation of Tyr^{66}, Phe^{93}, Phe^{101}, and His^{97} significantly impaired RAMP1 expression and CGRP binding, although a less pronounced effect on CGRP-stimulated cAMP production was seen.

In contrast, alanine substitution of Leu[94], while not altering total RAMP1 expression, significantly enhanced cell-surface expression, which suggested that Leu[94] may impair CLR association by steric hinderance.[75] The RAMP1 extracellular domain crystal structure shows that the side chain of Leu[94] faces away from the main helical bundle of the protein and it is therefore feasible that this face of the extracellular domain packs up against CLR. A more recent study concluded that Tyr[66] in helix 2 and His[97] were particularly important in promoting RAMP1 interaction with CLR, with Phe[93] and Phe[101] playing less significant roles.[78] All these residues cluster together and form an exposed surface, and are consistent with a role for them in defining an interaction site with CLR.[67] Mutation of residues in helix 1, which is on the other face of the protein, had little effect on CGRP responsiveness or its association with CLR.[78]

The crystal structure of the extracellular domain of RAMP1 also suggested that the binding site for ligands could be formed by a cluster of residues at the C-terminal end of helix 2 involving Arg[67], Asp[71], Trp[74], Asn[78], and Trp[84]. A recent mutagenesis study has identified Leu[69] and Thr[73], both of which lie within this region, as being important for CGRP binding.[78] However, mutation of Trp[74] does not affect CGRP binding,[79] and deletion of Asn[78] and surrounding amino acids does not impair CGRP signaling.[75]

Deletion mutants in the N-terminus of RAMP1 suggested that the region involved in AM binding in the CGRP receptor is its upper portion, which is adjacent to the N-terminus of RAMP1. Mutagenesis data provided firm evidence that position 74 in RAMP1/3, on the edge of this region, is a key determinant of AM pharmacology as substitution of Trp[74] in RAMP1 to its equivalent residue in RAMP3 (Glu) resulted in an increased affinity of the complex for AM.[23] Additionally, substitution of Glu[74] in RAMP3 to Trp resulted in a loss of the AM phenotype at AM_2 receptors.[23] Deletion experiments in RAMP2 identified at least one important region, consisting of residues 86–92, that plays a significant role in its interaction with CLR to form a functional AM receptor.[76] A deletion of the corresponding region in RAMP3 had a similar effect on generation of the AM phenotype.[76] There is little sequence identity in this region between RAMPs 2 and 3, which suggests this region has either an allosteric effect or an important structural role in determining the conformation of the CLR. Unlike the situation for AM binding, these deletion studies did not reveal any discrete area of RAMP1 that was specifically involved in CGRP recognition, rather than CLR recognition.[75] This suggests that the domains of RAMP1 that are involved in CLR recognition may be intimately associated with those that recognize CGRP. In support of this, Phe[93], which is close to the presumed CLR interaction site, appears to have a specific role in CGRP binding. The substitution of this residue in RAMP1 to its equivalent in RAMP3 (Ile), inhibited CGRP potency without altering cell-surface expression of the CLR/RAMP1 complex.[23]

Very little is known about the regions of any GPCR that are involved in RAMP recognition. The extreme N-terminus of CLR has been shown to be important (residues 23–60) for its association with RAMP1. Transfer of this region to the PTH1 receptor which does not normally associate with RAMP1 resulted in a receptor with an ability to associate with RAMP1.[15] A recent study that combined site-directed mutagenesis and molecular modeling techniques identified residues Ile^{32}, Gly^{35}, Thr^{37} Leu^{41}, Ala^{44}, Gln^{45}, and Tyr^{49} as being important for CGRP affinity and efficacy.[80] Of these, Gln^{45} and Tyr^{49} were predicted to mediate their effects by interacting with RAMP1 and the other residues were suggested to modulate CGRP affinity. In other members of family B GPCRs, the equivalent of this segment of the receptor N-terminus is, at least partially, an alpha helix and is in proximity to bound ligands.[49,81–83] In addition to this information on the CLR, there is also some evidence to support the structural basis of RAMP3 association with the SCTR. Using a combination of truncation constructs, peptide competition experiments, and chimeric secretin-GLP1 receptor constructs, it was established that this association is dependent on the intramembranous region of the RAMP and TM6 and TM7 of the SCTR.[11] This demonstrates that, in addition to the N-terminus, the TM regions of both the RAMP and the interacting GPCR are important for the association of some, if not all, of these complexes.

In the absence of a crystal structure depicting the interface of CLR with RAMPs and their interaction with their respective ligands, it is, at present, only possible to speculate how RAMPs interact with their respective partners. As indicated previously, a major unresolved issue is whether the ligands are in direct contact with the RAMPs or whether the RAMPs act indirectly by altering the conformation of the receptor. Cross-linking studies have demonstrated that the RAMPs are in close proximity to CGRP or AM when bound to their respective receptors, and it is therefore very much possible that ligand contact points do occur at the receptor/RAMP interface and may likely involve those residues identified by the mutagenesis studies.[9,34]

A small number of nonpeptide antagonists have been identified for the CGRP receptor; the best characterized of these is BIBN4096BS.[84] This is an allosteric antagonist that displays slow dissociation kinetics[84,85] and has high a selectivity toward primate CGRP receptors. The basis of this selectivity is the nature of the residue at position 74 of RAMP1. In primates, this residue is a tryptophan but in rodents and many other species it is a lysine. As with endogenous peptides, it is unclear whether the Trp^{74} of RAMP1 interacts with BIBN4096BS directly or indirectly or by modifying the conformation of CLR. Furthermore, RAMP2 and RAMP3 containing AM_1 and AM_2 receptors (CLR containing complexes) have only a low affinity for BIBN4096BS. Interestingly, substitution of the glutamic acid at position 74 in RAMP3 for tryptophan increases the sensitivity of AM_2 receptors for BIBN4096BS.[85]

In addition, chimeras of the CLR and the CTR suggest that BIBN4096BS-like compounds require the presence of CLR for high affinity, with CTR/RAMP1 complexes displaying only weak affinity for these antagonists.[86] Studies have implicated residues 37–63 of CLR in high-affinity RAMP binding, in addition to being implicated in antagonist binding,[86] further suggesting that it is this extreme N-terminus of the CLR that is in contact with RAMP1. A recent study using site-directed mutagenesis identified Met^{42} as being crucial for high BIBN4096BS affinity.[87] In addition, Met^{42} of the CLR was also shown to be essential for high-affinity interaction of another nonpeptide antagonist of the CGRP receptor, MK0974.[87] Furthermore, the substitution of Trp^{74} in RAMP1 by Lys impaired MK0974 affinity at the CGRP receptor to a degree that was similar to that observed for BIBN4096BS. When combined, these studies suggest that nonpeptide antagonists occupy a binding site close to the interface of the N-terminal domains of CLR and RAMP1 and this interface between the two proteins is likely to be responsible for the selectivity of these compounds.

C. Direct Targeting of RAMPs or the RAMP/Receptor Complex in Drug Development

Principally, whether targeting the RAMP by itself or the RAMP/receptor complexes, these proteins are viable targets for novel therapeutic agents. This is exemplified by BIBN4096BS and MK0974 which were developed for the treatment of migraine headaches (the pathology of which involves CGRP) and have shown great promise in clinical trials.[84,88] As noted above, these compounds act at the RAMP/receptor complex interface (Fig. 2). It is anticipated that the complex between the RAMP and the receptor would provide a unique structure containing binding pockets that could be exploited for drug discovery. In this regard, future work to elucidate the structure of RAMPs in complex with a GPCR would therefore be very valuable for the development of drug-like compounds. Inspection of the crystal structure of the extracellular N-terminus of RAMP1 has identified a potential "druggable" cavity within the RAMP protein that lies at the predicted interface of the complex between RAMP and CLR with Trp^{74} lining this cavity (Fig. 2). As mentioned above, Trp^{74} is responsible, at least in part, for the pharmacological specificity of the CGRP antagonists BIBN4096BS and MK0974 for RAMP1 containing complexes compared to the RAMP2/3 containing complexes.[85,87] This large cavity may accommodate both BIBN4096BS and MK0974 and therefore, it is possible that this forms a part of the binding pocket for these compounds. As this cavity lies at the predicted interface for CLR interaction, it is also possible that this may be the association interface for other GPCRs. The development of compounds targeted to this pocket combined with the capability to make

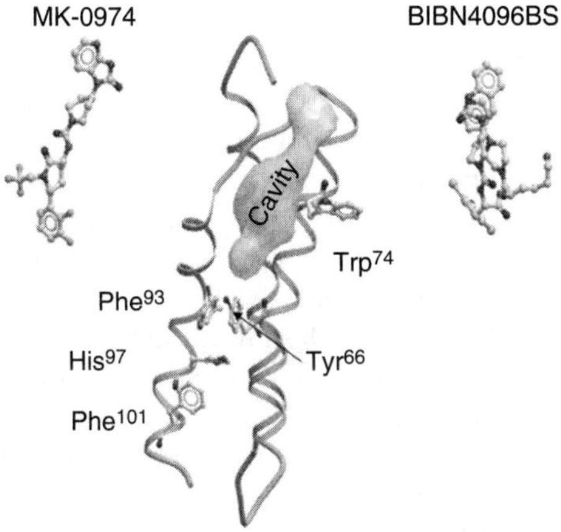

FIG. 2. Representation of the crystal structure of RAMP1. Depicted on the diagram are the positions of residues suggested through mutational studies to line the interface between RAMP1 and CLR (Tyr^{66}, Phe^{93}, His^{97}, and Phe^{101}). Also shown is the position of Trp^{74} that exhibits species and subtype specificity for the drugs BIBN4096BS and MK0974. The cavity prediction was obtained from the icmpocket macro as implemented in ICM Pro (MolSoft, La Jolla, CA). The orientation of BIBN0974BS and MK0974 represents a speculative docked conformation of these compounds in the cavity adjacent to Trp^{74} of RAMP1.

multiple contacts with both the RAMP protein and the interacting receptor partner may provide a feasible strategy for identifying selective drugs that target these complexes. Evidence for the success for this type of approach can be gained from the development of BIBN4906BS and MK0974, both of which form contacts with RAMP1 and CLR, thus producing selective CGRP antagonism (see previous). It is interesting that both these compounds are large when compared to most known drugs. This may be due to their need to form interactions with both partners of the complex.

For some GPCRs, such as the SCTR and the VPAC1R, the binding phenotype for their known ligands is not altered by their interaction with RAMPs.[10,11] Recent studies on the RAMP3/SCTR complex showed that the N-terminus of the RAMP is not required for its interaction with the receptor.[11] Instead, the basis of this interaction has been localized to the TM domain of the RAMP and TM6 and TM7 of the SCTR. Therefore, for GPCRs where the interaction of RAMPs does not alter the binding phenotype, peptides targeting the RAMP TM domain may provide a better therapeutic avenue than compounds directed at the extracellular N-terminal domain. This approach has been previously adopted by Yin *et al.*, wherein the TM helices of two closely related integrins were targeted in a sequence specific manner using computationally designed peptides.[89]

Modified versions of native peptides themselves may also provide an opportunity for development of selective drugs. This approach has shown considerable success with the development of Symlin, a modified form of AMY, which is used for the treatment of diabetes (reviewed in Ref. 90). In addition, a recent study using a similar approach with chimeric peptides has identified a series of novel nonselective AM_1/AM_2/CGRP receptor antagonists and modestly selective AM_2 and CGRP receptor antagonists.[91] Alternatively, designer peptides could be developed which consist of peptide fragments of the RAMP and/or GPCR of interest; these may be able to disrupt the RAMP/receptor complex. Antibody-based therapeutics may also provide a further avenue for drug development. As RAMPs are present at the cell surface in complex with a partner protein, this approach would involve the design of selective antibodies that target the RAMP/receptor complex by bridging the interface, thereby preventing ligand interaction.

Most small molecule drugs that act on GPCRs are believed to interact within the TM region of the receptor. Due to the molecular nature of receptors that have been shown to interact with RAMPs, it is anticipated that molecules acting at the GPCR would lack specificity for specific receptor phenotypes. However, if the interaction with RAMPs substantially altered the structure of the GPCR, then it may be possible to target this novel structure by itself for selective drug development. Attempts to develop allosteric modulators targeting specific RAMP receptor complexes, with the potential to alter the conformational bias of the complex either by disrupting or promoting RAMP or ligand interactions, may also be of use.

While compounds targeted to the extracellular face of the RAMP are most likely to be of therapeutic interest, development of low molecular weight drugs capable of passing the cell membrane should also be considered as a viable method of targeting RAMPs. For example, molecules that prevent RAMP-receptor association within the ER will inhibit their trafficking to the cell surface. In addition, development of compounds that promote an interaction between a RAMP and a receptor partner to enhance cell-surface expression may also be useful for conditions wherein naturally occurring mutations of a GPCR contribute to the disease phenotype. For example, the inherited disorders such as familial hypocalciuric hypercalcemia and neonatal severe hyperparathyroidism are caused by inactivating mutations in the CaSR gene that result in impaired posttranslational processing and subsequent trafficking of the CaSR to the plasma membrane.[92] In addition, compounds designed to interact with the C-terminus of the RAMP may alter trafficking or signaling profiles of the complexes.

RAMPs could also be targeted directly by small molecule inhibitors, RNAi or siRNA therapeutics, by designing drugs which mask the RAMP promoter which prevent or enhance RAMP production, or by using antibodies that bind to epitopes on the RAMP alone. As RAMPs have a widespread distribution and the full extent of RAMP interactions are unknown, targeting the RAMP by itself may have many unforeseen consequences. Additionally, as highlighted previously, in some disease states RAMP expression is altered in a tissue specific manner. Therefore targeted delivery of drug-like molecules to these tissues alone may limit some off-target effects. There is promise for such types of drug delivery therapies with the development of phage display technologies in which bacteriophages or retroviruses can be genetically and chemically modified to deliver drugs to specific targets within cells and tissues. For example, this technique may be applied to deliver siRNAs to specific tissues where RAMP expression has been shown to be upregulated in certain disease conditions or to bring RAMP expression back to normal levels. However, complexities may still arise when several RAMP and receptor partners are expressed in the same tissue, and in these cases, it may be desirable to target RAMP/receptor complexes in a tissue-specific manner rather than the RAMP itself.

D. Indirect Targeting of RAMPs

Under some conditions of disease, RAMP expression levels can be up- or downregulated (as discussed previously). In a number of cases, it has been demonstrated that use of drugs targeted at other receptor proteins can reverse these changes in RAMP expression. For example, an endothelin A/B receptor antagonist reduced the upregulation of RAMP2 in the atria and ventricles of rats with congestive heart failure.[50] There is therefore scope to indirectly target RAMPs by altering their expression levels in disease conditions using drugs directed toward other targets.

E. The Use of Computational Methods for Development of Drugs to Target RAMPs

To date, a number of crystal structures have been reported for the N terminal domain of Family B GPCRs and this, in addition to the RAMP1 structure, raises the possibility of using computational techniques to exploit the interface region. Conceivably, the interface region between each RAMP and a receptor partner would be unique for that pairing, raising the possibility of therapeutics that would target the potential interface. However, this area of computational biology is still in its infancy and faces a combinatorial problem in that all six degrees of freedom (three rotational combined with translation in three perpendicular axis) will have to be searched to find a native-like docking pose. This approach is computationally demanding but in combination with mutational data or other computational techniques to narrow the search space, this approach may be tractable for rational drug design across a protein interface.

XII. Concluding Remarks

Many aspects of RAMP function remain poorly understood and the full extent of their action remains to be explored. The challenge now is to gain a greater understanding of their function, their properties and the relevance of their interaction with other proteins, particularly outside cultured cell systems. These unique receptor complexes are promising drug targets and this potential needs to be explored for development of the selective pharmaceutical intervention that these RAMP complexes are likely to allow for the treatment and management of many different disease conditions.

References

1. Hay DL, Poyner DR, Sexton PM. GPCR modulation by RAMPs. *Pharmacol Ther* 2006;**109**:173–97.
2. Perez DM, Karnik SS. Multiple signaling states of G-protein-coupled receptors. *Pharmacol Rev* 2005;**57**:147–61.
3. Jacoby E, Bouhelal R, Gerspacher M, Seuwen K. The 7 TM G-protein-coupled receptor target family. *ChemMedChem* 2006;**1**:761–82.
4. Kolakowski Jr LF. GCRDb: a G-protein-coupled receptor database. *Recept Channels* 1994;**2**:1–7.
5. Milligan G, Ramsay D, Pascal G, Carrillo JJ. GPCR dimerisation. *Life Sci* 2003;**74**:181–8.
6. Fluhmann B, Muff R, Hunziker W, Fischer JA, Born W. A human orphan calcitonin receptor-like structure. *Biochem Biophys Res Commun* 1995;**206**:341–7.
7. Njuki F, Nicholl CG, Howard A, Mak JC, Barnes PJ, Girgis SI, et al. A new calcitonin-receptor-like sequence in rat pulmonary blood vessels. *Clin Sci (Lond)* 1993;**85**:385–8.
8. Poyner DR, Sexton PM, Marshall I, Smith DM, Quirion R, Born W, et al. International Union of Pharmacology. XXXII. The mammalian calcitonin gene-related peptides, adrenomedullin, amylin, and calcitonin receptors. *Pharmacol Rev* 2002;**54**:233–46.
9. McLatchie LM, Fraser NJ, Main MJ, Wise A, Brown J, Thompson N, et al. RAMPs regulate the transport and ligand specificity of the calcitonin-receptor-like receptor. *Nature* 1998;**393**:333–9.
10. Christopoulos A, Christopoulos G, Morfis M, Udawela M, Laburthe M, Couvineau A, et al. Novel receptor partners and function of receptor activity-modifying proteins. *J Biol Chem* 2003;**278**:3293–7.
11. Harikumar KG, Simms J, Christopoulos G, Sexton PM, Miller LJ. Molecular basis of association of receptor activity-modifying protein 3 with the family B G protein-coupled secretin receptor. *Biochemistry* 2009;**48**:11773–85.
12. Bouschet T, Martin S, Henley JM. Receptor-activity-modifying proteins are required for forward trafficking of the calcium-sensing receptor to the plasma membrane. *J Cell Sci* 2005;**118**:4709–20.
13. Bouschet T, Martin S, Henley JM. Regulation of calcium-sensing-receptor trafficking and cell-surface expression by GPCRs and RAMPs. *Trends Pharmacol Sci* 2008;**29**:633–9.
14. Tilakaratne N, Smyth K, Popp B, Summers RJ, Sexton PM. Beta-adrenergic receptor properties in SK-N-MC cells are altered by receptor activity modifying proteins (RAMPs). In: *Proceedings of Australian health and medical research congress 2002*. 2002; Abstract 2145.

15. Kunz TH, Mueller-Steiner S, Schwerdtfeger K, Kleinert P, Troxler H, Kelm JM, et al. Interaction of receptor-activity-modifying protein1 with tubulin. *Biochim Biophys Acta* 2007;**1770**:1145–50.
16. Hay DL, Howitt SG, Conner AC, Schindler M, Smith DM, Poyner DR. CL/RAMP2 and CL/RAMP3 produce pharmacologically distinct adrenomedullin receptors: a comparison of effects of adrenomedullin22-52, CGRP8-37 and BIBN4096BS. *Br J Pharmacol* 2003;**140**:477–86.
17. Roh J, Chang CL, Bhalla A, Klein C, Hsu SY. Intermedin is a calcitonin/calcitonin gene-related peptide family peptide acting through the calcitonin receptor-like receptor/receptor activity-modifying protein receptor complexes. *J Biol Chem* 2004;**279**:7264–74.
18. Sexton PM, Christopoulos G, Christopoulos A, Nylen ES, Snider Jr. RH, Becker KL. Procalcitonin has bioactivity at calcitonin receptor family complexes: potential mediator implications in sepsis. *Crit Care Med* 2008;**36**:1637–40.
19. Katafuchi T, Hamano K, Minamino N. Identification, structural determination, and biological activity of bovine and canine calcitonin receptor-stimulating peptides. *Biochem Biophys Res Commun* 2004;**313**:74–9.
20. Fraser NJ, Wise A, Brown J, McLatchie LM, Main MJ, Foord SM. The amino terminus of receptor activity modifying proteins is a critical determinant of glycosylation state and ligand binding of calcitonin receptor-like receptor. *Mol Pharmacol* 1999;**55**:1054–9.
21. Fitzsimmons TJ, Zhao X, Wank SA. The extracellular domain of receptor activity-modifying protein 1 is sufficient for calcitonin receptor-like receptor function. *J Biol Chem* 2003;**278**:14313–20.
22. Udawela M, Christopoulos G, Tilakaratne N, Christopoulos A, Albiston A, Sexton PM. Distinct receptor activity-modifying protein domains differentially modulate interaction with calcitonin receptors. *Mol Pharmacol* 2006;**69**:1984–9.
23. Qi T, Christopoulos G, Bailey RJ, Christopoulos A, Sexton PM, Hay DL. Identification of N-terminal receptor activity-modifying protein residues important for calcitonin gene-related peptide, adrenomedullin, and amylin receptor function. *Mol Pharmacol* 2008;**74**:1059–71.
24. Udawela M, Hay DL, Sexton PM. The receptor activity modifying protein family of G protein coupled receptor accessory proteins. *Semin Cell Dev Biol* 2004;**15**:299–308.
25. Steiner S, Muff R, Gujer R, Fischer JA, Born W. The transmembrane domain of receptor-activity-modifying protein 1 is essential for the functional expression of a calcitonin gene-related peptide receptor. *Biochemistry* 2002;**41**:11398–404.
26. Udawela M, Christopoulos G, Morfis M, Christopoulos A, Ye S, Tilakaratne N, et al. A critical role for the short intracellular C terminus in receptor activity-modifying protein function. *Mol Pharmacol* 2006;**70**:1750–60.
27. Flahaut M, Rossier BC, Firsov D. Respective roles of calcitonin receptor-like receptor (CRLR) and receptor activity-modifying proteins (RAMP) in cell surface expression of CRLR/RAMP heterodimeric receptors. *J Biol Chem* 2002;**277**:14731–7.
28. Hilairet S, Foord SM, Marshall FH, Bouvier M. Protein-protein interaction and not glycosylation determines the binding selectivity of heterodimers between the calcitonin receptor-like receptor and the receptor activity-modifying proteins. *J Biol Chem* 2001;**276**:29575–81.
29. Kuwasako K, Shimekake Y, Masuda M, Nakahara K, Yoshida T, Kitaura M, et al. Visualization of the calcitonin receptor-like receptor and its receptor activity-modifying proteins during internalization and recycling. *J Biol Chem* 2000;**275**:29602–9.
30. Kuwasako K, Cao YN, Chu CP, Iwatsubo S, Eto T, Kitamura K. Functions of the cytoplasmic tails of the human receptor activity-modifying protein components of calcitonin gene-related peptide and adrenomedullin receptors. *J Biol Chem* 2006;**281**:7205–13.
31. Bomberger JM, Parameswaran N, Hall CS, Aiyar N, Spielman WS. Novel function for receptor activity-modifying proteins (RAMPs) in post-endocytic receptor trafficking. *J Biol Chem* 2005;**280**:9297–307.

32. Cottrell GS, Padilla B, Pikios S, Roosterman D, Steinhoff M, Grady EF, et al. Post-endocytic sorting of calcitonin receptor-like receptor and receptor activity-modifying protein 1. *J Biol Chem* 2007;**282**:12260–71.
33. Bomberger JM, Spielman WS, Hall CS, Weinman EJ, Parameswaran N. Receptor activity-modifying protein (RAMP) isoform-specific regulation of adrenomedullin receptor trafficking by NHERF-1. *J Biol Chem* 2005;**280**:23926–35.
34. Hilairet S, Belanger C, Bertrand J, Laperriere A, Foord SM, Bouvier M. Agonist-promoted internalization of a ternary complex between calcitonin receptor-like receptor, receptor activity-modifying protein 1 (RAMP1), and beta-arrestin. *J Biol Chem* 2001;**276**:42182–90.
35. Morfis M, Tilakaratne N, Furness SG, Christopoulos G, Werry TD, Christopoulos A, et al. Receptor activity-modifying proteins differentially modulate the G protein-coupling efficiency of amylin receptors. *Endocrinology* 2008;**149**:5423–31.
36. Muff R, Buhlmann N, Fischer JA, Born W. An amylin receptor is revealed following co-transfection of a calcitonin receptor with receptor activity modifying proteins-1 or -3. *Endocrinology* 1999;**140**:2924–7.
37. Tilakaratne N, Christopoulos G, Zumpe ET, Foord SM, Sexton PM. Amylin receptor phenotypes derived from human calcitonin receptor/RAMP coexpression exhibit pharmacological differences dependent on receptor isoform and host cell environment. *J Pharmacol Exp Ther* 2000;**294**:61–72.
38. Christopoulos G, Perry KJ, Morfis M, Tilakaratne N, Gao Y, Fraser NJ, et al. Multiple amylin receptors arise from receptor activity-modifying protein interaction with the calcitonin receptor gene product. *Mol Pharmacol* 1999;**56**:235–42.
39. Prado MA, Evans-Bain B, Dickerson IM. Receptor component protein (RCP): a member of a multi-protein complex required for G-protein-coupled signal transduction. *Biochem Soc Trans* 2002;**30**:460–4.
40. Dickerson IM, Loiseau SC. CGRP-receptor: a multi-protein complex required for G protein-coupled signal transduction, Poster 19. In: *Fifth annual joint meeting Great Lakes GPCR retreat, Nov 5–7, 2004.* ; 2004 Poster 19.
41. Flahaut M, Pfister C, Rossier BC, Firsov D. N-Glycosylation and conserved cysteine residues in RAMP3 play a critical role for the functional expression of CRLR/RAMP3 adrenomedullin receptor. *Biochemistry* 2003;**42**:10333–41.
42. Choksi T, Hay DL, Legon S, Poyner DR, Hagner S, Bloom SR, et al. Comparison of the expression of calcitonin receptor-like receptor (CRLR) and receptor activity modifying proteins (RAMPs) with CGRP and adrenomedullin binding in cell lines. *Br J Pharmacol* 2002;**136**:784–92.
43. Tsujikawa K, Yayama K, Hayashi T, Matsushita H, Yamaguchi T, Shigeno T, et al. Hypertension and dysregulated proinflammatory cytokine production in receptor activity-modifying protein 1-deficient mice. *Proc Natl Acad Sci USA* 2007;**104**:16702–7.
44. Zhang Z, Winborn CS, Marquez de Prado B, Russo AF. Sensitization of calcitonin gene-related peptide receptors by receptor activity-modifying protein-1 in the trigeminal ganglion. *J Neurosci* 2007;**27**:2693–703.
45. Dackor R, Fritz-Six K, Smithies O, Caron K. Receptor activity-modifying proteins 2 and 3 have distinct physiological functions from embryogenesis to old age. *J Biol Chem* 2007;**282**:18094–9.
46. Fritz-Six KL, Dunworth WP, Li M, Caron KM. Adrenomedullin signaling is necessary for murine lymphatic vascular development. *J Clin Invest* 2008;**118**:40–50.
47. Ichikawa-Shindo Y, Sakurai T, Kamiyoshi A, Kawate H, Iinuma N, Yoshizawa T, et al. The GPCR modulator protein RAMP2 is essential for angiogenesis and vascular integrity. *J Clin Invest* 2008;**118**:29–39.
48. Cueille C, Pidoux E, de Vernejoul MC, Ventura-Clapier R, Garel JM. Increased myocardial expression of RAMP1 and RAMP3 in rats with chronic heart failure. *Biochem Biophys Res Commun* 2002;**294**:340–6.

49. Totsune K, Takahashi K, Mackenzie HS, Murakami O, Arihara Z, Sone M, et al. Increased gene expression of adrenomedullin and adrenomedullin-receptor complexes, receptor-activity modifying protein (RAMP)2 and calcitonin-receptor-like receptor (CRLR) in the hearts of rats with congestive heart failure. *Clin Sci (Lond)* 2000;**99**:541–6.
50. Oie E, Vinge LE, Yndestad A, Sandberg C, Grogaard HK, Attramadal H. Induction of a myocardial adrenomedullin signaling system during ischemic heart failure in rats. *Circulation* 2000;**101**:415–22.
51. Qi YF, Shi YR, Bu DF, Pang YZ, Tang CS. Changes of adrenomedullin and receptor activity modifying protein 2 (RAMP2) in myocardium and aorta in rats with isoproterenol-induced myocardial ischemia. *Peptides* 2003;**24**:463–8.
52. Nishikimi T, Wang X, Akimoto K, Tadokoro K, Mori Y, Ishikawa Y, et al. Alteration of renal adrenomedullin and its receptor system in the severely hypertensive rat: effect of diuretic. *Regul Pept* 2005;**124**:89–98.
53. Mishima K, Kato J, Kuwasako K, Imamura T, Kitamura K, Eto T. Angiotensin II modulates gene expression of adrenomedullin receptor components in rat cardiomyocytes. *Life Sci* 2003;**73**:1629–35.
54. Nishikimi T, Yoshihara F, Kanazawa A, Okano I, Horio T, Nagaya N, et al. Role of increased circulating and renal adrenomedullin in rats with malignant hypertension. *Am J Physiol Regul Integr Comp Physiol* 2001;**281**:R2079–87.
55. Cao YN, Kitamura K, Kato J, Kuwasako K, Ito K, Onitsuka H, et al. Chronic salt loading upregulates expression of adrenomedullin and its receptors in adrenal glands and kidneys of the rat. *Hypertension* 2003;**42**:369–72.
56. Stachniak TJ, Krukoff TL. Receptor activity modifying protein 2 distribution in the rat central nervous system and regulation by changes in blood pressure. *J Neuroendocrinol* 2003;**15**:840–50.
57. Tadokoro K, Nishikimi T, Mori Y, Wang X, Akimoto K, Matsuoka H. Altered gene expression of adrenomedullin and its receptor system and molecular forms of tissue adrenomedullin in left ventricular hypertrophy induced by malignant hypertension. *Regul Pept* 2003;**112**:71–8.
58. Wang X, Nishikimi T, Akimoto K, Tadokoro K, Mori Y, Minamino N. Upregulation of ligand, receptor system, and amidating activity of adrenomedullin in left ventricular hypertrophy of severely hypertensive rats: effects of angiotensin-converting enzyme inhibitors and diuretic. *J Hypertens* 2003;**21**:1171–81.
59. Nagae T, Mukoyama M, Sugawara A, Mori K, Yahata K, Kasahara M, et al. Rat receptor-activity-modifying proteins (RAMPs) for adrenomedullin/CGRP receptor: cloning and upregulation in obstructive nephropathy. *Biochem Biophys Res Commun* 2000;**270**:89–93.
60. Márquez-Rodas I, Longo F, Aras-López R, Blanco-Rivero J, Diéguez E, Tejerina T, et al. Aldosterone increases RAMP1 expression in mesenteric arteries from spontaneously hypertensive rats. *Regul Pept* 2006;**134**:61–6.
61. Hiragushi K, Wada J, Eguchi J, Matsuoka T, Yasuhara A, Hashimoto I, et al. The role of adrenomedullin and receptors in glomerular hyperfiltration in streptozotocin-induced diabetic rats. *Kidney Int* 2004;**65**:540–50.
62. Thota C, Gangula PR, Dong YL, Yallampalli C. Changes in the expression of calcitonin receptor-like receptor, receptor activity-modifying protein (RAMP) 1, RAMP2, and RAMP3 in rat uterus during pregnancy, labor, and by steroid hormone treatments. *Biol Reprod* 2003;**69**:1432–7.
63. Dong YL, Green KE, Vegiragu S, Hankins GD, Martin E, Chauhan M, et al. Evidence for decreased calcitonin gene-related peptide (CGRP) receptors and compromised responsiveness to CGRP of fetoplacental vessels in preeclamptic pregnancies. *J Clin Endocrinol Metab* 2005;**90**:2336–43.

64. Ono Y, Okano I, Kojima M, Okada K, Kangawa K. Decreased gene expression of adrenomedullin receptor in mouse lungs during sepsis. *Biochem Biophys Res Commun* 2000;**271**:197–202.
65. Kitamuro T, Takahashi K, Totsune K, Nakayama M, Murakami O, Hida W, et al. Differential expression of adrenomedullin and its receptor component, receptor activity modifying protein (RAMP) 2 during hypoxia in cultured human neuroblastoma cells. *Peptides* 2001;**22**:1795–801.
66. Qing X, Svaren J, Keith IM. mRNA expression of novel CGRP1 receptors and their activity-modifying proteins in hypoxic rat lung. *Am J Physiol Lung Cell Mol Physiol* 2001;**280**:L547–54.
67. Kusano S, Kukimoto-Niino M, Akasaka R, Toyama M, Terada T, Shirouzu M, et al. Crystal structure of the human receptor activity-modifying protein 1 extracellular domain. *Protein Sci* 2008;**17**:1907–14.
68. Simms J, Hay DL, Wheatley M, Poyner DR. Characterization of the structure of RAMP1 by mutagenesis and molecular modeling. *Biophys J* 2006;**91**:662–9.
69. Kuwasako K, Kitamura K, Uemura T, Nagoshi Y, Kato J, Eto T. The function of extracellular cysteines in the human adrenomedullin receptor. *Hypertens Res* 2003;**26**(Suppl.):S25–31.
70. Steiner S, Born W, Fischer JA, Muff R. The function of conserved cysteine residues in the extracellular domain of human receptor-activity-modifying protein. *FEBS Lett* 2003;**555**:285–90.
71. Koller D, Ittner LM, Muff R, Husmann K, Fischer JA, Born W. Selective inactivation of adrenomedullin over calcitonin gene-related peptide receptor function by the deletion of amino acids 14-20 of the mouse calcitonin-like receptor. *J Biol Chem* 2004;**279**:20387–91.
72. Heroux M, Hogue M, Lemieux S, Bouvier M. Functional calcitonin gene-related peptide receptors are formed by the asymmetric assembly of a calcitonin receptor-like receptor homo-oligomer and a monomer of receptor activity-modifying protein-1. *J Biol Chem* 2007;**282**:31610–20.
73. Qi T, Simms J, Bailey RJ, Wheatley M, Rathbone DL, Hay DL, et al. Structure-function analysis of RAMP1-RAMP3 chimeras. *Biochemistry* 2009;**49**:522–31.
74. Zumpe ET, Tilakaratne N, Fraser NJ, Christopoulos G, Foord SM, Sexton PM. Multiple ramp domains are required for generation of amylin receptor phenotype from the calcitonin receptor gene product. *Biochem Biophys Res Commun* 2000;**267**:368–72.
75. Kuwasako K, Kitamura K, Nagoshi Y, Cao YN, Eto T. Identification of the human receptor activity-modifying protein 1 domains responsible for agonist binding specificity. *J Biol Chem* 2003;**278**:22623–30.
76. Kuwasako K, Kitamura K, Ito K, Uemura T, Yanagita Y, Kato J, et al. The seven amino acids of human RAMP2 (86) and RAMP3 (59) are critical for agonist binding to human adrenomedullin receptors. *J Biol Chem* 2001;**276**:49459–65.
77. Kuwasako K, Kitamura K, Onitsuka H, Uemura T, Nagoshi Y, Kato J, et al. Rat RAMP domains involved in adrenomedullin binding specificity. *FEBS Lett* 2002;**519**:113–6.
78. Simms J, Hay DL, Bailey RJ, Konycheva G, Bailey G, Wheatley M, et al. Structure-function analysis of RAMP1 by alanine mutagenesis. *Biochemistry* 2009;**48**:198–205.
79. Mallee JJ, Salvatore CA, LeBourdelles B, Oliver KR, Longmore J, Koblan KS, et al. Receptor activity-modifying protein 1 determines the species selectivity of non-peptide CGRP receptor antagonists. *J Biol Chem* 2002;**277**:14294–8.
80. Barwell J, Miller PS, Donnelly D, Poyner DR. Mapping interaction sites within the N-terminus of the calcitonin gene-related peptide receptor; the role of residues 23-60 of the calcitonin receptor-like receptor. *Peptides* 2009;**31**:170–6.
81. Grace CR, Perrin MH, DiGruccio MR, Miller CL, Rivier JE, Vale WW, et al. NMR structure and peptide hormone binding site of the first extracellular domain of a type B1 G protein-coupled receptor. *Proc Natl Acad Sci USA* 2004;**101**:12836–41.

82. Grace CR, Perrin MH, Gulyas J, Digruccio MR, Cantle JP, Rivier JE, et al. Structure of the N-terminal domain of a type B1 G protein-coupled receptor in complex with a peptide ligand. *Proc Natl Acad Sci USA* 2007;**104**:4858–63.
83. Parthier C, Kleinschmidt M, Neumann P, Rudolph R, Manhart S, Schlenzig D, et al. Crystal structure of the incretin-bound extracellular domain of a G protein-coupled receptor. *Proc Natl Acad Sci USA* 2007;**104**:13942–7.
84. Doods H, Arndt K, Rudolf K, Just S. CGRP antagonists: unravelling the role of CGRP in migraine. *Trends Pharmacol Sci* 2007;**28**:580–7.
85. Hay DL, Christopoulos G, Christopoulos A, Sexton PM. Determinants of 1-piperidinecarboxamide, *N*-[2-[[5-amino-l-[[4-(4-pyridinyl)-l-piperazinyl]carbonyl]pentyl]amino]-1-[(3, 5-dibromo-4-hydroxyphenyl)methyl]-2-oxoethyl]-4-(1,4-dihydro-2-oxo-3(2*H*)-quinazolinyl) (BIBN4096BS) affinity for calcitonin gene-related peptide and amylin receptors—the role of receptor activity modifying protein 1. *Mol Pharmacol* 2006;**70**:1984–91.
86. Salvatore CA, Mallee JJ, Bell IM, Zartman CB, Williams TM, Koblan KS, et al. Identification and pharmacological characterization of domains involved in binding of CGRP receptor antagonists to the calcitonin-like receptor. *Biochemistry* 2006;**45**:1881–7.
87. Miller PS, Barwell J, Poyner DR, Wigglesworth MJ, Garland SL, Donnelly D. Non-peptidic antagonists of the CGRP receptor, BIBN4096BS and MK-0974, interact with the calcitonin receptor-like receptor via methionine-42 and RAMP1 via tryptophan-74. *Biochem Biophys Res Commun* 2009;**391**:437–42.
88. Benemei S, Nicoletti P, Capone JA, Geppetti P. Pain pharmacology in migraine: focus on CGRP and CGRP receptors. *Neurol Sci* 2007;**28**(Suppl 2):S89–93.
89. Yin H, Slusky JS, Berger BW, Walters RS, Vilaire G, Litvinov RI, et al. Computational design of peptides that target transmembrane helices. *Science* 2007;**315**:1817–22.
90. Ryan G, Briscoe TA, Jobe L. Review of pramlintide as adjunctive therapy in treatment of type 1 and type 2 diabetes. *Drug Des Devel Ther* 2009;**2**:203–14.
91. Robinson SD, Aitken JF, Bailey RJ, Poyner DR, Hay DL. Novel peptide antagonists of adrenomedullin and calcitonin gene-related peptide receptors: identification, pharmacological characterization, and interactions with position 74 in receptor activity-modifying protein 1/3. *J Pharmacol Exp Ther* 2009;**331**:513–21.
92. Pidasheva S, Canaff L, Simonds WF, Marx SJ, Hendy GN. Impaired cotranslational processing of the calcium-sensing receptor due to signal peptide missense mutations in familial hypocalciuric hypercalcemia. *Hum Mol Genet* 2005;**14**:1679–90.

Regulators of G Protein Signaling Proteins as Targets for Drug Discovery

BENITA SJÖGREN,[*]
LEVI L. BLAZER,[*] AND
RICHARD R. NEUBIG[*,†]

[*]*Department of Pharmacology, University of Michigan, Ann Arbor, Michigan, USA*
[†]*Department of Internal Medicine (Cardiovascular Medicine), University of Michigan, Ann Arbor, Michigan, USA*

Signaling via G-protein-coupled receptors (GPCRs) is central for the function of biological systems. Many clinically used drugs target GPCRs directly or target molecules involved in GPCR signaling. As an alternative to targeting receptors directly, one could modulate signaling cascades downstream of receptor activation. In recent years, there has been substantial interest in a family of proteins called regulators of G protein signaling (RGS) proteins. They modulate GPCR signaling by accelerating GTP hydrolysis on active Gα subunits, thereby reducing the amplitude and duration of signaling. Modulating RGS activity would be a useful strategy to control GPCR signaling. An RGS

Progress in Molecular Biology
and Translational Science, Vol. 91
DOI: 10.1016/S1877-1173(10)91004-1

1877-1173/10 $35.00

inhibitor would be expected to enhance GPCR signaling and could do so in a tissue- or pathway-specific manner. Apart from the central GAP (GTPase accelerating protein) activity, many RGS proteins also have other functions like regulating protein–protein interactions, subcellular localization of signaling molecules, and protein translation. It is clear that these proteins serve important functions in a number of physiological and pathophysiological processes, and they are emerging as potential drug targets. This chapter gives an overview of what is currently known about biological functions of RGS proteins based on *in vivo* and *in vitro* data. We also summarize the current status in targeting RGS proteins in drug discovery.

I. Introduction

G-protein-coupled receptors (GPCRs) represent one of the largest families of proteins in the human genome with more than 1000 members, including olfactory receptors. They are important therapeutically, as about 30% of Federal Drug Administration (FDA) approved drugs target these receptors.[1] GPCRs couple to heterotrimeric G proteins, consisting of an α and a $\beta\gamma$ subunit. When the G protein is in its inactive state, the α subunit is bound to guanosine diphosphate (GDP). When activated, the GDP is exchanged for guanosine triphosphate (GTP) and the α subunit dissociates from the $\beta\gamma$ subunits. Both entities can then stimulate downstream effectors. The G protein is inactivated when GTP is hydrolyzed back to GDP. Although G proteins possess the ability to hydrolyze GTP, the process is very slow and does not account for the rate at which GPCR signaling is terminated *in vivo*. In the mid-1990s, a family of proteins was identified that is responsible for the physiologically observed subsecond regulation of G protein signaling. These proteins, termed regulators of G protein signaling (RGS) proteins,[2,3] modulate GPCR signaling by accelerating the rate of GTP hydrolysis by Gα, thereby reducing amplitude and duration of G protein signaling (Fig. 1). These GTPase accelerating proteins (GAPs) can increase the rate of GTP hydrolysis $>$ 1000-fold.[3,4] RGS proteins were first discovered in *Saccharomyces cerevisiae*, where Sst2 was identified as a novel protein that could inhibit G protein function.[5] EGL-10[6] was then identified in *Caenorhabditis elegans* as a regulator of serotonin signaling that shared domains with high similarity to mammalian proteins. Since the initial reports on RGS proteins,[6–8] more than 30 human RGS-domain-containing proteins have been discovered with homologs in yeast and worms, and numerous reports have been published indicating the potential biological and clinical implications for RGS proteins as novel drug targets (e.g., Refs. 9–18). RGS proteins bind to and primarily modulate signals

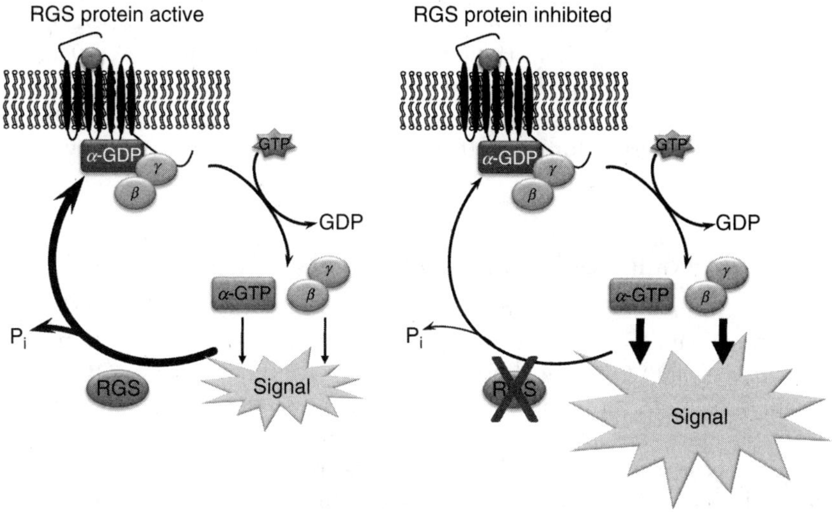

FIG. 1. A model of the rationale for using RGS inhibitors in therapeutics. Binding of an agonist to the GPCR induces a conformational change on the G protein which enables the exchange of GDP for GTP on the α subunit. The active α and $\beta\gamma$ subunits dissociate and can mediate downstream signals. The amplitude and duration of the signal is dependent on the rate of activation as well as the rate of inactivation through GTP hydrolysis. This process is accelerated by RGS proteins, returning the G protein to its heterotrimeric GDP-bound inactive state. In the absence of an RGS protein (in the presence of an RGS protein inhibitor), the rate of GTP hydrolysis is radically reduced. The G protein subunits stay active, and the signal is increased.

mediated through α subunits of G proteins of the Gq and Gi/o family. To date, there are no widely accepted reports of an RGS protein that binds to α subunits of Gs proteins.

Many review articles have been published on RGS proteins, their structure and function as well as their potential as targets in drug discovery. Their main focus has been on *in vitro* findings. In the present chapter, we focus on current knowledge on RGS protein function *in vivo*. We also give an update on strategies that have been or could be taken to screen for and characterize modulators of RGS protein function.

II. RGS Protein Families

More than 30 RGS-domain-containing proteins have been identified in humans. These are divided into subfamilies based on sequence homology. The classification of RGS protein subfamilies has been extensively reviewed, and there are some discrepancies in the nomenclature (e.g., Refs. 2,3,19,20); we will use that of Hollinger and Hepler.[2] Apart from their common ~120 amino acid RGS domain, many RGS proteins also contain domains for protein–protein interactions, subcellular localization, target specificity, and protein stability, which permit functions that go beyond the simple GAP activity for $G\alpha$ subunits.[21]

A. The R4 Family of RGS Proteins

The R4 family of RGS proteins (RGS1–5, 8, 13, 16, 18, and 21) is the largest RGS protein family. Besides the RGS domain, nearly all of its members include only small N- and C-terminal extensions. The N-terminus of most R4 family members contains an amphipathic helix, which is important for membrane and/or receptor association and can bind phospholipids directly.[22,23] The N-terminal can also serve as a site for protein–protein interactions. The exception to the short N-terminus characteristic of the R4 family is RGS3 which is expressed as three isoforms. RGS3S is similar in structure to the other members of the R4 family, but the two longer splice variants RGS3L and PDZ-RGS3 contain additional domains with functions involved in protein–protein interactions.[24]

Despite the lack of apparent targeting domains (with the exception of PDZ-RGS3), there are emerging data that suggest that R4 family members have physiological roles beyond their GAP activity. In contrast to the RGS domain, which is well conserved among members of the R4 family of RGS proteins, the N- and C-termini vary, enabling specificity of non-GAP functions of these proteins. RGS16, but not RGS4, binds to $G\alpha_{13}$ through its N-terminus, thereby inhibiting $G\alpha_{13}$-Rho signaling.[25] Also, in a recent report, RGS2 was shown to suppress protein translation through a 37-residue stretch in its N-terminus partly overlapping the RGS domain. That region interacts with and inhibits the eukaryotic initiation factor 2Bε (eIF2Bε).[26] Thus, despite their small size, R4 family RGS proteins have other functions besides GAP activity for $G\alpha$ protein subunits.

The G protein specificity of R4 family members is broad. Members of the R4 family of RGS proteins bind to and function as GAPs for all Gq or Gi/o family members ($G\alpha i_{1-3}$, $G\alpha_o$, $G\alpha_z$, transducin, $G\alpha_{11}$, $G\alpha_{14/15}$, and $G\alpha_{16}$).[27,28] One exception is RGS2, which is selective for $G\alpha_q$ at least *in vitro*.[29]

B. The R7 Family of RGS Proteins

The R7 family (consisting of RGS6, 7, 9, and 11) is closely related to the *C. elegans* homologs EGL-10 and EAT-16. These proteins have larger structures than the R4 family members including several domains in addition to the RGS domain. One of these is the disheveled-EGL10-Pleckstrin homology (DEP) domain, which is responsible for protein–protein interactions.[30,31] In addition to receptors,[32,33] two proteins, R9AP (for *R*GS9 *A*nchoring *P*rotein) in the retina and R7BP (for *R7 B*inding *P*rotein) in the brain, have been identified as key R7 family interacting proteins. They anchor R7 family RGS proteins to the plasma membrane via the DEP domain.[34] In the case of RGS9, this markedly increases protein stability, something that does not seem to apply to the other R7 family members.[35] Membrane targeting through binding to R7BP also increases the efficiency for R7 family RGS proteins to act as GAPs on G proteins.[36] All four members of the R7 family of RGS proteins also contain a G-γ-like (GGL) domain that binds to $G\beta_5$, an interaction that is necessary for stable expression of these proteins. Mice lacking Gβ5 have reduced expression levels of all R7 family RGS proteins, supporting the importance of the interaction.[37] The X-ray structure of the RGS9–$G\beta_5$ complex was recently solved, revealing a large interaction interface consisting of 54 residues in $G\beta_5$ and 39 residues in the GGL domain of RGS9, with a total buried surface area of 4700 $Å^2$.[38]

Members of the R7 family of RGS proteins show selectivity for $G\alpha_{i/o}$ proteins and do not bind $G\alpha_q$. However, within the R7 family the selectivity toward $G\alpha_{i/o}$ protein subtypes varies. While the efficacy toward $G\alpha_o$ is similar among the four proteins, RGS6 and RGS7 have lower potency to function as GAPs for $G\alpha_{i1}$, $G\alpha_{i2}$, and $G\alpha_{i3}$ than $G\alpha_o$ *in vitro*.[39] Hence, RGS6 and RGS7 are selective for $G\alpha_o$, while RGS9 and RGS11 are more promiscuous among Gi/o family members.

C. The R12 Family of RGS Proteins

The R12 family of RGS proteins consists of RGS10, RGS12, and RGS14. There has previously been some debate on whether RGS10 belongs to this family, due to its lack of structural domains present in the other two members[3]; however, the sequence of its RGS domain clearly places it in this family. RGS10 is a small protein consisting of the RGS domain and small N- and C-termini, similar to members of the R4 family. RGS12 and 14 are larger proteins with additional domains for protein–protein interactions. One of these is the $G\alpha_{i/o}$-Loco (GoLoco) motif, which also binds to Gα. It has GDI activity (for *G*uanine nucleotide *D*issociation *I*nhibitor) toward $G\alpha_{i1}$, $G\alpha_{i2}$, and $G\alpha_{i3}$.[40,41] Binding of the GoLoco motif to Gα inhibits the exchange of GDP for GTP and thereby prevents activation of the G protein. It also blocks association Gα with $G\beta\gamma$ potentially leading to prolonged $\beta\gamma$ signaling. This provides a more complex mode of regulating G protein signaling than just reducing amplitude and duration through GAP activity.

Similar to the R7 family of RGS proteins, RGS12 and RGS14 also contain membrane targeting protein–protein interaction domains. One splice variant of RGS12 includes an N-terminal PDZ domain that has been shown to interact with the C-terminus of GPCRs including the CXCR2 receptor.[42,43] RGS12 also contains a phosphotyrosine binding (PTB) domain shown to interact with the pore-forming unit of the N-type calcium channel in a phosphorylation-dependent manner.[44]

D. The RZ Family of RGS Proteins

The RZ family of RGS proteins is less well characterized than the R4 or R7 families. It consists of, RGS17 (also known as RGSZ2), RGS19 (also known as GAIP) and RGS20 (with several splice variants including RGSZ1 and Ret-RGS), all with an N-terminal cysteine string motif (reviewed in Ref. 45). The members of the RZ protein family are palmitoylated at several sites, most probably in the cysteine string. The function of this palmitoylation is not yet fully understood, but studies of other proteins with such modifications suggest that it could be a signal for subcellular trafficking or protein stability or to enable protein–protein interactions (reviewed in Ref. 46). The cysteine string could also function as a protein–protein interaction domain. RGS20 and RGS17 interact with the protein kinase C interacting protein 1 (PKCI-1) through their N-termini.[47] This interaction is specific to members of the RZ family since no interaction occurred between PKCI-1 and RGS7 or RGS2. PKCI-1 was also shown to phosphorylate RGS20 and RGS17, thereby inhibiting the ability of the RGS protein to function as a GAP on $G\alpha$ proteins.

Members of the RZ family of RGS proteins are the only RGS proteins that have selective GAP activity on $G\alpha_z$, a function that resulted in the name of the family. $G\alpha_z$ is a member of the Gi/o family with the distinction that it is insensitive to pertussis toxin (PTX).[48] Subsequent studies showed GAP activity for other members of the Gi/o family with varying selectivity among the RZ family members. RGS20 is relatively selective for $G\alpha_z$, although it does have activity toward other Gi/o proteins as well.[49,50] In contrast, RGS17 seems to be relatively nonselective, with similar efficiencies in GAP activity toward all members of the $G\alpha_{i/o}$ members although with some preference toward $G\alpha_z$ and $G\alpha_o$.[51] In contrast to the other RZ family members, RGS17 has been co-immunoprecipitated with $G\alpha_q$ but there are no reports on GAP activity for RGS17 on $G\alpha_q$.[51]

E. RGS Homology Domain-Containing Proteins

The 20 RGS proteins described above could be considered "canonical" or "classic" RGS proteins since they all bind to GTP-bound $G\alpha$ subunits and strongly stimulate GTPase activity. Consequently, a major role is likely to be turnoff of $G\alpha$ signaling and/or $G\alpha$-state-dependent recruitment to cellular sites. In addition to the 20 canonical RGS proteins, there are 19 other proteins that have a $\sim$ 120–130 amino acid long region that has substantial homology to

RGS proteins. These have been termed RGS homology (RH) domains. The RH-domain-containing proteins are the subject of a recent comprehensive review discussing their structure and function.[52] The seven families of RH-domain-containing proteins include G protein-coupled receptor kinases (GRKs), Rho guanine nucleotide exchange factors (RhoGEFs), Axin, an A-kinase anchoring protein (DAKAP-2), a subset of sorting nexins (SNXs), and two RGS-like (RGSL) proteins of unknown structure and function. Although a number of these are involved in signal transduction, they appear not to act via the same mechanism as RGS proteins.

Two of the RH-domain-containing protein families (GRK and RhoGEF) have been convincingly shown to bind Gα subunits. The RH domains from GRK2 and GRK3 bind to activated Gα_q and appear to inhibit its function, possibly by sterically blocking binding of the effector phospholipase C.[53] The three RH-domain-containing RhoGEFs (p115-RhoGEF, PDZ-RhoGEF, and leukemia-associated RhoGEF also known as. LARG) bind to $G_{12/13}$ family Gα subunits.[54] Instead of inhibiting $G_{12/13}$ function, the RhoGEFs appear to be the major effector of $G_{12/13}$ action.[55] Receptor-mediated $G_{12/13}$ activation enhances RhoGEF activity by causing a conformational change in the RhoGEF protein[54] and/or by recruitment of the RhoGEF to the plasma membrane[56] where inactive Rho-GDP resides. This leads to a unique signal switch from a heterotrimeric G protein through the RhoGEF to a small ras-superfamily G protein (RhoA or RhoC).

As detailed in the review by Tesmer,[52] crystal structures have been solved for both the GRK/Gα_q and the RhoGEF/Gα_{13} complexes. Interestingly, the mode of interaction of the RH domain with the Gα subunit in both cases is markedly different from that in the RGS/Gα subunit complexes with canonical RGS proteins. As described below, the Gα subunit bound to RGS proteins contacts the "A site" at the junction between the bundle subdomain and the terminal subdomain including contacts with loops between helix pairs $\alpha3-\alpha4$ and $\alpha5-\alpha6$ (see Pocket 1 in Fig. 2). In contrast, the Gα_q binding site on the GRK-RH is along the length of $\alpha5-\alpha6$ pair,[57–59] while Gα_{13} binds to a C-terminal region of the p115 and PDZ-RhoGEF RH domains.[60,61] Interestingly, in both cases the contact site on the Gα subunit involves a region of Gα that is important for effector binding and does not overlap with the site at which canonical RGS proteins would be expected to bind (Refs. 58,59 and see Ref. 52 for discussion). This sets up an interesting scenario (at least for Gα_q) in which an RGS protein (e.g., RGS4) could deactivate Gα_q and reverse the binding of GRK.

For the other RH-domain-containing proteins, there have not been definitively replicated studies to document binding to Gα subunits. Additional information about their structure and function can be found in the recent review from Tesmer.[52]

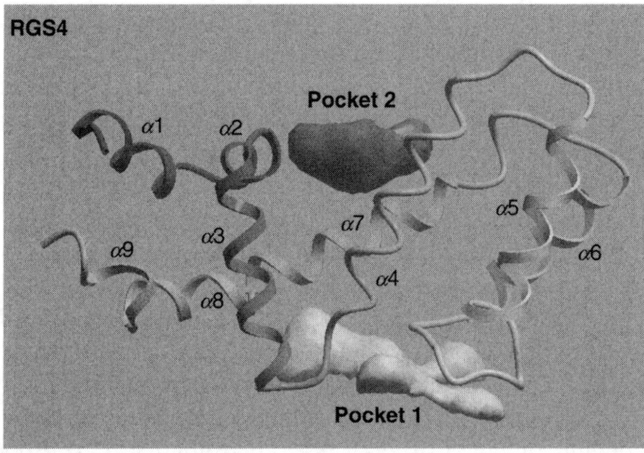

FIG. 2. Potential drug binding pockets on the surface of RGS4. The crystal structure of RGS4 (PDB: 1agr) was used to identify potential drug binding pockets on the surface of the protein. The pocketfinder algorithm in ICM Pro (Molsoft, San Diego, CA) was used, and the two most favorable pockets are illustrated. The α helices in the RGS structure are numbered 1 to 9 from N- to C-terminus. Pocket 1 (top) is the site of contact between the RGS4 and $G\alpha_{i1}$ subunit in the cocrystal (also known as the A site as defined by Zhong and Neubig[20]) and is adjacent to residues 85, 87, 88, 121–129, 134, 159, 163, and 164. It has a volume of 129 Å^3, area of 258 Å^2, and a nonsphericity of 2.09. Pocket 2 (bottom) is on the "back" of the molecule and may represent the "B site" of Zhong and Neubig. It is adjacent to residues 64, 67, 68, 92, 96, 99, 150, 153, 154, and 157. It has a volume of 151 Å^3, area of 177 Å^2, and a nonsphericity of 1.29. Pocket B is more similar in size and shape to "druggable" pockets on other proteins. Also, it is directly adjacent to cysteines 95 and 148, which may represent sites of allosteric sulfhydryl modification (see text).

III. RGS Protein Mechanisms of Actions

There is a wealth of structural and functional studies regarding the mechanism of action of RGS proteins. The activity of RGS proteins as Gα GAPs was first demonstrated *in vitro* by Berman et al. in 1996.[62] They and others[28,62–64] demonstrated that RGS proteins bind directly to GTP-bound Gα subunits and accelerate the rate of GTP hydrolysis. This mechanism seemed to explain previous cellular studies such as where the yeast protein, Sst2p, was able to negatively modulate signaling by the α factor pheromone—a G protein (Gpa1p)-dependent process.[5,65] Subsequent biochemical studies with Gpa1p and Sst2p confirmed that prediction.[66] From these studies sprang a wealth of publications that characterize the effects of RGS proteins both *in vitro* and *in vivo* (reviewed in Refs. 15,30,67).

The molecular mechanism underlying RGS GAP activity was further expounded by the publication of an RGS4–$G\alpha_{i1}$ complex structure (PDB ID 1AGR, Ref. 68). In this structure, $G\alpha_{i1}$ is bound with GDP and aluminum fluoride, with the planar AlF_4^- in the position normally occupied by the γ-phosphate in GTP. This is similar to the catalytic transition state for GTP hydrolysis. Since RGS proteins mediate catalysis in part by stabilizing this transition state, this structure is particularly useful in understanding the molecular mechanisms underlying the GAP activity of RGS proteins. A large number of structures of other RH-domain-containing proteins have been obtained (over 20 RGS or RH domain structures are present in the RCSB protein data bank), and the diversity of these structures and their Gα interactions have been thoroughly reviewed recently.[52]

The canonical RGS/RH domain structure (Fig. 2) is a series of nine α helices that form two lobes. The smaller lobe contains the N- and C-termini and α1, α2, α3, α8, and α9, while the larger lobe comprises a four-helix antiparallel bundle including helices α4, α5, α6, and α7. The contact site on RGS4 for $G\alpha_{i1}$ (1AGR), as well as that for other R4 and R7 family RGS proteins, encompasses regions of both lobes and covers greater than 1000 Å^2 of surface area. This interaction interface is comprised of the loops between α3/α4, α5/α6, and the residues at the ends of helices α7 and α8. This Gα contact surface is located on the "bottom" of the RGS protein in its usual display representation (Pocket 1 in Fig. 2) and has been termed the "A" site for RGS small molecule targeting.[69]

The RGS protein binding site on $G\alpha_{i1}$ is particularly informative in regard to the structural underpinnings of RGS GAP activity and the potential for small molecule targeting of this interface. In the RGS4–$G\alpha_{i1}$ complex structure, RGS4 interacts almost exclusively with the three switch regions of $G\alpha_{i1}$, sites which are critical for regulation of GTP hydrolysis and effector binding. Upon binding, RGS4 reduces the mobility of all three switch regions and restricts them into a conformation that promotes GTP hydrolysis. The binding is primarily driven by hydrogen bonding and electrostatic interactions, with only three hydrophobic side chain residues contributing. The hydrophilicity of the interaction also makes this site more difficult to target with standard "drug-like" small molecules.

Several reports in the literature suggest that RGS4 GAP activity can be modulated by endogenous phospholipids and calmodulin.[70–72] The binding sites for at least some of these regulators are in allosteric sites on the RGS protein. Relevant to allosteric modulation, several RGS proteins can be palmitoylated in a manner that affects GAP activity. Palmitoylation of RGS proteins was first observed for RGS19 and was subsequently expanded to include most of the canonical RGS protein families.[64,73–76] There are three sites for palmitoylation on RGS4, two located at the N-terminus (C2, C12) which are important for plasma membrane localization[76] and one (Cys^{95}) located within the RGS domain on helix 4.[77] In RGS4, palmitoylation of Cys^{95} inhibits GAP activity both in solution-phase single-turnover experiments and in receptor-stimulated steady-state GTPase experiments.[78] This cysteine residue is located on the "back" side of the RGS protein, far removed from the site of Gα interactions. This allosteric regulatory site (termed the "B site") contains a smaller, more hydrophobic pocket (Pocket 2 in Fig. 2) than does the Gα interface site. This site may be more amenable to traditional small molecule targeting approaches given the more favorable binding pocket geometry (Fig. 2). At least one member of every major family of GAP-active RH-domain-containing proteins has a cysteine residue on helix 4, which corresponds to Cys^{95} in RGS4. Interestingly, in certain RGS proteins (e.g., RGS10, RGS16), palmitoylation of the cysteine at this site actually potentiates GAP activity, at least in liposomes where this palmitoylation may assist in membrane localization or proper orientation at the membrane.[77,78] While the functionality of this cysteine has not been fully characterized in all RGS proteins, this level of conservation suggests that palmitoylation at this site may be physiologically relevant.

RGS4 is also reciprocally regulated by PIP_3 and calmodulin at a site close to the site of palmitoylation.[70–72] PIP_3 and phosphatidic acid, but not PIP_2, IP_3, or diacylglycerol (DAG), inhibit the activity of RGS4 in purified protein single-turnover GAP assays.[72] The binding site of PIP_3 appears to include regions of helix 4 and 5, which are close to the Cys^{95} palmitoylation site. It is also near the site at which calmodulin binds to RGS4. Calmodulin binding to RGS4 is calcium-dependent and noncompetitive with G protein binding or RGS GAP activity. Furthermore, calmodulin is able to inhibit the effect of PIP_3 on RGS4 GAP activity. This phospholipid/calmodulin-mediated regulation is believed to be a mechanism for feedback regulation of RGS activity and has been proposed to describe the RGS-dependent calcium oscillations observed after activation of Gq-coupled receptors.[79,80] Since this site has been designed by nature to provide allosteric regulation of an RGS protein by an organic molecule, we propose that this site will be more amenable to small molecule targeting than the expansive Gα–RGS interaction interface. In fact, the small molecule RGS inhibitor CCG-4986 appears to bind at this site (Ref. 81 and 82), and other small molecule RGS inhibitors that have been developed also appear to bind directly to this portion of the molecule (Ref. 83).

IV. Biological Functions of RGS Proteins

Signaling through GPCRs regulates functions in every aspect of physiology and is the single most prominent pathway in drug discovery. Given their role in regulating GPCR function, it is not surprising that RGS proteins have been hypothesized to regulate diverse physiological processes.[20,69] Potential pathophysiological roles have also been suggested for RGS proteins in disorders of the central nervous system (CNS) and cardiovascular system as well as in cancer and regulation of the immune system.

To understand the role of RGS proteins as potential drug targets, it would be helpful to consider new data on mouse models with perturbed RGS function (summarized in Table I). Although some of the phenotypes could be masked by compensatory effects from other RGS proteins, these knock-out models have greatly increased our knowledge of the biological function of several RGS proteins. Several RGS knock-out models have strong phenotypes such as RGS2 knock-out mice, which are hypertensive, and RGS9 knock-out mice, which show motor deficiencies and markedly altered pharmacological responses (see below). In parallel with ongoing developments with knock-out mice for specific RGS proteins, our group has taken another approach using knock-in mice expressing RGS-insensitive mutants of Gα proteins. These mice express a Gα subunit with a mutation in the Switch I region (G184S) that disrupts the interaction between Gα and RGS, making it resistant to the GAP activity of the RGS protein.[4] In these models, the overall effects of RGS proteins can be studied on a specific pathway. They provide strong phenotypes since RGS redundancy cannot overcome the effect, but they cannot immediately identify the specific RGS protein responsible for any phenotype. To date, RGS-insensitive mutants of Gα_q, Gα_o, and Gα_{i2} have all been characterized.[12,16,84] Despite these recent advances outlined below, a full understanding of the *in vivo* functions of RGS proteins will require additional data from other knock-out models and other approaches.

A. RGS Proteins in the Cardiovascular System

Many signals regulating cardiovascular function are mediated via GPCRs. Numerous hormones and neurotransmitters, such as noradrenaline, angiotensin II, serotonin (5-hydroxytryptamine; 5-HT), endothelin, and acetylcholine have strong cardiovascular actions. Hypertension is a common problem in industrialized countries and is a risk factor for developing further complications, such as heart disease and stroke. Hypertension is strongly associated with increased GPCR signaling, and many drugs for treating this condition target these receptors (e.g., α- and β-adrenergic receptor and angiotensin II receptor antagonists) or related processes (e.g., angiotensin-converting enzyme (ACE) inhibitors). Among the many genes that have been implicated in regulating blood pressure, several RGS proteins, including RGS2, RGS4, and RGS5, have been proposed to play a role.

The importance of RGS proteins for cardiovascular functions *in vivo* was first demonstrated in the RGS2 knock-out mice[18,85] which are hypertensive. Further studies revealed a role for RGS2 in regulating smooth muscle cell contraction by modulating signaling via the PAR-1 receptor.[86] RGS2 is a

TABLE I

BIOLOGICAL FUNCTIONS OF RGS PROTEINS

Model	Cardiovascular phenotype	Neuronal phenotype	Other phenotypes	References
$RGS1^{(-/-)}$			B cell response to chemokines ↑	173,174
$RGS2^{(-/-)}$	Hypertension[a]; prolonged response to vasoconstrictors	Long-term potentiation (LTP) ↑	Impaired immune response	18,26,78,85,88,175
		Anxiety response ↑	(T-cell proliferation, IL-2 ↓)	
		Male aggression ↓	Protein translation ↑	
$RGS4^{(-/-)}$	Muscarinic-induced bradycardia ↑	Opiate dependence ↑ (Local KO in NAc)	Circulating FFA ↑	11,96,101,113,120
	Renal blood flow ↓	Audiogenic seizures ↓	Glucose tolerance ↓	
	Sensitivity to renal toxicity ↑	*Note: Whole-body KO without effect on opiate function*		
$RGS5^{(-/-)}$	Hypotension		Tumor vessel normalization	10,99
$RGS9^{(-/-)}$		Adaption to light changes[a] ↓ (RGS9-1)		9,14,106,115,117
		Morphine response ↑		
		Motor coordination and working memory ↓		
		Drug-induced dyskinesia ↑		
$RGS10^{(-/-)}$		Microglial inflammatory response ↑	Osteoclast differentiation ↓	118,176
$RGS13^{(-/-)}$			CREB-dependent transcription (B cells) ↑	142
$G\beta_5^{(-/-)}$			Expression of R7 family of RGS proteins ↓	37
			Reduced body weight during weaning	
			Increased mortality during weaning	

R7BP$^{(-/-)}$		Morphine response ↑		35,177
		Motor coordination and working memory ↓		
		RGS9-2 expression ↓		
R9AP		Mutations related to bradyopsia[a]		104
$G\alpha_{i2}$G184S	Muscarinic-induced bradycardia ↑	Antidepressant	Insulin sensitivity, glucose tolerance ↑	16,84,178
$G\alpha_{o}$G184S		Sensitivity to Epinephrine-mediated reduction of CA3 epileptiform bursts ↑		119

KO, knock-out; R7BP, R7 binding protein; NAc, nucleus accumbens; IL-2, interleukin 2; FFA, free fatty acids; CREB, cAMP response element binding protein.
[a]Phenotypes of human RGS mutations, RGS knock-out mice, and RGS protein-related knock-out and transgenic models.

relatively specific and highly potent inhibitor of Gα_q signaling.[29,87] It suppresses signaling via multiple receptors (noradrenaline, angiotensin II, vasopressin, and endothelin) in vascular smooth muscle cells (VMSCs); all represent potent vasoconstrictor agonists. Furthermore, RGS2 appears to be an effector of NO-mediated vasodilation[88] as well as directly binding and modulating certain adenylyl cyclase subtypes,[89] which could be important in signaling via dopamine and vasopressin receptors in the kidney. Several rare ($\sim$ 1–5/1000) nonsynonymous human single nucleotide polymorphisms (SNPs) have been identified in RGS2 in a Japanese population study.[90] One of these (RGS2 Q2L) was found only in hypertensive patients, and we have shown that the RGS2-Q2L mutation results in destabilized RGS2 expression *in vitro*.[91] Another mutation (RGS2-R44H) is found predominantly in hypertensive patients and it disrupts membrane localization and function of RGS2.[92] Other noncoding polymorphisms in both Japanese and African-American hypertensive patients may modify RGS2 expression.[90,93] Together, all of these data provide strong evidence for a role of RGS2 in regulating blood pressure homeostasis in mice, and in rare cases, it may be a potential contributor to the pathophysiology of hypertension. As with other rare genetic causes of disease, this information may provide insights into approaches that extend to the broader population who do not have that specific genetic cause of the disorder.

Gi-protein-mediated activation of GIRK (G protein-coupled inward rectifying potassium) channels via M_2 muscarinic receptors is a key mechanism in parasympathetic signaling in the heart.[94] The importance of proper RGS function in cardiac automaticity has been demonstrated in studies using knock-in mice expressing an RGS-insensitive mutant of Gα_{i2} (G184S).[12] Embryonic-stem-cell-derived myocytes[13] as well as isolated hearts[12] from these mice show an increased response to cholinergic stimulation. *In vivo*, they are markedly more sensitive to muscarinic-induced bradycardia.[13] In contrast, adenosine A_1 receptor-induced bradycardia was unaffected,[12,13] suggesting a selective function for RGS proteins to regulate cardiac automaticity. Although these studies did not define which RGS protein(s) was responsible for the effect, subsequent work from the Heximer lab identified RGS4 as the key RGS type. RGS4 is expressed at low levels in ventricular and atrial myocytes but much more strongly in sinoatrial nodal cells.[11] Similar to the Gα_{i2} knock-in, RGS-insensitive mice, RGS4 knock-out mice show an increased response to agonists at the M_2 muscarinic receptor. In addition, they showed decreased GIRK channel desensitization[95] and altered kinetics of acetylcholine-activated K^+ currents.[11] Consequently, RGS4 is a key regulator of cardiac automaticity.

A recent study identified a role for RGS4 in regulating renal blood flow via actions on Gα_q signaling by endothelin type A (ET-A) receptors.[96] Calcineurin inhibitors, such as cyclosporine A (CyA), are used as immunomodulators to prevent transplant rejection but they can cause renal toxicity. One hypothesis for this is an increase in endothelin-1 release, which promotes vasoconstriction. In this study, RGS4 knock-out mice displayed increased sensitivity to CyA-induced renal toxicity compared to wild-type (WT) controls and this could be rectified by treatment with an endothelin-1 receptor antagonist. These results support a role for RGS4 in regulating the control of renal blood flow by endothelin.

RGS5 is highly expressed in vascular tissue and has been shown to regulate signaling via several Gi- and Gq-coupled receptors including angiotensin, sphingosine-1-phosphate, and endothelin-1. Specifically, knock-down of RGS5 in the A10 rat aortic cultured VSMC line leads to increased AngII-mediated ERK activation.[97] In the same cell system, knock-down of RGS3 enhanced ERK phosphorylation induced by carbachol, suggesting differential regulation of $G\alpha_q$-mediated signaling for these two proteins. *In vivo*, RGS5 mRNA expression seems to be greatest in pericytes in the developing vasculature, suggesting a role for RGS5 in regulating neovascularization.[98] Despite this notion, RGS5-deficient mice show no alterations in vascular development, possibly due to compensatory mechanisms. However, RGS5 knock-out mice show vascular stabilization in a genetic model of pancreatic cancer, suggesting a role for RGS5 to inhibit angiogenesis in tumor formation.[99] In contrast to the RGS2 knock-out mice, which are hypertensive, RGS5 knock-out mice were recently shown to have reduced blood pressure compared to WT controls.[10,100] The molecular mechanism for this phenotype is yet to be determined.

Another proposed role for RGS4 is in regulating endocrine functions. RGS4 knock-out mice have increases in serum levels of catecholamines with resulting increases in circulating free fatty acids and reduced glucose tolerance and insulin sensitivity.[101] They also have increased release of norepinephrine and epinephrine from the adrenal gland in response to acetylcholine.[101] In contrast, mice overexpressing RGS-insensitive $G\alpha_{i2}$ ($G\alpha_{i2}{}^{G184S}$) also have increased catecholamine levels but they are leaner than WT littermate controls on both low- and high-fat diets. They also show increased insulin sensitivity and glucose tolerance,[16] an effect that could be due to in part to a reduction in body weight. These data suggest that the effects of RGS4 on cardiac function may be by actions on $G\alpha_{i2}$, whereas the function in regulating metabolic events is probably due to actions on another $G\alpha$ subunit.

B. RGS Proteins in the Central Nervous System

Many RGS proteins bind to members of the $G\alpha_{i/o}$ family with high affinity, and several RGS proteins are highly (and some exclusively) expressed in the CNS. Also, $G\alpha_o$ is the most highly expressed G protein α subunit expressed in the brain. As a result, there have been numerous studies on the importance of RGS proteins regulating physiological and pathophysiological functions in the CNS.

RGS9 is a member of the R7 family of RGS proteins and is expressed as two splice variants, RGS9-1 in the retina and the longer form RGS9-2 primarily postsynaptic in dopamine receptor-rich brain regions, such as the striatum.[102,103] RGS9-1 is exclusively found in the outer segment of photoreceptor cells of the retina where it regulates signaling through transducin α subunits in rods and cones. This protein seems to be essential for normal visual perception as demonstrated by the RGS9 knock-out mice and humans with mutations that disrupt RGS9 expression or function,[104,105] where there are decreases in their ability to adapt to changes in light intensity and difficulty in seeing moving low-contrast objects.[106] The human syndrome is called bradyopsia. The RGS9-2 splice variant is highly expressed in the striatum, which is an important brain region for movement and reward. Among other RGS proteins, RGS9-2 regulates the effects of many drugs of abuse (reviewed in Refs. 17,107,108). The reward system in the striatum utilizes dopaminergic, serotonergic, and opioid signaling. Many drugs, like amphetamine and cocaine, as well as morphine and other opiates, mediate their effects directly or indirectly via GPCRs for opioids, dopamine, and serotonin in the brain. Morphine is well known for its potent, short-term analgesic effects, but it is also associated with dependence and tolerance after long-term treatment. Several RGS proteins have been identified as possible regulators of opioid effects. Chronic morphine treatment increases both mRNA and protein expression of RGS2 and RGS4 in the rat locus coeruleus (LC).[109] In both cases, the level of expression returns to normal after 24 h of withdrawal. Overexpression of WT but not a GAP-deficient mutant of RGS4 inhibits morphine-induced electrophysiological responses in LC neurons.[109]

Several studies have focused on the role of RGS4 in regulating signaling via opioid receptors. Endogenous RGS4 mRNA has been shown to be upregulated in PC-12 cells in response to several opioid receptor agonists.[110] This upregulation was blocked by antagonists at opioid receptors and also by pretreatment with PTX, linking the effect to a Gi/o pathway effect. RGS4 was later shown to directly interact with the C-terminus of both μ- and δ-opioid receptors as well as with active Gα, forming a heterotrimeric complex.[111] This notion was partly contradicted in a recent study showing direct interaction between RGS4 and the C-terminus of endogenously expressed δ-opioid but not μ-opioid receptors (MORs) in the human neuronal SH-SY5Y cell line.[112] Thus, the latter study suggests specificity for RGS4 to regulate signaling via δ-opioid receptors over MORs. Two responses to the selective δ-opioid receptor agonist SNC80 were increased in cells when RGS4 expression was knocked down.[112] Both inhibition of cAMP accumulation and induction of ERK phosphorylation were enhanced. The effects of morphine, which acts on MORs, were unaltered. Despite these *in vitro* studies showing RGS4 regulation of opioid signaling, the first report on RGS4 knock-out mice did not show alterations in either acute or chronic responses to opioids.[113] However, a recent study showed that local knock-down of RGS4 expression in the nucleus accumbens (NAc) increased morphine reward.[113] Also, targeted knock-down of RGS4 in the LC increased physical dependence on morphine. Thus a clear picture of the role of RGS4 in drug abuse *in vivo* will require further study.

Another RGS protein that has very pronounced effects on drugs of abuse is RGS9. Its long splice variant (RGS9-2) is highly expressed in the striatum, where it modulates $G\alpha_{i/o}$ signaling via dopamine and opioid receptors. As a member of the R7 family of RGS proteins, RGS9 has a more complex structure with several domains in addition to the RGS domain as well as the $G\beta_5$ and R7BP binding partners. RGS9-2 expression in the striatum is increased by acute and decreased by chronic morphine administration.[114] RGS9 knock-out mice show enhanced behavioral effects after both acute and chronic morphine treatment, consistent with a role of RGS9 in suppressing the function of MORs. This may involve a direct interaction between MOR and the DEP domain of RGS9, an effect that is independent of the GAP activity.[115]

RGS9 has also been implicated in regulating motor activity via the dopamine system. Early studies showed decreased locomotor activity in response to the selective dopamine D_2 receptor agonist quinpirole after overexpression of RGS9 in the nucleus accumbens (NAc).[116] Subsequently, RGS9 knock-out mice were shown to be deficient in motor coordination and working memory.[9] Compared to WT mice, RGS9 knock-out mice show increased locomotor responses to drugs targeting opioidergic and dopaminergic signaling, such as morphine or cocaine.[117] Furthermore, in a primate model of Parkinson's disease (PD), overexpression of RGS9 in the striatum suppressed L-DOPA-induced dyskinesias.[14] These results, together with the high expression of RGS9 in the striatum and its selectivity for dopamine D_2 over dopamine D_1 receptors, suggest that a compound that enhances RGS9 function could be useful in treating the side effects of L-DOPA treatment in PD.

In addition to RGS9, RGS10 has been proposed to play a role, perhaps indirect, in the pathophysiology of PD. RGS10 is expressed in both neurons and microglia of the hippocampus, striatum, and dorsal raphe but its function in these regions is poorly understood. In one study, Lee et al. demonstrated that RGS10 knock-out mice display dysregulated, inflammation-related gene responses.[118] Furthermore, that study showed that a dopaminergic cell line was more susceptible to microglial-derived inflammatory cell death after knock-down of RGS10. This suggests that using a compound that enhances RGS10 function and/or expression could emerge as a possible therapeutic strategy to prevent neuronal cell death in PD.

RGS proteins may also be relevant to the therapy of epilepsy. Mice expressing RGS-insensitive $G\alpha_o$ protein showed enhanced α_{2a} adrenergic suppression of hippocampal CA3 epileptiform bursts.[119] In contrast, mice expressing

RGS-insensitive $G\alpha_{i2}$ protein showed no difference in sensitivity compared to WT mice. Although it is not yet established which RGS protein is responsible for this effect, the members of the R7 family (RGS6, RGS7, RGS9, and RGS11) seem to be good candidates based on their expression pattern and selectivity for $G\alpha_o$. In addition, a recent study of RGS4 knock-out mice showed suppression of seizure activity in a mouse model of the Fragile X mental retardation syndrome.[120]

RGS4 has been implicated in schizophrenia, although recent results seem contradictory. The role of RGS4 in schizophrenia was initially based on several studies of gene expression. RGS4 was found to be downregulated in the prefrontal cortex of schizophrenic patients at both the mRNA and protein level.[121] Other studies have seen similar results in this and other important brain regions, such as the dentate gyrus,[122–124] yet others have failed to see a correlation between RGS4 expression and the incidence of schizophrenia.[125,126] A number of genetic association studies have provided support for a linkage to RGS4[124] but this has also been disputed.[127] Thus, human studies of a role for RGS4 in schizophrenia are tantalizing but still need clarification. Analysis of RGS4 knock-out mice also failed to demonstrate altered prepulse inhibition in behavioral studies, raising further questions about the role of RGS4 as a vulnerability gene for schizophrenia.[128] Thus, further clarification of the involvement of RGS4 in schizophrenia is necessary.

Depression is a rapidly growing problem in the industrialized world. The disease is commonly treated with selective serotonin reuptake inhibitors (SSRIs) that increase synaptic levels of serotonin (5-HT). Which 5-HT receptor signals that mediate this effect is not fully understood. Major problems with SSRI treatment include the late onset of the effect (weeks or even months) as well as various side effects.[129] For many patients, SSRI treatment has proven to be ineffective, which increases the need to better understand the disease and develop more effective treatments. We have recently identified a potential role for RGS proteins as a therapeutic target in depression using RGS-insensitive $G\alpha_{i2}$G184S mutant mice.[178] Both heterozygous and homozygous mice showed an antidepressant-like effect in both tail-suspension and forced swim tests, with no effects on basal locomotor activity. $G\alpha_{i2}$ knock-out mice show the opposite effect. The selective 5-HT_{1A} receptor antagonist, WAY-100635, reversed the anti-depressant phenotype of the $G\alpha_{i2}$ G184S mutants. This suggests that enhanced 5-HT_{1A} signaling especially through $G\alpha_{i2}$ may be beneficial in depression. Achieving this therapeutically could be accomplished either with 5-HT_{1A} agonists that are functionally selective for $G\alpha_{i2}$ activation or by inhibition of an appropriate RGS protein to enhance endogenous 5-HT_{1A}–$G\alpha_{i2}$ signaling. Interestingly, other 5-HT_{1A}-mediated responses, as well as responses due to opioidergic or dopaminergic activation, were unaffected, suggesting specificity for this anti-depressant-like effect via 5-HT_{1A} receptors. A key benefit of targeting RGS proteins rather than, for example, positive allosteric modulators of the receptor is that the response to a specific receptor in different tissues may be differentially enhanced by RGS modulation promoting therapeutic effects while potentially avoiding side effects.

C. RGS Proteins in Cancer

The role of GPCR signaling in cancer has been extensively reviewed.[67,130–133] GPCRs mediate several signals coupled to cancer progression, such as cell proliferation and survival, invasion, and metastasis. Many GPCR agonists are found at high concentration in cancerous tissues and can activate their cognate receptors to induce cancer progression. These include lysophosphatidic acid (LPA), thrombin, angiotensin, and chemokines. Many of the corresponding receptors for these ligands have also been shown to be overexpressed in these tissues, leading to overactive signaling. Lately, there has been much focus on receptor tyrosine kinase (RTK) receptors in mediating cancer progressive signals, and as a result several drugs have emerged for cancer treatment that target these receptors.[134,135] Transactivation of RTKs by G proteins is another factor that has been associated with cancer progression. In this aspect, RGS proteins may prove to play a key role in cancer therapeutics.

As with many other physiological and pathophysiological states, several RGS proteins have been implicated in regulating cancer progression. In many cases, the link is related to changes in expression. RGS2 and RGS4 mRNA are both downregulated in ovarian cancer cell lines.[136] In contrast, RGS6 mRNA is upregulated, indicating different functional roles for these proteins. A recent study also identified RGS4 downregulation as a mechanism for breast cancer metastasis.[137] Overexpression of RGS4 in breast cancer cell lines led to reduced invasion and migration both in *in vitro* assays and in a mouse xenograft model. The tumor incidence and size were not affected by RGS4, and so the effect seems to be specific to the invasion aspects of breast cancer cell models. This effect on migration and invasion was connected to the GAP activity of RGS4, as a GAP-deficient mutant was ineffective at inhibiting migration and invasion. It was also found that RGS4 protein expression was severely reduced in breast cancer tissue. In contrast, RGS4 mRNA levels were greatly increased compared to healthy tissue, leading to the hypothesis that increased protein degradation is responsible for the reduced RGS4 protein expression. RGS4 has indeed been previously suggested to be rapidly degraded in cells,[91] as mentioned above. This result emphasizes the point that, to understand functional consequences, the analysis of RGS expression must be done at the protein level.

The most studied RGS protein in regulating cancer progression is RGS2, which has been linked to ovarian cancer (as mentioned above), prostate cancer, breast cancer, acute myeloid leukemia, mantle cell lymphoma, and bladder cancer (reviewed in, e.g., Ref. 67). In most cases of prostate cancer, the disease progresses from being androgen-dependent to an androgen-independent form, rendering hormone treatment ineffective.[138] This has been suggested to be due to ligand-independent activation of the androgen receptor (AR), which could be via a GPCR-mediated pathway. RGS2 mRNA is downregulated in androgen-independent prostate cancer cells, and overexpression of RGS2 inhibits androgen-independent signaling via the androgen receptor.[139] As is the case for RGS4 in breast cancer cells, this effect is due, at least in part, to the GAP activity of RGS2, as a GAP-deficient mutant was not as effective at inhibiting the AR signaling.

Also, RGS5 has been suggested to play a role in vascular remodeling during angiogenesis associated with tumor formation in a model of pancreatic cancer,[99] making enhancement of its function a possible therapeutic approach to inhibiting angiogenesis and improving sensitivity to immunotherapy. However, this is still under investigation, and another study found no differences in vascular stabilization between WT and RGS5 knock-out mice in subcutaneous fibrosarcoma T241 tumors. Nor could investigators detect a difference in angiogenesis in oxygen-induced retinopathies.[100] However, RGS5 seems to be interesting with respect to tumor development, although it seems that the results are dependent on the model studied.

D. Recent Discoveries—RGS Functions Beyond GAP Activity

Although the GAP activity mediated by the RGS domain is the central function for RGS proteins, there is emerging evidence that other functions can be mediated by this large group of proteins. These include actions such as scaffolding proteins, regulating transcription, and subcellular targeting of other proteins involved in signal transduction. It is clear that many RGS-domain-containing proteins can play other roles than merely serving as GAPs for Gα subunits. One of these is the direct interaction between the DEP domain of R7 family RGS proteins and receptors. Although these proteins are specific for members of the Gi/o family of G proteins, other functions have been discovered that involve this domain and others. RGS7 inhibits Ca^{2+} signaling via the Gq-coupled receptor 5-HT_{2C}[140] and can directly interact with Gα_q as demonstrated with fluorescence resonance energy transfer (FRET).[141] This interaction resulted in a modest inhibition of Gα_q-mediated increase in Ca^{2+} levels, but it is not known whether this is due to RGS7 working as a GAP at the G protein.

RGS13 has been shown to directly bind to the transcription factor CREB (cAMP response element binding protein) in B lymphocytes, thereby acting as a nuclear repressor which reduces the affinity of CREB for binding to promoter regions.[142] In that study, activation of adrenergic β_2 receptors led to increases in intracellular Ca^{2+} and cAMP levels, and this increase was connected to nuclear accumulation of RGS13.

In a recent study, RGS2 was shown to suppress protein synthesis by direct binding to eIF2ε (for *e*ukaryotic *i*nitiation *f*actor *2B* ε subunit).[26] This interaction was mapped to a stretch of 37 residues partially, but not exclusively, located in the RGS domain of the protein. An RGS2 "minigene" made up of the 37 residues alone was sufficient to suppress protein translation in stably transfected cells. Also, in RGS2 knock-out mice the investigators found increased total protein synthesis as compared to WT mice. In addition, RGS2 has been reported to modulate $G\alpha_s$-mediated increases in intracellular cAMP levels,[89,143] although this is probably due at least in part to a direct interaction with certain subtypes of adenylyl cyclase (ACI, II, V, and VI). Such an interaction was identified using bioluminescence resonance energy transfer (BRET) in HEK293 cells. Furthermore, that investigation used a blocking antibody RGS2 in olfactory neurons where it was observed to increase cAMP levels. The interaction was mapped to the N-terminal region of RGS2 and was independent of the RGS domain. This further strengthens the hypothesis that, even though members of the R4 family are small with not much more than an RGS domain, they can have functions beyond serving as GAPs on Gα.

V. Reaching Specificity—Modulating Specific Receptors and Signaling Pathways

How does the large group of RGS proteins reach specificity in regulating GPCR signaling? GAP activity, the main function of RGS proteins, modulates the actions of most Gq and Gi/o protein α subunits. As previously mentioned, there are more than 30 RGS-domain-containing proteins in mammalian systems. Several factors, such as cell and tissue distribution, selectivity for certain Gα subtypes, and protein–protein interactions with receptors or other signaling components, can determine the action of a specific RGS protein on a certain pathway. One obvious source of specificity is the tissue- or cell-specific expression pattern. Needless to say, in order to modulate a specific signaling pathway, the RGS protein must be expressed in the same cell as the receptor, G protein, and other effectors. A great example of this is the distinct localization of the splice variants of RGS9. RGS9-1 is exclusively expressed in the outer segment of photoreceptor cells, where it regulates the α subunit of transducin ($G\alpha_t$).[144,145] RGS9-2, in contrast, is highly expressed in the striatum where it can regulate opioidergic and dopaminergic responses.[102] Although these receptors are expressed outside of this brain region, the striatum is the only site where they are regulated by RGS9; consequently, RGS9 actions will control only a subset of opioid and dopaminergic functions. Other RGS proteins also show a localized expression pattern in certain brain regions or cell types.[102] RGS9 is almost exclusively expressed in the striatum, whereas RGS7 is also expressed at high levels in the hippocampus. The expression of RGS4 and RGS8 in the CNS is widespread, whereas other members of the R4 family of RGS proteins show a much more discrete expression pattern. Specifically, RGS5 mRNA is detected only in regions of the thalamus and to some degree in the medial mammillary nucleus in the hypothalamus and the LC in the brainstem. In the cardiovascular system, the expression of RGS4 is mainly restricted to the SA and AV nodes in the heart, whereas RGS2 and RGS5 have a broader expression in cardiac myocytes and VMSCs. Often, several different RGS proteins are expressed in the same cell but they can still display distinct selectivity for separate pathways. Selectivity can be obtained by differential actions on specific Gα subunits. As mentioned above, the selectivity for different Gα subunits can vary between or even within a subfamily of RGS proteins. The R7 family RGS proteins act only on members of the G$\alpha_{i/o}$ protein family and may have differential GAP activity even with that family. While the efficacy toward Gα_o is similar among R7 family members, RGS9 and RGS11 are more promiscuous with similar GAP activity on Gα_{i1}, Gα_{i2}, and Gα_{i3} as on Gα_o *in vitro*. In contrast, RGS6 and RGS7 are more selective, with 5–10 fold higher activity for Gα_o than on the Gα_i subtypes.[39,146] In the R4 family of RGS proteins, several studies have shown that RGS2 is a selective GAP for Gα_q, whereas other members show strong activity toward G$\alpha_{i/o}$ as well as Gα_q. In a recombinant system with either the M_2 or M_3 muscarinic receptor co-expressed with members of the R4 family, Anger et al. observed that Gα_q-mediated Akt phosphorylation was only blocked by RGS3, whereas activation of ERK was also blocked by RGS2 and RGS5.[87] In addition, G$\alpha_{i/o}$-mediated activation of Akt was blocked by all tested RGS proteins (RGS2-5), but RGS2 was not capable of inhibiting ERK phosphorylation via G$\alpha_{i/o}$ proteins. However, this selectivity may be mediated at the effector level, since Bodenstein et al.[91] found no difference in the activity of five RGS proteins to suppress Ca^{2+} signals from M1, M3, and M5 receptors once RGS expression levels were taken into account. These data suggest that affinity toward different Gα subtypes is not the only determinant for selectivity of RGS proteins targeting specific signaling pathways.

Apart from serving as GAPs for G protein α subunits RGS proteins also have other functions including protein–protein interactions, as previously discussed. Although selective for G$\alpha_{i/o}$, RGS7 can regulate Gα_q-mediated signaling via the M_3 muscarinic receptor through a direct interaction between the third intracellular loop of the receptor and the DEP domain of RGS7.[32] Another example of RGS proteins regulating signaling pathways via protein–

protein interactions is the regulation of adenylyl cyclase by RGS2, as discussed above.[89,143] Although prominent reports on GAP activity of RGS proteins on $G\alpha_s$ have appeared,[147,148] these have generally not been confirmed and the regulation of Gs-mediated pathways by RGS2 through direct effects on adenlyl cyclase shows another means of reaching specificity for RGS proteins as regulators of GPCR-mediated pathways.

Finally, subcellular co-localization is a prerequisite for an interaction to occur. Targeting the RGS protein to the plasma membrane is crucial for its GAP activity in G protein α subunits. For the R7 family of RGS proteins, two membrane targeting proteins have been characterized (reviewed in Ref. 34). R9AP (for *R*GS9 *A*nchoring *P*rotein) is exclusively expressed in the retina, where it anchors RGS9-1 to the membrane, and the site of action for GAP activity on $G\alpha_t$. Its counterpart R7BP (for *R7 B*inding *P*rotein) is expressed in the CNS and recruits all R7 family RGS proteins to the plasma membrane. Another example is the long splice variant of RGS3, PDZ-RGS3, which has an N-terminal PDZ domain. This is a common protein–protein interaction motif found in many classes of proteins. PDZ-RGS3 was shown to bind the Eph receptor tyrosine kinase B (Ephrin-B), which mediates signals related to axon guidance and neuronal migration.[149] This interaction targets PDZ-RGS3 to the plasma membrane and also enables crosstalk between GPCR- and receptor tyrosine kinase-mediated signaling pathways. The membrane targeting signal could also be through a posttranslational modification of the RGS protein itself. Several groups have studied the mechanisms of subcellular targeting, identifying the N-terminus as a key structure for targeting signals, such as palmitoylation (e.g., Refs. 76,92,150–152). Taken together, there are many ways that RGS proteins can obtain specificity for regulating specific GPCR-mediated signaling events, many of which have yet to be discovered.

VI. Targeting RGS Proteins in Drug Discovery—From Increased Knowledge to Increased Throughput

Considering the many proposed biological functions of RGS proteins, it is clear that they are intriguing targets for drug discovery. In drug development, it is more straightforward to screen for inhibitors than enhancers and therefore many more studies have used this approach. The theoretical rationale for an RGS inhibitor could be to reduce the dose of the primary drug and/or to increase tissue selectivity (discussed in Refs. 20,69,153). Two recent studies have indicated the possible clinical uses of RGS inhibitors in the treatment of epilepsy and depression (Ref. 119 and 178). Both studies utilized knock-in mice, which express an RGS-insensitive mutant of $G\alpha$ protein ($G\alpha^{G184S}$).

Goldenstein et al. showed that hippocampal CA3 epileptiform activity was mediated via a $G\alpha_o$-mediated pathway and the $mG\alpha_o^{G184S}$-expressing mice showed increased epinephrine-induced suppression of hippocampal seizure activity.[119] Mice expressing $G\alpha_{i2}^{G184S}$ have a spontaneous antidepressant-like phenotype, which can be directly connected to 5-HT_{1A}-receptor-mediated signaling.[178] Interestingly, other 5-HT_{1A}-mediated effects were unaffected in these mice, suggesting that an RGS inhibitor could be useful in the development of more selective antidepressants.

Enhancers of RGS protein function could also be useful therapeutically. RGS2 enhancers could be used to treat hypertension, as several groups have confirmed that RGS2 knock-out mice are hypertensive.[85] The importance of RGS4 in breast cancer was recently demonstrated by Xie et al.,[137] who found that RGS4 protein expression was severely reduced in models of metastatic breast cancer despite a dramatic increase in the level of mRNA. Finally, RGS9 enhancers could be useful in the treatment of PD, as overexpression of RGS9 in both rodent and primate models of PD dramatically reduces L-DOPA-induced involuntary movements.[14]

The primary strategy for targeting RGS proteins in drug discovery is to inhibit the RGS–Gα interaction, thereby inhibiting the GTPase activity. However, the other functions of RGS proteins in addition to the GAP activity could contribute when studying these proteins. Consequently, choosing the right model system is crucial when attempting to screen for either inhibitors or enhancers of RGS protein function. There are several possible mechanisms by which the function of an RGS protein could modulated, including direct inhibition of the RGS–Gα interaction or allosteric modulation by compounds binding to a site on the RGS protein outside of the RGS–Gα interface. Another approach could be to modulate yet other functions, such as stability, membrane targeting, or blocking other protein–protein interaction sites. In the development of drugs targeting RGS protein function, it may not prove to be necessary to target GAP function. One example of this would be the interaction between the members of the R7 family of RGS proteins and $G\beta_5$, which is necessary for stable protein expression. Disruption of the interaction could be another way to inhibit function by downregulating RGS protein expression. Nevertheless, choosing the assay system will be a key in finding compounds that have the desired effect. This section summarizes some strategies that have been or could be taken to assay for novel RGS protein modulators.

A. Biochemical Methods

Disrupting protein–protein interactions with small molecules has proven challenging and the RGS–Gα protein–protein interaction is no exception.[153–155] As discussed above (and see Fig. 2), the RGS–Gα interface is large and relatively devoid of "druggable" pockets, as traditionally defined. Despite this challenge, several groups have reported the development of either peptides or small molecules that inhibit RGS activity (Table II).[81,156–161] Most of these inhibitors have been developed from either random or focused compound/peptide libraries and are still in the early development stages.

To date, the primary methods of screening for inhibitors of RGS proteins use measurements of the physical RGS–Gα interaction since the biochemical measurement of GAP activity has been very cumbersome. We have developed spectroscopic biochemical assays in our lab,[156,158,159] and the laboratory of Kathleen Young at Wyeth Pharmaceuticals has utilized a yeast two-hybrid assay method.[160,161] The first small molecules identified from an RGS–Gα screen were from the yeast two-hybrid screen reported by Young and colleagues in 2004.[161] Unfortunately, the structures were never disclosed, so further independent research on these compounds has not been possible. In the same year, our laboratory published the first of a series of papers that characterized the RGS inhibitory effects of peptides corresponding to the Switch I region of $G\alpha_i$.[157–159,162] This original cyclic octa-peptide inhibitor (YJ34) was used as the basis for a constrained peptide library of approximately 2.5 million peptides that were screened for the ability to bind a fluorescently tagged, purified RGS4. From this, a number of novel peptides were discovered that possess inhibitory activity against several RGS proteins.[159] Recently, Young's group also reported a peptide inhibitor identified from a random peptide library in a yeast two-hybrid format.[160] This assay, which used RGS4 as the bait, identified a peptide containing a positively charged residue (R), followed by a series of predominantly hydrophobic residues, a sequence that is not known to correspond to any RGS interacting protein. It is intriguing to postulate that the amphipathic nature of the peptide might allow it to function in a manner similar to the allosteric regulation of RGS4 by endogenous acidic phospholipids. At present, however, there are no direct data to address this hypothesis.

The first publicly disclosed small molecule inhibitors of an RGS protein were identified in our lab using the flow cytometry protein interaction assay (FCPIA).[83] This methodology is described in detail in the literature (Refs. 81,156 and 179). Briefly, the binding of Gα to an RGS protein is measured by immobilizing the purified RGS protein on a polystyrene bead and then incubating it with a fluorescently tagged G protein in the presence or absence of a potential inhibitor. These beads are then analyzed in a flow cytometer for bead-associated fluorescence. While flow cytometry has traditionally not been thought of as a high-throughput method, the advent of the Luminex plate reader with 96-well capabilities and, more recently, the HyperCyt system developed by Larry Sklar[163] have changed this landscape. One advantage of the FCPIA method is that it can be multiplexed; different RGS proteins are immobilized on beads that can be optically differentiated by the

TABLE II
IDENTIFIED RGS PROTEIN INHIBITORS FROM BIOCHEMICAL AND YEAST TWO-HYBRID SCREENS

Name	Class	Structure	Target	Assay(s)	References
CCG-4986	Small molecule	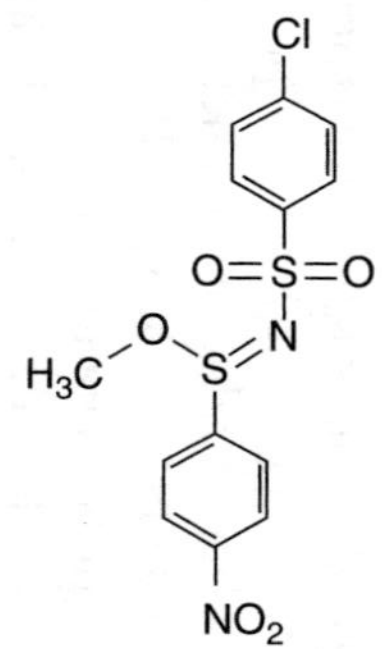 Methyl *N*-[(4-chlorophenyl)-sulfonyl]-4-nitrobenzenesulfinimidoate	RGS4	FCPIA Single-turnover GAP Permeabilized cell cAMP assay	81
YJ34	Cyclic peptide	Ac-VK-c[CTGIC]-E-NH_2 (S-S)	RGS4 (RGS8)	Steady-state GAP Single-turnover GAP	162
YJ33	Cyclic peptide	Ac-VK-c[CTGIC]-E-NH_2 (S-CH_2-S)	RGS4 (RGS8)	Single-turnover GAP GIRK currents in atrial myocytes	157

CCG-63802	Small Molecule	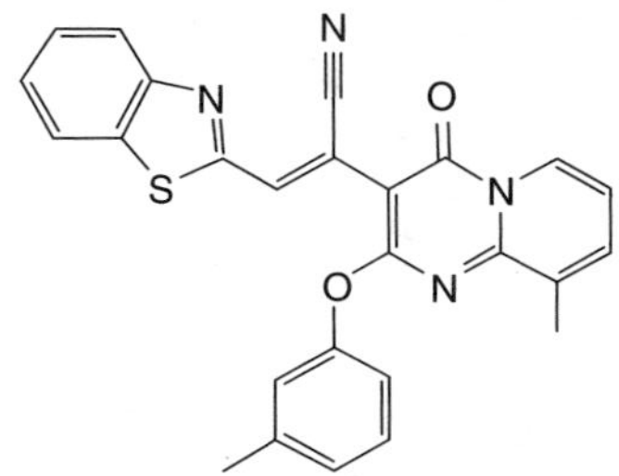 (2*E*)-2-(1,3-Benzothiazol-2-yl)-3-[9-methyl-2-(3-methylphenoxy)-4-oxo-4*H*-pyrido[1,2-*a*]pyrimidin-3-yl]prop-2-enenitrile	RGS4	TR-FRET RGS4/Gαo FCPIA Single-turnover GAP Ligand-induced thermal stability	83
P17	Peptide	VRHVAVEVGGVVVVVG	RGS4	Yeast two-hybrid Single-turnover GAP GIRK currents in oocytes	160
WAY-243	Small molecule	Not disclosed	RGS20	Yeast two-hybrid Single-turnover GAP	161

Amino acids are given in standard one-letter code; FCPIA, flow cytometry protein interaction assay; GAP, GTPase activation protein; TR-FRET, time-resolved fluorescence resonance energy transfer.

flow cytometer. They are mixed and placed in a single well and probed with a single labeled Gα subunit. This allowed several RGS proteins to be screened against in a single high-throughput assay.[156] We have also utilized a high-throughput time-resolved FRET assay[164] using purified RGS4 and $G\alpha_o$ that are labeled with AlexaFluor 488 and the LanthaScreen terbium probe, respectively. To date, we have screened over 200,000 small molecules using this approach and are currently characterizing the lead compounds.[83] CCG-63802 (Table II) was identified in the TR-FRET screen. It inhibits RGS4 binding and GAP activity with an IC_{50} value of $\sim 10\ \mu M$ and is the first example of a reversible small molecule RGS inhibitor. CCG-63802 binds specifically to the RGS protein and not the Gα and shows specificity for RGS4 > 19 > 16 > 8 > > 7. Using mutagenesis and protein NMR, we are currently determining the binding site of this family of compounds; however, the data to date suggest that they bind at the "B" site on RGS4.[83]

Recently, an assay for RGS function (i.e., Gα GAP activity) that is amenable to high-throughput screening (HTS) was reported by Bellbrook labs in collaboration with the laboratory of David Siderovski.[165] It utilizes a mutant Gα with rapid GDP release. This makes the GTP hydrolysis step rate-limiting, so that an RGS protein can significantly accelerate steady-state GTPase activity. Consequently, it does not require either a single-turnover approach or a GEF such as receptor to produce rapid GDP → GTP exchange. A second innovation was the use of a high-affinity GDP antibody and a fluorescence polarization method to sensitively detect the released GDP. This assay is designed for HTS and should provide a useful approach to developing small molecule inhibitors of RGS function.

B. Cell-Based Approaches

Cell-based studies of RGS protein function are usually based on observing the activity of a signaling pathway in the absence and presence of a specific RGS protein. Given the importance of GPCR signaling in therapeutics, many methods have been developed that are easily adapted to HTS assays and they may be used to identify novel compounds modulating RGS function. There are, however, several considerations when developing a cell-based assay for RGS protein function. First, overexpression of receptors and RGS proteins leads to altered stoichiometry, which could change the mode of action. Furthermore, in complex cell systems, alterations in signaling by a compound could be due to effects on steps other than the RGS protein. Thus any compounds identified from a cell-based screen must be verified for mechanism of action with direct biochemical assays (see above). Also, given the non-GAP functions of RGS proteins, different cellular assays might reflect different modes of RGS action, so the identified compounds could be targeting different aspects of RGS function. This section summarizes approaches that could be taken when screening for modulators of RGS protein function. To date, there are no published studies on mammalian-cell-based drug screening in high throughput for modulators of RGS function.

One could directly assess the RGS/Gα protein–protein interaction in live cells by means of bioluminescence or fluorescence resonance energy transfer (BRET or FRET, respectively). This methodology has been widely used and is extensively reviewed elsewhere (e.g., Refs. 166–169). These methods generally employ fluorescent or bioluminescent donor (e.g., CFP for FRET or Rluc for BRET) and acceptor (e.g., YFP, citrine, or venus) proteins coupled to the proteins whose interaction is being studied. This approach has been utilized for RGS proteins to investigate interactions with Gβ_5.[170] A similar approach could be taken to investigate the interaction between an RGS protein and Gα in a cell-based system, similar to what already has been done biochemically (see above). An advantage of this study of protein–protein interactions in living cells is that it would immediately identify compounds with access to intracellular sites at which RGS/Gα interactions normally occur.

Enzyme fragment complementation assays has also been used to investigate protein–protein interactions in cells (for review see Ref. 169 and references therein). This could be utilized in a similar manner as FRET or BRET. In brief, two fragments of an enzyme which have no catalytic activity on their own are fused to the proteins of interest.[171] When these fusion proteins are brought into close proximity by an interaction between the two target proteins, the enzyme fragments will reform an active enzyme. In the presence of the enzyme substrate, a product is produced that emits a signal (luminescence or fluorescence) which is proportional to the amount of interaction. This approach can also be integrated with FRET or BRET to study interactions with more than two proteins. All of the above-mentioned techniques are easily adapted to screening setups. They also have the advantage that the signal can be imaged and subcellular location information can be obtained. However, caution should be taken when interpreting the results, as these assays, like many high-throughput assays, are generally performed in recombinant systems that may not reflect the endogenous signaling events.

Ultimately, when screening for modulators of RGS protein function, studies of the interaction with Gα may not be the optimal choice. Although informative, merely inhibiting the RGS–Gα interaction may not give sufficient information on the functional properties of an RGS modulator compound. RGS effects on functional GPCR signaling would directly replicate the physiological actions that would be desired of an RGS modulator drug. Thus, robust measures of RGS effects on standard signal transduction outputs would be useful. GPCR function measures commonly involve effects on intracellular Ca^{2+} or cAMP levels. These readouts are typically controlled by Gq and Gi family G proteins, which are the best established targets of RGS proteins. There are many commercially available assays for these systems and they are very adaptable to HTS platforms. In this context, calcium assays give a robust response to activation of both $G\alpha_q$- and $G\alpha_i$-mediated pathways. Functional effects of RGS proteins on intracellular calcium mobilization have been examined by several groups (e.g., Refs. 32,51,140,141,172) and this could be a good system for HTS. Utilizing cAMP measurements to look at RGS effects might not be the best option given that most RGS proteins act on $G\alpha_{i/o}$, leading to a decrease in cAMP levels which is generally not a robust assay for HTS purposes. Alternative functional readouts of Gi/o signaling pathways would be desirable.

In addition to inhibiting RGS function, there are a number of situations in which it would be desirable to increase RGS function. For example, enhancing the action of RGS2 could be useful in hypertension (e.g., Refs. 85,86,90). Recent data suggest that regulation of RGS protein expression might be another way to target these proteins in drug discovery. RGS4 protein expression seems to be important in suppressing breast cancer metastasis, and RGS2 has been shown to suppress protein translation (Refs. 26,137 and mentioned above). Furthermore, there is evidence that RGS4 is rapidly degraded in cells via the N-end rule pathway.[91] Indeed, in the study of RGS4 expression in breast cancer, it was found that, although protein levels were undetectable in invasive cancer cells, mRNA levels were greatly upregulated.[137] Targeting RGS proteins in a way that stabilizes expression might prove to be an additional effective way to increase RGS protein function apart from modulating the action of binding to Gα and/or increasing or decreasing GAP activity.

VII. Future Prospects

With the discovery of the many roles that RGS proteins play in the regulation of GPCR-mediated signaling pathways, it is becoming clear that modulation of these proteins could be an interesting target for drug discovery in a number of pathophysiological states. There are good arguments for the utility of either inhibiting or enhancing the activity of RGS proteins as a way to selectively alter GPCR signaling. Key breakthroughs in GPCR research came with the development of selective agonists and antagonists at specific receptors. This greatly enhanced the knowledge of the functions of these receptors but also set the stage for important therapeutic advances. This is true also for other protein classes, such as kinases, phosphodiesterases, and ion channels, enabling both functional characterization and the development clinically important drugs. Initial work has identified a number of inhibitors of RGS proteins in biochemical systems, but a critical next step is to obtain activity in cellular and *in vivo* systems. The development of such modulators of RGS protein function would provide important pharmacological tools to increase our understanding of this large protein family and could also lead to novel clinical therapeutics.

References

1. Williams C, Hill SJ. GPCR signaling: understanding the pathway to successful drug discovery. G protein-coupled receptors in drug discovery. *Methods Mol Biol* 2009;**552**:39–50.
2. Hollinger S, Hepler JR. Cellular regulation of RGS proteins: modulators and integrators of G protein signaling. *Pharmacol Rev* 2002;**54**:527–59.
3. Ross EM, Wilkie TM. GTPase-activating proteins for heterotrimeric G proteins: regulators of G protein signaling (RGS) and RGS-like proteins. *Annu Rev Biochem* 2000;**69**:795–827.
4. Lan KL, Sarvazyan NA, Taussig R, Mackenzie RG, DiBello PR, Dohlman HG, et al. A point mutation in Galphao and Galphai1 blocks interaction with regulator of G protein signaling proteins. *J Biol Chem* 1998;**273**:12794–7.
5. Dohlman HG, Apaniesk D, Chen Y, Song J, Nusskern D. Inhibition of G-protein signaling by dominant gain-of-function mutations in Sst2p, a pheromone desensitization factor in *Saccharomyces cerevisiae*. *Mol Cell Biol* 1995;**15**:3635–43.
6. Koelle MR, Horvitz HR. EGL-10 regulates G protein signaling in the *C. elegans* nervous system and shares a conserved domain with many mammalian proteins. *Cell* 1996;**84**:115–25.
7. Druey KM, Blumer KJ, Kang VH, Kehrl JH. Inhibition of G-protein-mediated MAP kinase activation by a new mammalian gene family. *Nature* 1996;**379**:742–6.
8. Siderovski DP, Hessel A, Chung S, Mak TW, Tyers M. A new family of regulators of G-protein-coupled receptors? *Curr Biol* 1996;**6**:211–2.
9. Blundell J, Hoang CV, Potts B, Gold SJ, Powell CM. Motor coordination deficits in mice lacking RGS9. *Brain Res* 2008;**1190**:78–85.
10. Cho H, Park C, Hwang IY, Han SB, Schimel D, Despres D, et al. Rgs5 targeting leads to chronic low blood pressure and a lean body habitus. *Mol Cell Biol* 2008;**28**:2590–7.
11. Cifelli C, Rose RA, Zhang H, Voigtlaender-Bolz J, Bolz SS, Backx PH, et al. RGS4 regulates parasympathetic signaling and heart rate control in the sinoatrial node. *Circ Res* 2008;**103**:527–35.
12. Fu Y, Huang X, Piao L, Lopatin AN, Neubig RR. Endogenous RGS proteins modulate SA and AV nodal functions in isolated heart: implications for sick sinus syndrome and AV block. *Am J Physiol Heart Circ Physiol* 2007;**292**:H2532–9.
13. Fu Y, Huang X, Zhong H, Mortensen RM, D'Alecy LG, Neubig RR. Endogenous RGS proteins and Galpha subtypes differentially control muscarinic and adenosine-mediated chronotropic effects. *Circ Res* 2006;**98**:659–66.
14. Gold SJ, Hoang CV, Potts BW, Porras G, Pioli E, Kim KW, et al. RGS9-2 negatively modulates L-3, 4-dihydroxyphenylalanine-induced dyskinesia in experimental Parkinson's disease. *J Neurosci* 2007;**27**:14338–48.
15. Gu S, Cifelli C, Wang S, Heximer SP. RGS proteins: identifying new GAPs in the understanding of blood pressure regulation and cardiovascular function. *Clin Sci (Lond)* 2009;**116**:391–9.
16. Huang X, Charbeneau RA, Fu Y, Kaur K, Gerin I, MacDougald OA, et al. Resistance to diet-induced obesity and improved insulin sensitivity in mice with a regulator of G protein signaling-insensitive G184S Gnai2 allele. *Diabetes* 2008;**57**:77–85.
17. Lomazzi M, Slesinger PA, Luscher C. Addictive drugs modulate GIRK-channel signaling by regulating RGS proteins. *Trends Pharmacol Sci* 2008;**29**:544–9.
18. Oliveira-Dos-Santos AJ, Matsumoto G, Snow BE, Bai D, Houston FP, Whishaw IQ, et al. Regulation of T cell activation, anxiety, and male aggression by RGS2. *Proc Natl Acad Sci U S A* 2000;**97**:12272–7.
19. Dohlman HG, Thorner J. RGS proteins and signaling by heterotrimeric G proteins. *J Biol Chem* 1997;**272**:3871–4.
20. Zhong H, Neubig RR. Regulator of G protein signaling proteins: novel multifunctional drug targets. *J Pharmacol Exp Ther* 2001;**297**:837–45.

21. Hepler JR. Emerging roles for RGS proteins in cell signalling. *Trends Pharmacol Sci* 1999;**20**:376–82.
22. Bernstein LS, Grillo AA, Loranger SS, Linder ME. RGS4 binds to membranes through an amphipathic alpha-helix. *J Biol Chem* 2000;**275**:18520–6.
23. Saitoh O, Masuho I, Terakawa I, Nomoto S, Asano T, Kubo Y. Regulator of G protein signaling 8 (RGS8) requires its NH_2 terminus for subcellular localization and acute desensitization of G protein-gated K + channels. *J Biol Chem* 2001;**276**:5052–8.
24. Kehrl JH, Srikumar D, Harrison K, Wilson GL, Shi CS. Additional 5' exons in the RGS3 locus generate multiple mRNA transcripts, one of which accounts for the origin of human PDZ-RGS3. *Genomics* 2002;**79**:860–8.
25. Johnson EN, Seasholtz TM, Waheed AA, Kreutz B, Suzuki N, Kozasa T, et al. RGS16 inhibits signalling through the G alpha 13-Rho axis. *Nat Cell Biol* 2003;**5**:1095–103.
26. Nguyen CH, Ming H, Zhao P, Hugendubler L, Gros R, Kimball SR, et al. Translational control by RGS2. *J Cell Biol* 2009;**186**:755–65.
27. Berman DM, Wilkie TM, Gilman AG. GAIP and RGS4 are GTPase-activating proteins for the Gi subfamily of G protein alpha subunits. *Cell* 1996;**86**:445–52.
28. Watson N, Linder ME, Druey KM, Kehrl JH, Blumer KJ. RGS family members: GTPase-activating proteins for heterotrimeric G-protein alpha-subunits. *Nature* 1996;**383**:172–5.
29. Heximer SP, Watson N, Linder ME, Blumer KJ, Hepler JR. RGS2/G0S8 is a selective inhibitor of Gqalpha function. *Proc Natl Acad Sci U S A* 1997;**94**:14389–93.
30. Anderson GR, Posokhova E, Martemyanov KA. The R7 RGS protein family: multi-subunit regulators of neuronal G protein signaling. *Cell Biochem Biophys* 2009;**54**:33–46.
31. Chen S, Hamm HE. DEP domains: more than just membrane anchors. *Dev Cell* 2006;**11**:436–8.
32. Sandiford SL, Slepak VZ. The Gbeta5-RGS7 complex selectively inhibits muscarinic M3 receptor signaling via the interaction between the third intracellular loop of the receptor and the DEP domain of RGS7. *Biochemistry* 2009;**48**:2282–9.
33. Ballon DR, Flanary PL, Gladue DP, Konopka JB, Dohlman HG, Thorner J. DEP-domain-mediated regulation of GPCR signaling responses. *Cell* 2006;**126**:1079–93.
34. Jayaraman M, Zhou H, Jia L, Cain MD, Blumer KJ. R9AP and R7BP: traffic cops for the RGS7 family in phototransduction and neuronal GPCR signaling. *Trends Pharmacol Sci* 2009;**30**:17–24.
35. Anderson GR, Semenov A, Song JH, Martemyanov KA. The membrane anchor R7BP controls the proteolytic stability of the striatal specific RGS protein, RGS9-2. *J Biol Chem* 2007;**282**:4772–81.
36. Drenan RM, Doupnik CA, Jayaraman M, Buchwalter AL, Kaltenbronn KM, Huettner JE, et al. R7BP augments the function of RGS7°Gbeta5 complexes by a plasma membrane-targeting mechanism. *J Biol Chem* 2006;**281**:28222–31.
37. Chen CK, Eversole-Cire P, Zhang H, Mancino V, Chen YJ, He W, et al. Instability of GGL domain-containing RGS proteins in mice lacking the G protein beta-subunit Gbeta5. *Proc Natl Acad Sci USA* 2003;**100**:6604–9.
38. Cheever ML, Snyder JT, Gershburg S, Siderovski DP, Harden TK, Sondek J. Crystal structure of the multifunctional Gbeta5-RGS9 complex. *Nat Struct Mol Biol* 2008;**15**:155–62.
39. Hooks SB, Waldo GL, Corbitt J, Bodor ET, Krumins AM, Harden TK. RGS6, RGS7, RGS9, and RGS11 stimulate GTPase activity of Gi family G-proteins with differential selectivity and maximal activity. *J Biol Chem* 2003;**278**:10087–93.
40. Kimple RJ, De Vries L, Tronchere H, Behe CI, Morris RA, Gist Farquhar M, et al. RGS12 and RGS14 GoLoco motifs are G alpha(i) interaction sites with guanine nucleotide dissociation inhibitor activity. *J Biol Chem* 2001;**276**:29275–81.

41. Siderovski DP, Willard FS. The GAPs, GEFs, and GDIs of heterotrimeric G-protein alpha subunits. *Int J Biol Sci* 2005;**1**:51–66.
42. Snow BE, Brothers GM, Siderovski DP. Molecular cloning of regulators of G-protein signaling family members and characterization of binding specificity of RGS12 PDZ domain. *Methods Enzymol* 2002;**344**:740–61.
43. Snow BE, Hall RA, Krumins AM, Brothers GM, Bouchard D, Brothers CA, et al. GTPase activating specificity of RGS12 and binding specificity of an alternatively spliced PDZ (PSD-95/Dlg/ZO-1) domain. *J Biol Chem* 1998;**273**:17749–55.
44. Schiff ML, Siderovski DP, Jordan JD, Brothers G, Snow B, De Vries L, et al. Tyrosine-kinase-dependent recruitment of RGS12 to the N-type calcium channel. *Nature* 2000;**408**:723–7.
45. Nunn C, Mao H, Chidiac P, Albert PR. RGS17/RGSZ2 and the RZ/A family of regulators of G-protein signaling. *Semin Cell Dev Biol* 2006;**17**:390–9.
46. Linder ME, Deschenes RJ. Palmitoylation: policing protein stability and traffic. *Nat Rev Mol Cell Biol* 2007;**8**:74–84.
47. Ajit SK, Ramineni S, Edris W, Hunt RA, Hum W-T, Hepler JR, et al. RGSZ1 interacts with protein kinase C interacting protein PKCI-1 and modulates mu opioid receptor signaling. *Cell Signal* 2007;**19**:723–30.
48. Wong YH, Conklin BR, Bourne HR. Gz-mediated hormonal inhibition of cyclic AMP accumulation. *Science* 1992;**255**:339–42.
49. Glick JL, Meigs TE, Miron A, Casey PJ. RGSZ1, a Gz-selective regulator of G protein signaling whose action is sensitive to the phosphorylation state of Gzalpha. *J Biol Chem* 1998;**273**:26008–13.
50. Wang J, Ducret A, Tu Y, Kozasa T, Aebersold R, Ross EM. RGSZ1, a Gz-selective RGS protein in brain. Structure, membrane association, regulation by Galphaz phosphorylation, and relationship to a Gz gtpase-activating protein subfamily. *J Biol Chem* 1998;**273**:26014–25.
51. Mao H, Zhao Q, Daigle M, Ghahremani MH, Chidiac P, Albert PR. RGS17/RGSZ2, a novel regulator of Gi/o, Gz, and Gq signaling. *J Biol Chem* 2004;**279**:26314–22.
52. Tesmer JJG. Structure and function of regulator of G protein signaling homology domains. In: Fisher R, editor. *Molecular biology of RGS proteins.* Elsevier, Amsterdam, The Netherlands; 2009;**86**:75–113.
53. Carman CV, Parent JL, Day PW, Pronin AN, Sternweis PM, Wedegaertner PB, et al. Selective regulation of Galpha(q/11) by an RGS domain in the G protein-coupled receptor kinase, GRK2. *J Biol Chem* 1999;**274**:34483–92.
54. Hart MJ, Jiang X, Kozasa T, Roscoe W, Singer WD, Gilman AG, et al. Direct stimulation of the guanine nucleotide exchange activity of p115 RhoGEF by Galpha13. *Science* 1998;**280**:2112–4.
55. Meigs TE, Juneja J, DeMarco CT, Stemmle LN, Kaplan DD, Casey PJ. Selective uncoupling of G alpha 12 from Rho-mediated signaling. *J Biol Chem* 2005;**280**:18049–55.
56. Meyer BH, Freuler F, Guerini D, Siehler S. Reversible translocation of p115-RhoGEF by G (12/13)-coupled receptors. *J Cell Biochem* 2008;**104**:1660–70.
57. Day PW, Tesmer JJ, Sterne-Marr R, Freeman LC, Benovic JL, Wedegaertner PB. Characterization of the GRK2 binding site of Galphaq. *J Biol Chem* 2004;**279**:53643–52.
58. Tesmer VM, Kawano T, Shankaranarayanan A, Kozasa T, Tesmer JJ. Snapshot of activated G proteins at the membrane: the Galphaq-GRK2-Gbetagamma complex. *Science* 2005;**310**:1686–90.
59. Shankaranarayanan A, Thal DM, Tesmer VM, Roman DL, Neubig RR, Kozasa T, et al. Assembly of high order G alpha q-effector complexes with RGS proteins. *J Biol Chem* 2008;**283**:34923–34.
60. Chen Z, Singer WD, Sternweis PC, Sprang SR. Structure of the p115RhoGEF rgRGS domain-Galpha13/i1 chimera complex suggests convergent evolution of a GTPase activator. *Nat Struct Mol Biol* 2005;**12**:191–7.

61. Chen Z, Singer WD, Danesh SM, Sternweis PC, Sprang SR. Recognition of the activated states of Galpha13 by the rgRGS domain of PDZRhoGEF. *Structure* 2008;**16**:1532–43.
62. Berman DM, Kozasa T, Gilman AG. The GTPase-activating protein RGS4 stabilizes the transition state for nucleotide hydrolysis. *J Biol Chem* 1996;**271**:27209–12.
63. Hunt TW, Fields TA, Casey PJ, Peralta EG. RGS10 is a selective activator of G alpha i GTPase activity. *Nature* 1996;**383**:175–7.
64. De Vries L, Elenko E, Hubler L, Jones TL, Farquhar MG. GAIP is membrane-anchored by palmitoylation and interacts with the activated (GTP-bound) form of G alpha i subunits. *Proc Natl Acad Sci USA* 1996;**93**:15203–8.
65. Dohlman HG, Song J, Ma D, Courchesne WE, Thorner J. Sst2, a negative regulator of pheromone signaling in the yeast Saccharomyces cerevisiae: expression, localization, and genetic interaction and physical association with Gpa1 (the G-protein alpha subunit). *Mol Cell Biol* 1996;**16**:5194–209.
66. Apanovitch DM, Slep KC, Sigler PB, Dohlman HG. Sst2 is a GTPase-activating protein for Gpa1: purification and characterization of a cognate RGS–Galpha protein pair in yeast. *Biochemistry* 1998;**37**:4815–22.
67. Hurst JH, Hooks SB. Regulator of G-protein signaling (RGS) proteins in cancer biology. *Biochem Pharmacol* 2009;**78**:1289–97.
68. Tesmer JJ, Berman DM, Gilman AG, Sprang SR. Structure of RGS4 bound to AlF4–activated G(i alpha1): stabilization of the transition state for GTP hydrolysis. *Cell* 1997;**89**:251–61.
69. Neubig RR, Siderovski DP. Regulators of G-protein signalling as new central nervous system drug targets. *Nat Rev Drug Discov* 2002;**1**:187–97.
70. Ishii M, Fujita S, Yamada M, Hosaka Y, Kurachi Y. Phosphatidylinositol 3, 4, 5-trisphosphate and Ca2+/calmodulin competitively bind to the regulators of G-protein-signalling (RGS) domain of RGS4 and reciprocally regulate its action. *Biochem J* 2005;**385**:65–73.
71. Ishii M, Ikushima M, Kurachi Y. *In vivo* interaction between RGS4 and calmodulin visualized with FRET techniques: possible involvement of lipid raft. *Biochem Biophys Res Commun* 2005;**338**:839–46.
72. Popov SG, Krishna UM, Falck JR, Wilkie TM. Ca2+/Calmodulin reverses phosphatidylinositol 3, 4, 5-trisphosphate-dependent inhibition of regulators of G protein-signaling GTPase-activating protein activity. *J Biol Chem* 2000;**275**:18962–8.
73. Castro-Fernandez C, Janovick JA, Brothers SP, Fisher RA, Ji TH, Conn PM. Regulation of RGS3 and RGS10 palmitoylation by GnRH. *Endocrinology* 2002;**143**:1310–7.
74. Druey KM, Ugur O, Caron JM, Chen CK, Backlund PS, Jones TL. Amino-terminal cysteine residues of RGS16 are required for palmitoylation and modulation of Gi- and Gq-mediated signaling. *J Biol Chem* 1999;**274**:18836–42.
75. Rose JJ, Taylor JB, Shi J, Cockett MI, Jones PG, Hepler JR. RGS7 is palmitoylated and exists as biochemically distinct forms. *J Neurochem* 2000;**75**:2103–12.
76. Srinivasa SP, Bernstein LS, Blumer KJ, Linder ME. Plasma membrane localization is required for RGS4 function in Saccharomyces cerevisiae. *Proc Natl Acad Sci USA* 1998;**95**:5584–9.
77. Jones TL. Role of palmitoylation in RGS protein function. *Methods Enzymol* 2004;**389**:33–55.
78. Tu Y, Popov S, Slaughter C, Ross EM. Palmitoylation of a conserved cysteine in the regulator of G protein signaling (RGS) domain modulates the GTPase-activating activity of RGS4 and RGS10. *J Biol Chem* 1999;**274**:38260–7.
79. Luo X, Popov S, Bera AK, Wilkie TM, Muallem S. RGS proteins provide biochemical control of agonist-evoked [Ca2+]i oscillations. *Mol Cell* 2001;**7**:651–60.
80. Muallem S, Wilkie TM. G protein-dependent Ca2+ signaling complexes in polarized cells. *Cell Calcium* 1999;**26**:173–80.
81. Roman DL, Talbot JN, Roof RA, Sunahara RK, Traynor JR, Neubig RR. Identification of small-molecule inhibitors of RGS4 using a high-throughput flow cytometry protein interaction assay. *Mol Pharmacol* 2007;**71**:169–75.

82. Roman, D., Blazer LL, Monroy CA, Neubig RR. Allosteric Inhibition of the RGS-G Protein-Protein Interaction by CCG-4986, Mol Pharm 2010 in press.
83. Blazer LL, Roman DL, Chung A, Larsen M, Greedy B, Husbands S, Neubig RR. Reversible, allosteric, small-molecule inhibitors of RGS proteins. Mol Pharm 2010. Accepted for publication.
84. Fu Y, Zhong H, Nanamori M, Mortensen RM, Huang X, Lan K, et al. RGS-insensitive G-protein mutations to study the role of endogenous RGS proteins. *Methods Enzymol* 2004;**389**:229–43.
85. Heximer SP, Knutsen RH, Sun X, Kaltenbronn KM, Rhee MH, Peng N, et al. Hypertension and prolonged vasoconstrictor signaling in RGS2-deficient mice. *J Clin Invest* 2003;**111**:445–52.
86. Tang KM, Wang GR, Lu P, Karas RH, Aronovitz M, Heximer SP, et al. Regulator of G-protein signaling-2 mediates vascular smooth muscle relaxation and blood pressure. *Nat Med* 2003;**9**:1506–12.
87. Anger T, Klintworth N, Stumpf C, Daniel WG, Mende U, Garlichs CD. RGS protein specificity towards Gq- and Gi/o-mediated ERK 1/2 and Akt activation, *in vitro*. *J Biochem Mol Biol* 2007;**40**:899–910.
88. Sun X, Kaltenbronn KM, Steinberg TH, Blumer KJ. RGS2 is a mediator of nitric oxide action on blood pressure and vasoconstrictor signaling. *Mol Pharmacol* 2005;**67**:631–9.
89. Salim S, Sinnarajah S, Kehrl JH, Dessauer CW. Identification of RGS2 and type V adenylyl cyclase interaction sites. *J Biol Chem* 2003;**278**:15842–9.
90. Yang J, Kamide K, Kokubo Y, Takiuchi S, Tanaka C, Banno M, et al. Genetic variations of regulator of G-protein signaling 2 in hypertensive patients and in the general population. *J Hypertens* 2005;**23**:1497–505.
91. Bodenstein J, Sunahara RK, Neubig RR. N-terminal residues control proteasomal degradation of RGS2, RGS4, and RGS5 in HEK293 cells. *Mol Pharmacol* 2007;**71**(4):1040–50.
92. Heximer SP, Lim H, Bernard JL, Blumer KJ. Mechanisms governing subcellular localization and function of human RGS2. *J Biol Chem* 2001;**276**:14195–203.
93. Riddle EL, Rana BK, Murthy KK, Rao F, Eskin E, O'Connor DT, et al. Polymorphisms and haplotypes of the regulator of G protein signaling-2 gene in normotensives and hypertensives. *Hypertension* 2006;**47**:415–20.
94. Kurachi Y, Nakajima T, Sugimoto T. On the mechanism of activation of muscarinic K+ channels by adenosine in isolated atrial cells: involvement of GTP-binding proteins. *Pflugers Arch* 1986;**407**:264–74.
95. Bender K, Nasrollahzadeh P, Timpert M, Liu B, Pott L, Kienitz MC. A role for RGS10 in beta-adrenergic modulation of G-protein-activated K+ (GIRK) channel current in rat atrial myocytes. *J Physiol* 2008;**586**:2049–60.
96. Siedlecki A, Anderson JR, Jin X, Garbow JR, Lupu TS, Muslin AJ. RGS4 controls renal blood flow and inhibits cyclosporine-mediated nephrotoxicity. *Am J Transplant* 2010;**10**(2):231–41.
97. Wang Q, Liu M, Mullah B, Siderovski DP, Neubig RR. Receptor-selective effects of endogenous RGS3 and RGS5 to regulate mitogen-activated protein kinase activation in rat vascular smooth muscle cells. *J Biol Chem* 2002;**277**:24949–58.
98. Cho H, Kozasa T, Bondjers C, Betsholtz C, Kehrl JH. Pericyte-specific expression of Rgs5: implications for PDGF and EDG receptor signaling during vascular maturation. *FASEB J* 2003;**17**:440–2.
99. Hamzah J, Jugold M, Kiessling F, Rigby P, Manzur M, Marti HH, et al. Vascular normalization in Rgs5-deficient tumours promotes immune destruction. *Nature* 2008;**453**:410–4.
100. Nisancioglu MH, Mahoney Jr WM, Kimmel DD, Schwartz SM, Betsholtz C, Genove G. Generation and characterization of rgs5 mutant mice. *Mol Cell Biol* 2008;**28**:2324–31.

101. Iankova I, Chavey C, Clape C, Colomer C, Guerineau NC, Grillet N, et al. Regulator of G protein signaling-4 controls fatty acid and glucose homeostasis. *Endocrinology* 2008;**149**:5706–12.
102. Gold SJ, Ni YG, Dohlman HG, Nestler EJ. Regulators of G-protein signaling (RGS) proteins: region-specific expression of nine subtypes in rat brain. *J Neurosci* 1997;**17**:8024–37.
103. Mancuso JJ, Qian Y, Long C, Wu GY, Wensel TG. Distribution of RGS9-2 in neurons of the mouse striatum. *J Neurochem* 2010;**112**(3):651–61.
104. Michaelides M, Li Z, Rana NA, Richardson EC, Hykin PG, Moore AT, et al. Novel mutations and electrophysiologic findings in RGS9- and R9AP-associated retinal dysfunction (bradyopsia). *Ophthalmology* 2009;**117**:120–127.e121.
105. Nishiguchi KM, Sandberg MA, Kooijman AC, Martemyanov KA, Pott JWR, Hagstrom SA, et al. Defects in RGS9 or its anchor protein R9AP in patients with slow photoreceptor deactivation. *Nature* 2004;**427**:75–8.
106. Chen CK, Burns ME, He W, Wensel TG, Baylor DA, Simon MI. Slowed recovery of rod photoresponse in mice lacking the GTPase accelerating protein RGS9-1. *Nature* 2000;**403**:557–60.
107. Hooks SB, Martemyanov K, Zachariou V. A role of RGS proteins in drug addiction. *Biochem Pharmacol* 2008;**75**:76–84.
108. Traynor JR, Neubig RR. Regulators of G protein signaling & drugs of abuse. *Mol Interv* 2005;**5**:30–41.
109. Gold SJ, Han MH, Herman AE, Ni YG, Pudiak CM, Aghajanian GK, et al. Regulation of RGS proteins by chronic morphine in rat locus coeruleus. *Eur J Neurosci* 2003;**17**:971–80.
110. Nakagawa T, Minami M, Satoh M. Up-regulation of RGS4 mRNA by opioid receptor agonists in PC12 cells expressing cloned mu- or kappa-opioid receptors. *Eur J Pharmacol* 2001;**433**:29–36.
111. Georgoussi Z, Leontiadis L, Mazarakou G, Merkouris M, Hyde K, Hamm H. Selective interactions between G protein subunits and RGS4 with the C-terminal domains of the mu- and delta-opioid receptors regulate opioid receptor signaling. *Cell Signal* 2006;**18**:771–82.
112. Wang Q, Liu-Chen LY, Traynor JR. Differential modulation of mu- and delta-opioid receptor agonists by endogenous RGS4 protein in SH-SY5Y cells. *J Biol Chem* 2009;**284**:18357–67.
113. Han MH, Renthal W, Ring RH, Rahman Z, Psifogeorgou K, Howland D, et al. Brain region specific actions of regulator of G protein signaling 4 oppose morphine reward and dependence but promote analgesia. *Biol Psychiatry* 2010;**67**(8):761–9.
114. Zachariou V, Georgescu D, Sanchez N, Rahman Z, DiLeone R, Berton O, et al. Essential role for RGS9 in opiate action. *Proc Natl Acad Sci U S A* 2003;**100**:13656–61.
115. Psifogeorgou K, Papakosta P, Russo SJ, Neve RL, Kardassis D, Gold SJ, et al. RGS9-2 is a negative modulator of mu-opioid receptor function. *J Neurochem* 2007;**103**:617–25.
116. Rahman Z, Schwarz J, Gold SJ, Zachariou V, Wein MN, Choi KH, et al. RGS9 modulates dopamine signaling in the basal ganglia. *Neuron* 2003;**38**:941–52.
117. Kovoor A, Seyffarth P, Ebert J, Barghshoon S, Chen CK, Schwarz S, et al. D2 dopamine receptors colocalize regulator of G-protein signaling 9-2 (RGS9-2) via the RGS9 DEP domain, and RGS9 knock-out mice develop dyskinesias associated with dopamine pathways. *J Neurosci* 2005;**25**:2157–65.
118. Lee J-K, McCoy MK, Harms AS, Ruhn KA, Gold SJ, Tansey MG. Regulator of G-protein signaling 10 promotes dopaminergic neuron survival via regulation of the microglial inflammatory response. *J Neurosci* 2008;**28**:8517–28.
119. Goldenstein BL, Nelson BW, Xu K, Luger EJ, Pribula JA, Wald JM, et al. Regulator of G protein signaling protein suppression of Galphao protein-mediated alpha2A adrenergic receptor inhibition of mouse hippocampal CA3 epileptiform activity. *Mol Pharmacol* 2009;**75**:1222–30.

120. Pacey LK, Heximer SP, Hampson DR. Increased GABA(B) receptor-mediated signaling reduces the susceptibility of fragile X knockout mice to audiogenic seizures. *Mol Pharmacol* 2009;**76**:18–24.
121. Mirnics K, Middleton FA, Stanwood GD, Lewis DA, Levitt P. Disease-specific changes in regulator of G-protein signaling 4 (RGS4) expression in schizophrenia. *Mol Psychiatry* 2001;**6**:293–301.
122. Bowden NA, Scott RJ, Tooney PA. Altered expression of regulator of G-protein signalling 4 (RGS4) mRNA in the superior temporal gyrus in schizophrenia. *Schizophr Res* 2007;**89**:165–8.
123. Erdely HA, Tamminga CA, Roberts RC, Vogel MW. Regional alterations in RGS4 protein in schizophrenia. *Synapse* 2006;**59**:472–9.
124. Hon-Cheong S, Ronald YLC, Eric YHC, Eric FCC, Tao L, Pak CS. An association study of RGS4 polymorphisms with clinical phenotypes of schizophrenia in a Chinese population. *Am J Med Genet B Neuropsychiatr Genet* 2008;**147B**:77–85.
125. Ishiguro H, Horiuchi Y, Koga M, Inada T, Iwata N, Ozaki N, et al. RGS4 is not a susceptibility gene for schizophrenia in Japanese: association study in a large case-control population. *Schizophr Res* 2007;**89**:161–4.
126. Kampman O, Illi A, Hanninen K, Katila H, Anttila S, Rontu R, et al. RGS4 genotype is not associated with antipsychotic medication response in schizophrenia. *J Neural Transm* 2006;**113**:1563–8.
127. Mie AR, Andrew M, Vinay P, Khalid C, Susmita D, Srinivasa T, et al. Failure to confirm genetic association between schizophrenia and markers on chromosome 1q23.3 in the region of the gene encoding the regulator of G-protein signaling 4 protein (< I > RGS4</I>). *Am J Med Genet B Neuropsychiatr Genet* 2006;**141B**:296–300.
128. Grillet N, Pattyn A, Contet C, Kieffer BL, Goridis C, Brunet JF. Generation and characterization of Rgs4 mutant mice. *Mol Cell Biol* 2005;**25**:4221–8.
129. Whittington CJ, Kendall T, Pilling S. Are the SSRIs and atypical antidepressants safe and effective for children and adolescents? *Curr Opin Psychiatry* 2005;**18**:21–5.
130. Dorsam RT, Gutkind JS. G-protein-coupled receptors and cancer. *Nat Rev Cancer* 2007;**7**:79–94.
131. Li S, Huang S, Peng SB. Overexpression of G protein-coupled receptors in cancer cells: involvement in tumor progression. *Int J Oncol* 2005;**27**:1329–39.
132. Pierce KL, Premont RT, Lefkowitz RJ. Seven-transmembrane receptors. *Nat Rev Mol Cell Biol* 2002;**3**:639–50.
133. Spiegelberg BD, Hamm HE. Roles of G-protein-coupled receptor signaling in cancer biology and gene transcription. *Curr Opin Genet Dev* 2007;**17**:40–4.
134. Bennasroune A, Gardin A, Aunis D, Cremel G, Hubert P. Tyrosine kinase receptors as attractive targets of cancer therapy. *Crit Rev Oncol Hematol* 2004;**50**:23–38.
135. Zwick E, Bange J, Ullrich A. Receptor tyrosine kinases as targets for anticancer drugs. *Trends Mol Med* 2002;**8**:17–23.
136. Hurst JH, Mendpara N, Hooks SB. Regulator of G-protein signalling expression and function in ovarian cancer cell lines. *Cell Mol Biol Lett* 2009;**14**:153–74.
137. Xie Y, Wolff DW, Wei T, Wang B, Deng C, Kirui JK, et al. Breast cancer migration and invasion depend on proteasome degradation of regulator of G-protein signaling 4. *Cancer Res* 2009;**69**:5743–51.
138. Feldman BJ, Feldman D. The development of androgen-independent prostate cancer. *Nat Rev Cancer* 2001;**1**:34–45.
139. Cao X, Qin J, Xie Y, Khan O, Dowd F, Scofield M, et al. Regulator of G-protein signaling 2 (RGS2) inhibits androgen-independent activation of androgen receptor in prostate cancer cells. *Oncogene* 2006;**25**:3719–34.

140. Shuey DJ, Betty M, Jones PG, Khawaja XZ, Cockett MI. RGS7 attenuates signal transduction through the G(alpha q) family of heterotrimeric G proteins in mammalian cells. *J Neurochem* 1998;**70**:1964–72.
141. Witherow DS, Tovey SC, Wang Q, Willars GB, Slepak VZ. G beta 5.RGS7 inhibits G alpha q-mediated signaling via a direct protein-protein interaction. *J Biol Chem* 2003;**278**:21307–13.
142. Xie Z, Geiger TR, Johnson EN, Nyborg JK, Druey KM. RGS13 acts as a nuclear repressor of CREB. *Mol Cell* 2008;**31**:660–70.
143. Roy AA, Baragli A, Bernstein LS, Hepler JR, Hebert TE, Chidiac P. RGS2 interacts with Gs and adenylyl cyclase in living cells. *Cell Signal* 2006;**18**:336–48.
144. Cowan CW, Fariss RN, Sokal I, Palczewski K, Wensel TG. High expression levels in cones of RGS9, the predominant GTPase accelerating protein of rods. *Proc Natl Acad Sci USA* 1998;**95**:5351–6.
145. He W, Cowan CW, Wensel TG. RGS9, a GTPase accelerator for phototransduction. *Neuron* 1998;**20**:95–102.
146. Lan KL, Zhong H, Nanamori M, Neubig RR. Rapid kinetics of regulator of G-protein signaling (RGS)-mediated Galphai and Galphao deactivation. Galpha specificity of RGS4 AND RGS7. *J Biol Chem* 2000;**275**:33497–503.
147. Castellone MD, Teramoto H, Williams BO, Druey KM, Gutkind JS. Prostaglandin E2 promotes colon cancer cell growth through a Gs-axin-beta-catenin signaling axis. *Science* 2005;**310**:1504–10.
148. Zheng B, Ma YC, Ostrom RS, Lavoie C, Gill GN, Insel PA, et al. RGS-PX1, a GAP for GalphaS and sorting nexin in vesicular trafficking. *Science* 2001;**294**:1939–42.
149. Lu Q, Sun EE, Flanagan JG, David PS. Analysis of PDZ-RGS3 function in Ephrin-B reverse signaling. *Methods Enzymol* 2004;120–8. (Academic Press).
150. Hiol A, Davey PC, Osterhout JL, Waheed AA, Fischer ER, Chen CK, et al. Palmitoylation regulates regulators of G-protein signaling (RGS) 16 function. I. Mutation of amino-terminal cysteine residues on RGS16 prevents its targeting to lipid rafts and palmitoylation of an internal cysteine residue. *J Biol Chem* 2003;**278**:19301–8.
151. Chatterjee TK, Fisher RA. Cytoplasmic, nuclear, and golgi localization of RGS proteins. Evidence for N-terminal and RGS domain sequences as intracellular targeting motifs. *J Biol Chem* 2000;**275**:24013–21.
152. Chen C, Seow KT, Guo K, Yaw LP, Lin SC. The membrane association domain of RGS16 contains unique amphipathic features that are conserved in RGS4 and RGS5. *J Biol Chem* 1999;**274**:19799–806.
153. Blazer LL, Neubig RR. Small molecule protein-protein interaction inhibitors as CNS therapeutic agents: current progress and future hurdles. *Neuropsychopharmacology* 2009;**34**:126–41.
154. Berg T. Modulation of protein-protein interactions with small organic molecules. *Angew Chem Int Ed Engl* 2003;**42**:2462–81.
155. Berg T. Small-molecule inhibitors of protein-protein interactions. *Curr Opin Drug Discov Devel* 2008;**11**:666–74.
156. Roman DL, Ota S, Neubig RR. Polyplexed flow cytometry protein interaction assay: a novel high-throughput screening paradigm for RGS protein inhibitors. *J Biomol Screen* 2009;**14**(6):610–9.
157. Roof RA, Jin Y, Roman DL, Sunahara RK, Ishii M, Mosberg HI, et al. Mechanism of action and structural requirements of constrained peptide inhibitors of RGS proteins. *Chem Biol Drug Des* 2006;**67**:266–74.
158. Roof RA, Roman DL, Clements ST, Sobczyk-Kojiro K, Blazer LL, Ota S, et al. A covalent peptide inhibitor of RGS4 identified in a focused one-bead, one compound library screen. *BMC Pharmacol* 2009;**9**:9.
159. Roof RA, Sobczyk-Kojiro K, Turbiak AJ, Roman DL, Pogozheva ID, Blazer LL, et al. Novel peptide ligands of RGS4 from a focused one-bead, one-compound library. *Chem Biol Drug Des* 2008;**72**:111–9.

160. Wang Y, Lee Y, Zhang J, Young KH. Identification of peptides that inhibit regulator of G protein signaling 4 function. *Pharmacology* 2008;**82**:97–104.
161. Young KH, Wang Y, Bender C, Ajit S, Ramirez F, Gilbert A, et al. Yeast-based screening for inhibitors of RGS proteins. *Methods Enzymol* 2004;**389**:277–301.
162. Jin Y, Zhong H, Omnaas JR, Neubig RR, Mosberg HI. Structure-based design, synthesis, and pharmacologic evaluation of peptide RGS4 inhibitors. *J Pept Res* 2004;**63**:141–6.
163. Edwards BS, Young SM, Ivnitsky-Steele I, Ye RD, Prossnitz ER, Sklar LA. High-content screening: flow cytometry analysis. *Methods Mol Biol* 2009;**486**:151–65.
164. Leifert WR, Bailey K, Cooper TH, Aloia AL, Glatz RV, McMurchie EJ. Measurement of heterotrimeric G-protein and regulators of G-protein signaling interactions by time-resolved fluorescence resonance energy transfer. *Anal Biochem* 2006;**355**:201–12.
165. Zielinski T, Kimple AJ, Hutsell SQ, Koeff MD, Siderovski DP, Lowery RG. Two G{alpha}i1 rate-modifying mutations act in concert to allow receptor-independent steady-state measurements of RGS protein activity. *J Biomol Screen* 2009;**14**(10):1195–206.
166. Pfleger KD, Eidne KA. Illuminating insights into protein-protein interactions using bioluminescence resonance energy transfer (BRET). *Nat Methods* 2006;**3**:165–74.
167. Gales C, Rebois RV, Hogue M, Trieu P, Breit A, Hebert TE, et al. Real-time monitoring of receptor and G-protein interactions in living cells. *Nat Methods* 2005;**2**:177–84.
168. Bacart J, Corbel C, Jockers R, Bach S, Couturier C. The BRET technology and its application to screening assays. *Biotechnol J* 2008;**3**:311–24.
169. Lohse MJ, Nikolaev VO, Hein P, Hoffmann C, Vilardaga JP, Bunemann M. Optical techniques to analyze real-time activation and signaling of G-protein-coupled receptors. *Trends Pharmacol Sci* 2008;**29**:159–65.
170. Yost EA, Mervine SM, Sabo JL, Hynes TR, Berlot CH. Live cell analysis of G protein beta5 complex formation, function, and targeting. *Mol Pharmacol* 2007;**72**:812–25.
171. Vidi PA, Watts VJ. Fluorescent and bioluminescent protein-fragment complementation assays in the study of G protein-coupled receptor oligomerization and signaling. *Mol Pharmacol* 2009;**75**:733–9.
172. Xu X, Zeng W, Popov S, Berman DM, Davignon I, Yu K, et al. RGS proteins determine signaling specificity of Gq-coupled receptors. *J Biol Chem* 1999;**274**:3549–56.
173. Han SB, Moratz C, Huang NN, Kelsall B, Cho H, Shi CS, et al. Rgs1 and Gnai2 regulate the entrance of B lymphocytes into lymph nodes and B cell motility within lymph node follicles. *Immunity* 2005;**22**:343–54.
174. Moratz C, Hayman JR, Gu H, Kehrl JH. Abnormal B-cell responses to chemokines, disturbed plasma cell localization, and distorted immune tissue architecture in Rgs1−/− mice. *Mol Cell Biol* 2004;**24**:5767–75.
175. Hutchison RM, Peter C, Leung LS. Hippocampal long-term potentiation is enhanced in urethane-anesthetized RGS2 knockout mice. *Hippocampus* 2009;**19**:687–91.
176. Yang S, Li Y-P. RGS10-null mutation impairs osteoclast differentiation resulting from the loss of [Ca2+]i oscillation regulation. *Genes Dev* 2007;**21**:1803–16.
177. Anderson GR, Cao Y, Davidson S, Truong HV, Pravetoni M, Thomas MJ, et al. R7BP complexes with RGS9-2 and RGS7 in the striatum differentially control motor learning and locomotor responses to cocaine. *Neuropsychopharmacology* 2010;**35**(4):1040–50.
178. Talbot JN, Jutkiewicz EM, Graves SM, Clemans, CF, Nicol MR, Mortensen RM, et al. RGS inhibition at Gαi2 selectively potentiates 5-HT1A–mediated antidepressant effects, PNAS 2010; in press.
179. Blazer LL, Roman DL, Muxlow MR, Neubig RR. Use of Flow Cytometric Methods to Quantify Protein-Protein Interactions. *Curr Protocol Cytom* 2010; in press.

Escorts Take the Lead: Molecular Chaperones as Therapeutic Targets

Dumaine Williams and
Lakshmi A. Devi

Department of Pharmacology and Systems Therapeutics, Mount Sinai School of Medicine, New York, USA

The functional and physiological diversity of transmembrane receptors results from factors that influence the pharmacology, signaling, and trafficking of these receptors. Receptor mutations and other modifications may lead to misfolding, intracellular retention, and ineffective signaling of transmembrane receptors. The importance of such mutations is highlighted by the fact that various diseases have been linked to mutations that lead to ineffective signaling of these receptors, resulting from the retention of receptors in intracellular compartments. Studies focused on understanding the regulation of trafficking and cell surface expression of newly synthesized receptors have highlighted molecular chaperones as key regulators of receptor maturation and sorting. In this chapter, we discuss the functions of molecular chaperones in the regulation of seven-transmembrane-containing G-protein-coupled receptor function and trafficking and explore ways in which chaperones can serve as novel therapeutic targets.

Progress in Molecular Biology
and Translational Science, Vol. 91
DOI: 10.1016/S1877-1173(10)91005-3

1877-1173/10 $35.00

I. Introduction

A large family of over 800 genes encodes receptor proteins that are characterized by a signature seven-transmembrane structure. Members of this family include sensory receptors for taste, odorants, and light as well as receptors for many hormones, neurotransmitters, chemokines, and ions. These seven-transmembrane receptors (also known as heptahelical receptors) are commonly referred to as G-protein-coupled receptors (GPCRs) because they mediate their effects through the activation of a variety of heterotrimeric (α, β, γ-subunits) guanine nucleotide-binding G proteins. These GPCRs regulate many physiological processes, and the mechanism by which GPCRs translate extracellular signals into cellular changes has been an area of active research for many years.

Initial theories of GPCR signaling involved agonist binding leading to the activation of the receptor, resulting in dissociation of the G protein into an α subunit and a $\beta\gamma$ subunit. Both these subunits have been shown to activate or inhibit various downstream effector molecules. However, further advances in the field of GPCR signaling have demonstrated that the mechanisms by which cell surface receptors orchestrate cellular changes are more complex. The recognition of the importance of GPCR oligomerization, the discovery of regulators of G protein signaling (RGS) proteins, and the identification of accessory/chaperone molecules are just some of the factors that have contributed to the expansion of the role and function of GPCRs.

Not only do the GPCRs regulate a plethora of physiological processes but drugs that target these receptors account for most of the medicines sold worldwide. These drugs target these seven-transmembrane receptors directly or target other proteins that are crucial for signaling through these receptors. This chapter focuses on the functions of molecular chaperones in the regulation of transmembrane receptor function and trafficking and explores ways in which chaperones can serve as novel therapeutic targets.

II. Molecular Chaperones and Accessory Proteins

The molecular chaperone concept was first proposed by John Ellis in 1987; he proposed that the term molecular chaperone be used to describe "a class of cellular proteins whose function is to ensure that the folding of certain other polypeptide chains and their assembly into oligomeric structures occur correctly."[1] There is a commonly held misconception that molecular chaperones are solely involved in ensuring proper protein folding. While many chaperones are involved in stabilizing unfolded protein folding and are involved in protein unfolding and degradation, chaperones also play a crucial role in the assembly of folded subunits into oligomeric structures.

With regard to GPCR function, some GPCRs may require the specific assistance of chaperones for proper folding during maturation. In addition, recent findings have highlighted various cytoplasmic and membrane-associated proteins that interact with GPCRs as they traffic through intracellular compartments and facilitate the cell surface expression of these GPCRs. While many of these chaperone proteins have additional biological roles, it is clear that they are necessary for proper functional expression of the receptors with which they interact.

III. GPCR Maturation and Postendoplasmic Reticulum Trafficking

GPCRs are synthesized by ribosomes attached at the cytosolic face of the endoplasmic reticulum (ER). During biosynthesis, these proteins are targeted by their hydrophobic signal sequences to the translocation complex which facilitates cotranslational entry into the ER lumen. Insertion of transmembrane domains into the membrane is driven by the translocation complex and orientation signals contained in the protein's polypeptidic chain. This membrane insertion is assisted by molecular chaperones and folding factors.[2,3] Most GPCR proteins fold properly with the aid of a conventional chaperone system. This conventional chaperone system comprises classical and lectin chaperones as well as enzymes that catalyze disulfide-bond formation or peptidyl–prolyl cis–trans isomerization.[4,5]

Once transmembrane proteins have achieved their native conformation, they leave the ER and are transported through the secretory pathway to their subcellular destination. This complex ER machinery constitutes the most important quality-control system for crosschecking newly synthesized proteins. In the event that newly synthesized proteins are defective in their folding, these misfolded polypeptides are exported across the ER membrane into the cytosol and destroyed by the ER-associated degradation pathway (ERAD).[6]

The general chaperone system that is commonly utilized for membrane proteins in the secretory pathway involves the use of heat-shock proteins (HSPs). HSPs have been implicated as central components of the chaperone-mediated protein folding mechanism. Several physical and chemical conditions that are potentially harmful to cells (such as elevated temperature) result in inappropriate protein folding. This increase in inappropriate protein folding is accompanied by a concomitant increase in the levels and/or activity of HSPs whose job is to deal with misfolded proteins and get the protein folding machinery back on track.[7] Many HSPs are constitutively expressed in all cells. There are many families of HSPs and individual members within each family differ in their expression pattern and functions.

The Hsp70 family ER luminal protein BiP (also known as GRP78) is a key regulator of the quality-control system of the ER. BiP, working with Hsp40 family cofactors, facilitates translocation of nascent chains in the ER lumen, participates in protein folding and oligomerization and contributes to the retro-translocation of misfolded proteins to the ER-associated degradation pathway.[6] Some GPCRs, such as the luteinizing hormone receptor, have been shown to interact with BiP.[8,9] The luteinizing hormone receptor was also found to interact in the ER with GRP94, a member of the Hsp90 family and a known cofactor for BiP.[9]

Numerous reports have demonstrated the interaction between GPCRs and carbohydrate-binding chaperones.[8–10] It is a common feature that many mature GPCRs are *N*-glycosylated at the N-terminal region and/or at the extracellular loops. These regions are luminal when the immature receptors are present in the ER. Cotranslational addition of sugar chains to asparagine residues by oligosaccharyltransferases provides binding sites for lectin chaperones that are able to bind to carbohydrate molecules. Such lectin chaperones include the proteins calnexin and calreticulin. After the removal of the two terminal glucoses by a glucosidase, the resulting monoglucosylated newly synthesized GPCR proteins interact with lectin chaperones. The interaction between the lectin chaperone and the immature receptor protein is terminated by the cleavage of the last glucose by glucosidase II. Once released, correctly folded glycoproteins can exit the ER. By contrast, incorrectly or incompletely folded glycoproteins are reglucosylated by glucosyltransferase, resulting in another round of association with calnexin and calreticulin. Cycles of glucosylation and deglucosylation continue until the glycoprotein has either reached its proper, folded conformation or is targeted for degradation.

Exit of proteins from the ER occurs at ER-exit sites, where buds are formed and coated with the COPII coat protein. Proteins released from the ER quality-control machinery accumulate in these COPII-coated buds.[11] Signals in the cytoplasmically exposed C-terminal tails of transmembrane proteins, such as GPCRs, that are being transported in the secretory pathway are likely to be involved in direct binding with components of COPII.[12] These signals comprise diacidic motifs and pairs of aromatic or bulky hydrophobic amino acid residues.[11] Cargo receptors for soluble secretory proteins present in ER buds also possess these motifs in their C-terminal tail. Many GPCRs contain these ER-export motifs, indicating that they might interact directly with COPII complex proteins.[13]

The next step in the maturation of GPCRs involves trafficking to the Golgi apparatus. In mammalian cells, protein traffic moving from ER-exit sites to the Golgi complex passes through the ER–Golgi intermediate compartment. The ER–Golgi intermediate compartment is a site of anterograde and retrograde sorting under the control of COPI coat proteins, Rab and Arf GTPases.[14] Vesicles exiting from this compartment are directed either to the Golgi or back to the ER, depending on a series of factors. These factors include the nature of the cargo protein, the Arf GTPase isoform involved in coat recruitment and the Rab effectors.

The ER–Golgi intermediate compartment is known to utilize at least two Rab effector proteins to promote the formation of two separate pools of transport vesicles. It is believed that Rab1 isoforms are involved in the transport from the ER to the ER–Golgi intermediate compartment and transport from the ER–Golgi intermediate compartment to the *cis*-Golgi complex.[15] Conversely, Rab2 is believed to promote the formation of vesicles returning from the ER–Golgi intermediate compartment back to the ER.[16]

In order to investigate the involvement of these Rab effector proteins in GPCR trafficking through the ER–Golgi intermediate compartment, various studies have investigated the effect of selective small interfering RNA (siRNA)-mediated knockdown of Rab1 and Rab2 and overexpression of dominant-negative Rab1 and Rab2 on GPCR maturation and proper receptor export. Overexpression of dominant-negative Rab1a and siRNA-mediated knockdown of Rab1b perturbed trafficking of both the angiotensin AT_1 receptor and the β_2-adrenoceptor.[17] Such effects appear to be receptor-specific as, in the same report, trafficking of the α_{2B}-adrenoceptor was not affected by these same Rab protein manipulations. Additionally, overexpression of Rab2 mutants and siRNA-mediated knockdown of Rab2 resulted in an inhibition of cell surface expression of both the β_2- and α_{2B}-adrenoceptors.[18]

Transport vesicles that bud from the rough ER carry proteins to the luminal cavity of the Golgi complex. Three-dimensional reconstructions from serial sections of the Golgi complex show a series of flattened membrane sacs. These series of flattened membrane sacs have three defined regions, termed the *cis*-, *medial*-, and *trans*-Golgi. Transport vesicles from the rough ER fuse with the *cis*-Golgi region of the complex, where these cargo proteins are deposited. Proteins then progress from the cis to the medial to the trans region of the Golgi complex. Within each region are different enzymes (glycosyltransferases) that modify proteins differently depending on their structures and final destinations.[19]

Once in the Golgi, cargo proteins can be sorted to the endosomal system, the plasma membrane or back to the ER. Retrograde transport to the ER is likely to be involved in retargeting misfolded proteins to the ER-associated degradation pathway.[20] Other defective proteins are targeted to lysosomal degradation after sorting to endosomes.[20] As is the case with investigating GPCR transport out of the ER, studies examining GPCR trafficking through the Golgi stacks have looked at the effects of Rab effector protein perturbation. The Rab6 GTPase regulates vesicular transport in the Golgi. Perturbing the function of the Rab6 GTPase inhibits the anterograde transport of *Drosophila* rhodopsin and of mammalian GPCRs.[18,21] Additionally, studies of δ-opioid receptor glycosylation demonstrated that O-glycosylation (on serine or threonine residues) and final processing of N-linked oligosaccharides occur in different compartments of the Golgi.[22] Studies using a α_{2B}-adrenoceptor mutant in which adjacent tyrosine and serine amino acid residues were substituted provided evidence that the N-terminus of the α_{2B}-adrenoceptor might contain a signal that directs exit of the receptor from the Golgi. When this mutant α_{2B}-adrenoceptor protein was expressed, the mutant receptor was primarily trapped in the Golgi complex.[23]

IV. Regulated Translocation to Intracellular Compartments and/or the Plasma Membrane

In the most simplified model of GPCR trafficking, GPCRs are expressed on the cell surface following biosynthetic sorting and are then endocytosed in response to activation by agonists. Increasingly, there is evidence that for some GPCRs, this might not always be the case. This is demonstrated by studies performed studying the trafficking of the protease-activated receptor (PAR) family. Irreversible activation of the PAR1 and PAR2 thrombin receptors by cleavage results in internalization and degradation of the receptors in lysosomes. It has been demonstrated that large pools of PAR1 and PAR2 receptors reside in intracellular compartments. Most of these receptors are localized in the Golgi complex, where they are protected from activation by thrombin and subsequent degradation. Upon activation of the cell surface receptors, these intracellular thrombin receptors translocate to the plasma membrane and replenish the pool of plasma membrane receptors that have been degraded following receptor activation. In this way, replenishment of plasma membrane thrombin receptors is correlated with recovery of thrombin responsiveness.[24]

As is the case with the thrombin receptors, regulated pools of intracellular dopamine D_1 receptors exist in tubular renal cells. In these cells, agonist activation of cell surface dopamine receptors results in recruitment of the intracellular receptors to the cell surface.[25] It has also been reported that intracellular D_1 receptors can be recruited to the plasma membrane by atrial natriuretic peptide-dependent heterologous activation.[26] A similar translocation profile has been demonstrated for α_{1A}-adrenoceptors that translocate to the plasma membrane in response to stimulation by neuropeptide Y.[26] Additionally, nerve growth factor treatment was shown to lead to the formation of intracellular pools of the δ-opioid receptor in PC12 neurosecretory cells.[27] These intracellular δ-opioid receptors exhibited depolarization-dependent translocation to the plasma membrane. A similar intracellular distribution of δ-opioid receptors has been demonstrated in primary cells from the central nervous system of the rat.[28] Also, it has been shown that prolonged exposure of neurons to morphine leads to the recruitment of intracellular δ-opioid receptors to the cell surface, both *in vitro* and *in vivo*.[29,30]

The estrogen receptor GPR30 is unique in that it is an intracellular transmembrane estrogen receptor.[31] GPR30 is localized in the ER where it has the ability to specifically bind estrogen and estrogen derivatives. Activation of GPR30 with estrogen resulted in downstream signaling effects such as intracellular calcium mobilization, indicating that the intracellular receptors are functional. Similarly, studies investigating the function of cannabinoid receptor 1 (CB_1) have demonstrated that these receptors are localized at the plasma membrane as well as in intracellular vesicles.[32] Additionally, it was demonstrated that intracellular CB_1 receptors do not have an endocytic origin. However, these receptors were shown to interact with the adaptor protein AP-3 and traffic to endolysosomes. These CB_1 receptors in the endolysosomes associate with heterotrimeric G proteins and have the ability to mediate signaling, illustrating that the intracellular receptors are in fact functional. Hence, proteins that bind and retain these receptors modulate the spatial dynamics of signaling.

V. GPCR Oligomerization

GPCRs physically associate with other cellular proteins, including a large variety of soluble intracellular proteins such as β-arrestins, membrane proteins such as receptor-activity-modifying proteins (RAMPs) as well as other GPCRs. The recognition of GPCR homo- and heteromerization has generated numerous possibilities for expanding the roles and functions of GPCRs. Dimerization of GPCRs has been implicated in modulating a number of functional properties of the receptors, including ligand binding, receptor signaling, and receptor trafficking. Data supporting the association of GPCRs was first presented for muscarinic cholinergic and α_2- and β_2-adrenergic receptors (ARs).[33] However, despite published reports on β_2-AR homomerization,[34] the concept of GPCR dimerization remained highly controversial until the cloning of the γ-aminobutyric $acid_B$ ($GABA_B$) receptor.

After the cloning of the $GABA_BR1$ receptor it was quickly discovered that expression of a fully functional receptor in mammalian cells was not possible because the receptor was constitutively trapped in the ER and did not traffic to the cell surface.[35] In an attempt to find novel genes related to the $GABA_BR1$ gene, two independent groups identified a related protein by homology cloning. This protein exhibited 35% identity to $GABA_BR1$, and was named $GABA_BR2$.[36,37] When expressed in either *Xenopus* oocytes or HEK293 cells, neither $GABA_BR1$ nor $GABA_BR2$ receptors were cell surface localized or functional when expressed alone.[35,36] However, coexpression of the $GABA_BR1$ and R2 receptors in heterologous systems resulted in the expression of these receptors at the cell surface.[36]

Coexpression of the two receptors also resulted in acquisition of receptor recognition properties characteristic of native $GABA_B$ receptors, coupling to K^+ currents, and inhibition of adenylyl cyclase.[36–40] The $GABA_BR1$ and $GABA_BR2$ receptors were also shown to physically associate with each other. Yeast two-hybrid screens demonstrated interaction of the C-terminus of $GABA_BR1$ with cDNA-encoded $GABA_BR2$ fragments.[37,38] GST-pull down experiments showed interactions of the C-termini of $GABA_BR1$ and $GABA_BR2$,[38] and these two receptors were co-immunoprecipitated from the rat brain cortex and from heterologous cells expressing both proteins.[36,37,40]

Further investigation demonstrated that heteromerization of the $GABA_B$ receptor appears to promote cell surface localization by masking an ER retention motif, identified as RXR(R), which exists in the C-terminus of $GABA_BR1$.[41] It is important to note, however, that $GABA_BR2$ is not merely a chaperone for $GABA_BR1$. In fact, $GABA_BR2$ also contributed to signaling and hence is necessary for the formation of fully functional $GABA_B$ receptors. In addition, other domains of the two subunits were shown to interact, as C-terminally truncated constructs also form heteromeric complexes, suggesting that this interaction is not due entirely to interactions between the C-terminal domains.[42]

Unlike the case of the $GABA_B$ receptor in which dimerization of the R1 and R2 receptors is required for proper receptor cellular localization and function, opioid receptors subtypes which are properly localized and fully functional when expressed alone have been shown to form both homomers and heteromers.[43,44] For example, while both κ-and δ-opioid receptors show high affinity for agonists and antagonists that are subtype-selective, the κ–δ heteromers have no significant affinity for any of the subtype-selective agents. However, the κ–δ-opioid receptor heteromers demonstrate a high affinity for certain partially subtype-selective ligands and display synergistic agonist binding. These opioid receptor heteromers therefore demonstrate potentiated functional responses.[43] It is hypothesized that heteromerization results in the formation of novel recognition sites for previously characterized receptors which may provide functionally relevant receptors for endogenous opioid peptides.

In addition to the κ–δ heteromers, δ-opioid receptors also heteromerize with μ-opioid receptors.[45] Treatment of cells coexpressing μ- and δ-opioid receptors with very low concentrations of certain δ-opioid agonists significantly potentiates the binding of μ receptor-selective agonists and vice versa. Subsequently, it has been proposed that this μ–δ heteromer interaction may represent the molecular basis for the observed phenomenon of opiate synergy. Opiate synergy is the term used to describe the enhancement of the analgesic effects of one opioid receptor subtype through the treatment with the agonist that is selective for another receptor subtype.

Not only does heteromerization of κ- and δ-opioid receptors lead to the generation of completely distinct pharmacological profiles, heteromerization of these two opioid receptor subtypes modulates receptor trafficking. When these opioid receptors are expressed alone, stimulation of δ-opioid receptors with etorphine results in substantial internalization of these receptors; however, κ receptors do not internalize following stimulation with this agonist.[43] However, when these δ and κ receptors are coexpressed and then stimulated with etorphine, internalization of the δ receptor is substantially reduced, indicating that heteromerization of these two opioid receptors significantly changes the trafficking profiles of the receptors.[43]

In addition to investigating heteromerization between receptor subtypes in the same receptor families, there are reports of GPCR heteromerization between receptors from two distinct subfamilies. For example, it was reported that the β_2-adrenergic receptors (β_2AR) (which couple to the G_s protein) and κ- and δ-opioid receptors (which couple to G_i proteins) can also form heteromers. This heteromerization was demonstrated by co-immunoprecipitation of independently epitope-tagged receptors. As an important control, it was shown that this co-immunoprecipitation does not occur if the β_2AR and opioid receptors are expressed independently and the cells are mixed before extraction and immunoisolation. Heteromerization of these receptor pairs does not alter ligand binding or functional coupling of the receptors; however, the trafficking properties of the receptors was affected by dimerization. When expressed alone, β_2ARs undergo robust internalization following agonist stimulation. However, it was demonstrated that when β_2AR is coexpressed with κ receptor followed by stimulation with the β-adrenergic agonist isoproterenol the β_2AR receptors fail to internalize.[46] In contrast, coexpression of δ-opioid and β_2ARs leads to agonist-mediated internalization following treatment with either etorphine or isoproterenol.[46]

In addition to the aforementioned studies looking at β_2AR heteromerization, significant effort has been made looking at the ability of β_2AR to form homomers. Initial reports of negative cooperativity of ligand binding at β_2ARs provided the first indications that β_2AR may form homomers.[47] Evidence for the existence of β_2AR homomers include biochemical data in which differentially epitope-tagged β_2AR receptors were coexpressed in the same cell and subsequently co-immunoprecipitated.[48] Based on homology to a dimerization motif found in glycophorin A, sequence analysis of β_2AR revealed a putative dimerization motif in the sixth transmembrane-spanning segment. Hebert et al. synthesized a peptide corresponding to this motif in the sixth transmembrane-spanning segment of β_2AR and found that it was capable of disrupting β_2AR dimer formation, as assessed by SDS-PAGE analysis.[48] Additionally, researchers found that this peptide diminished β_2AR-stimulated adenylyl cyclase activity in a concentration-dependent manner. The functional relevance of β_2AR dimerization is further supported by the observation that isoproterenol enhances dimer formation and confers protection from interference by the sixth transmembrane-spanning segment peptide; consistent with the interpretation that agonist occupancy stabilizes β_2AR dimer formation.

The conclusion that β_2AR oligomerization occurs in living cells is further supported by bioluminescence resonance energy transfer (BRET) studies.[49] BRET relies on the transfer of energy from one tagged receptor (the donor) to a differentially tagged receptor (the acceptor). The energy transfer requires that the energy donor receptor and acceptor receptor are in very close proximity. In BRET studies, evidence of two receptors being in close proximity is used as an indication of protein–protein interactions and subsequently receptor oligomerization.

For studies investigating homomerization of the β_2AR, fusion proteins of β_2AR-luciferase and β_2AR-GFP were created and then coexpressed in cells. It was observed that BRET energy transfer could be detected in cells coexpressing the two differentially tagged receptors. In addition, when these cells were activated with the receptor agonist isoproterenol there was a dose-dependent increase in the BRET signal achieved. It was also demonstrated that no BRET signal is detected when the β_2AR-luciferase fusion protein is coexpressed with other receptors (including the melatonin 1a receptor-GFP or the chemokine CCR5 receptor-GFP fusion proteins).[50] These results are consistent with a model in which the β_2AR homomerization is promoted by the receptor agonist.[50]

But what are the roles of GPCR homomerization in promoting receptor maturation or plasma membrane localization? As discussed earlier, in the case of GABA receptors, heteromerization may help promote exit from the ER by masking retention signals present in the receptors. There is less evidence of specific examples of mechanisms by which homomerization of GPCRs can help promote receptor maturation during biosynthesis. One emerging hypothesis is that homomerization might assist in receptor folding. The membrane-spanning domains of GPCRs contain hydrophobic regions that are particularly prone to nonspecific aggregation. In fact, chaperones are known to associate with newly synthesized polypeptides in order to prevent unproductive interactions with the environment that could result in protein aggregation.[51] Thus, one hypothesis is that individual GPCR receptors can potentially serve as folding chaperones for each other. Therefore, the ordered association of two newly synthesized GPCR polypeptides via their hydrophobic transmembrane regions could hide a substantial proportion of the exposed hydrophobic surface and facilitate correct folding. In some cases these hydrophobic transmembrane regions constitute the dimerization interface, supporting such a hypothesis.

This model that GPCRs can act as folding chaperones for each other is supported by the fact that functional GPCR heteromers have been obtained in reconstituted cell models containing receptors that do not encounter each other under normal physiological conditions. In these artificial conditions, in which two distinct GPCR polypeptides are forced to enter simultaneously into the ER, if they display sufficient structure-driven propensity to assemble, they might form heteromers.

Another theory is that homomerization of GPCRs might also contribute to quality control. This theory is based on the fact that dimeric receptors are likely to be structurally symmetric and checking for symmetry between two receptors in a homomer might be one mechanism by which mutant receptor can be identified in the ER. Random mutations affecting the overall structure of one receptor might generate asymmetry within the homomer. If asymmetry is found between receptors in a homomer, this could lead to the recognition of mutant receptors by the ER quality-control system. Once recognized, these mutant receptors can then be retained for disposal via the ER-associated degradation pathway. Consistent with this model, mutant ER-retained GPCRs have been shown to display a dominant-negative effect over wild-type GPCRs in reconstituted cellular models and in heterozygous individuals.[52]

Studies on receptor homomerization have also provided useful information on how and where dimerization occurs in cells. In a study investigating the biogenesis of serotonin 5-HT_{2C} homomers, confocal microscopy and fluorescence resonance energy transfer (FRET) were used to monitor the proximity of differentially tagged serotonin 5-HT_{2C} receptors in intact living cells.[53] These differentially tagged receptors were tracked during biosynthesis in the ER, trafficking through the Golgi complex and subsequent localization at the plasma membrane. The FRET results from these studies provide evidence of oligomerization of these receptors in the ER and the Golgi of living cells. Similarly, a study using photoaffinity labeling to investigate the dimerization of wild-type and mutant dopamine D_2 receptors provided evidence for dimerization of these receptors in intracellular compartments.[54] These results also highlighted the fact that dimerization is a naturally occurring step in receptor maturation and indicated that, in some cases, dimerization might be a prerequisite for normal receptor trafficking and cell surface localization.

VI. Specific Molecular Chaperones for GPCRs

It has become evident that, in addition to using general chaperones for the folding and sorting of newly synthesized proteins, some proteins may require more specialized assistance and utilize specific chaperones. Some of these specialized chaperones or enzymes directly participate in the folding of their cognate proteins. In the case of GPCRs, a few specialized chaperones have been reported to be involved in GPCR folding. The functions of these chaperones are summarized in Table I. NinaA (*n*either *i*nactivation *n*or *a*fterpotential A) is known to be a chaperone for the visual pigment receptor rhodopsin 1 in *Drosophila melanogaster.* NinaA mediates the cell surface expression of Rhodopsin 1. In vertebrates, the rhodopsin 1 chaperone is RAN-binding protein 2. Both these chaperones are cyclophilin type-II proteins displaying peptidyl–prolyl cis–trans isomerase activity.[55,56]

NinaA was discovered when it was demonstrated that mutation in the gene encoding this protein leads to reduced expression of rhodopsin in R1–6 photoreceptor cells of the *Drosophila* eye.[57] Further investigation indicated that in the absence of NinaA, rhodopsin 1 accumulates in the ER, where it is degraded instead of being transported to rhabdomeres.[58] Unlike other cyclophilins which are usually cytosolic proteins, NinaA has a membrane permeating signal sequence and a single transmembrane-spanning domain and therefore functions as an integral membrane protein.[55] It has been demonstrated that NinaA forms a stable complex with rhodopsin and, functioning more stoichiometrically than catalytically, ensures proper folding and trafficking of the newly synthesized rhodopsin receptors to the plasma membrane.

Rhodopsin also interacts with human neuron-specific DnaJ-like protein (HSJ1). HSJ1 is a member of the DnaJ and Hsp40 chaperone family.[59] In neurons, HSJ1 proteins, which function at the cytosolic face of the ER, facilitate the transfer of newly synthesized proteins onto Hsc70 chaperones. This is followed by ubiquitination of the proteins and subsequent sorting to the proteasome.[60] Thus, HSJ1 isoforms probably participate in the ER-associated degradation pathway and protect neurons against cytotoxic protein aggregation. Other specialized chaperones for GPCRs are escort proteins.[61] These escort proteins bind newly synthesized proteins in the ER and escort them to the Golgi complex and to the plasma membrane.

One novel group of specialized escort proteins that work with GPCRs is represented by the receptor-activity-modifying protein (RAMP) family of protein.[62] RAMPs are type-I single transmembrane domain proteins with a large N-terminal extracellular domain and a short C-terminal domain. RAMPs were initially described as interacting partners for the calcitonin receptor-like receptor (CRLR). In the case of CRLR, RAMPs act as chaperones to assist in the transport of the CRLR receptor to the cell surface.[63]

TABLE I
THE FUNCTIONS OF KNOWN MOLECULAR CHAPERONES FOR GPCRS

GPCR	Molecular chaperone	Function(s)	Reference(s)
Luteinizing hormone receptor	BiP (GRP78)	Folding	8,9
	GPR94	Folding	9
Rhodopsin I	NinaA (*D. melanogaster*)	Cell surface targeting	55
	RAN-binding protein 2 (vertebrates)	Cell surface targeting	56
Calcitonin receptor-like receptor	RAMP 1, 2, 3	Cell surface targeting terminal glycosylation and maturation	63
Olfactory receptors	ODR4	Folding	65
		Cell surface targeting	
	RTPI,2	Cell surface targeting	66
	REEP I	Cell surface targeting	66
μ–δ-Opioid receptor	RTP4	Cell surface targeting heteromer protection from ubiquitination/ degradation	67
Vomeronasal receptors.	Ml0	Cell surface targeting	68
Dopamine D_1 receptors	DRiP78	Cell surface targeting	70
Angiotensin AT_2 receptors	ATBP5O	Cell surface targeting	73
Thromboxane A_2 receptor	RACK 1	Cell surface targeting	74
CXCR4 chemokine receptor	RACK1	cell surface targeting	74
κ-Opioid receptors	GEC1	Cell surface targeting	75
		Glycosylation	
μ-Opioid receptor	Ribophorin I	Cell surface targeting	76
Metabotropic glutamate receptor 5	Norbin	Cell surface targeting	77
Serotonin 5-HT_{1B} receptor	p11	Cell surface targeting	78
Melanocortin-2 receptor	MRAP	Cell surface targeting	81,82
		Glycosylation	
AdenosineA_{2a} receptor	Usp4	Cell surface targeting protection from ubiquitination/ degradation	90

RAMP1 was initially identified by expression cloning in Xenopus oocytes for cDNAs encoding the human calcitonin gene related peptide (CGRP) receptor. Coexpression of RAMP1 with the cAMP-activated cystic fibrosis transmembrane conductance regulator (CFTR) chloride channel leads to cAMP production following stimulation of CGRP and activation of the CFTR chloride channel. RAMP1 also plays an important role in the trafficking of CRLR receptor to the cell surface. When expressed alone, neither RAMP1 nor CRLR was found to be expressed at the cell surface. However, when the CRLR receptor was coexpressed with the RAMP1 chaperone protein, both proteins showed cell surface localization.[63] Such studies demonstrate that RAMPs remain associated with their cognate receptors at the cell surface.

In addition to its role in assisting the transport of the CRLR receptor to the cell surface, RAMP1 also helps define the glycosylation state and recognition properties of the receptor. Using SDS-PAGE analysis it was shown that when expressed alone CRLR is sensitive to endoglycosidase H treatment.[63] This result indicates that the CRLR receptors have not been terminally glycosylated. Conversely, CRLR when coexpressed with RAMP1, migrates as a larger molecular weight protein and is resistant to endoglycosidase H treatment. Additionally, when coexpressed with RAMP1, CRLR shows sensitivity to endoglycosidase F (which cleaves all N-linked carbohydrates). These results indicate that while RAMP1 assists in the trafficking of CRLR to the cell surface, the chaperone is also required for terminal glycosylation and maturation of the CRLR receptor through the Golgi complex on its way to the plasma membrane.

Two additional RAMP proteins, RAMP2 and RAMP3 have been identified.[63] These three RAMPs have different mRNA expression profiles. Interestingly, when different RAMP proteins act as the chaperone for the CRLR receptor they result in differential ligand binding properties for the receptor. While chaperoning by RAMP1 was shown to increase the CRLR-mediated cAMP production in response to CGRP stimulation, when RAMP2 is used as a chaperone for CRLR the receptor behaves pharmacologically like an adrenomedullin receptor.[63]

Even though it has been demonstrated that coexpression of CRLR with different RAMP proteins results in differences in the glycosylation state of CRLR, it has been found that the chaperone protein-dependent ligand binding selectivity observed for the CRLR receptor is determined by protein–protein interactions and not by these differences in the glycosylation state of the receptor.[64] The realization that CRLR can exhibit two distinct pharmacological profiles depending on which chaperone protein it interacts with expands the already extensive biological diversity of signaling responses that GPCR receptors can generate in cells. These GPCRs can exhibit differential responses to various stimuli based not only on the specific expression pattern of the receptors, but also the specific expression pattern of the cognate accessory proteins.

Another example of GPCRs that require specific chaperones for proper cell surface expression is the group of olfactory or odorant receptors. The necessity of specific chaperones for cell surface expression of olfactory receptors is highlighted by the fact that functional expression of these receptors is essentially impossible in heterologous cell systems where these olfactory neuron-specific chaperones are absent. In fact, when olfactory receptors are expressed in nonolfactory neuronal cells the receptors are retained in the ER and rarely pass into the Golgi complex. When a genetic screen was performed in *Caenorhabditis elegans* for mutants with olfactory defects the odorant response protein 4 (ODR4) was identified as a chaperone for olfactory receptors.[65] ODR4 is expressed specifically in chemosensory neurons where it promotes the folding and cell surface localization of olfactory receptors. In addition, receptor-transporting protein (RTP) 1 and 2 and receptor-expression-enhancing protein 1 (REEP1) were found to also act as chaperones for olfactory receptors and enable functional cell surface targeting of these receptors in fibroblasts.[66] Similarly, RTP4 acts as a chaperone for opioid receptors (this is discussed in more detail in Section VIII; Ref. 67).

Other sensory GPCRs are involved in the detection of pheromones. Mammals possess a small sensory organ located near the base of the nasal septum that is involved in the detection of pheromones called the vomeronasal organ. The vomeronasal organ contains specific GPCRs that function as pheromone receptors. As is the case with other GPCRs, these vomeronasal receptors require the association with specific chaperone proteins for proper trafficking to the cell surface. It has been demonstrated that proper cell surface localization of vomeronasal receptors is aided by the action of M10 chaperone.[67] M10s belong to the superfamily of major histocompatability complex (MHC) class I molecules and are exclusively expressed in the vomeronasal organ. As is the case with RAMP proteins, results have indicated that in addition to their role as chaperones, M10s might modulate the ligand specificity of vomeronasal receptors and might participate in neuronal plasticity.[68]

Another area of active research involves understanding the roles of molecular chaperones in modulating the cell surface expression of neurotransmitter receptors. The dopamine receptor-interacting protein of 78 kDa (DRiP78) is a known chaperone for the dopamine receptor. DRiP78 is a putative two trans-membrane-spanning domain protein and is ER membrane-associated. When DRiP78 is over expressed or knocked down this results in ER retention of dopamine D_1 receptors as well as reduced ligand binding of the receptors and modulation of the receptor glycosylation profile.[69] It was hypothesized that DRiP78's role is not limited to being a chaperone for dopamine receptors and that DRiP78 is capable of functioning as a chaperone for several GPCRs. This hypothesis was developed based on the fact that DRiP78 binds to a Phe-Xaa-Xaa-Xaa-Phe-Xaa-Xaa-Xaa-Phe motif on dopamine receptors and this motif is found in the C-terminus of various GPCRs. This hypothesis has been supported by a study demonstrating a role for DRiP78 in the maturation of the angiotensin II AT_1 receptor.[70] Expanding its role even further, it has also been demonstrated that DRiP78 specifically interacts with Gγ subunits of heterotrimeric G proteins. This interaction protects the Gγ subunits from degradation until a stable partner (the cognate Gβ subunit) is provided. Such results indicate an additional chaperone role for DRiP78 in the assembly of Gβγ subunits.[71]

As demonstrated by DRiP78, intracellular membrane proteins are capable of functioning as molecular chaperones and regulating the targeting of GPCR receptors to the cell surface without the chaperone itself being localized at the cell surface. Another example of such chaperone activity is the angiotensin AT_2 receptor-binding protein of 50 kDa (ATBP50) which is a membrane-associated Golgi protein. ATBP50 binds to the cytoplasmic C-terminal tail of the angiotensin AT_2 receptor and controls the cell surface expression of the receptor. It was demonstrated that inhibition of ATBP50 expression results in retention of the angiotensin AT_2 receptor within intracellular compartments.[72]

Similarly, the receptor for activated C-kinase 1 (RACK1) is an ER chaperone that regulates the cell surface localization of the thromboxane A_2 receptor. RACK1 constitutively binds to the C-terminal tail and also the first intracellular loop of the thromboxane A_2 receptor.[73] In cells where the levels of RACK1 were reduced by siRNA-mediated knockdown of the chaperone protein, the thromboxane A_2 receptor was retained in the ER. Interestingly, reducing the expression level of RACK1 is also able to affect the cell surface distribution of the chemokine receptor CXCR4 but has no effect of the cell surface localization of the β_2-adrenoceptor or prostanoid DP receptors, indicating that RACK1 shows specificity for different GPCRs.

Another chaperone protein that demonstrates comparable specificity for GPCRs is GEC1. GEC1 is a member of the microtubule-associated family of proteins.[74] It was demonstrated that in the ER and the Golgi, GEC1 interacts with the C-terminus of κ-opioid receptors, but not with that of μ- or δ-opioid receptor subtypes. Additionally, GEC1 expression enhances the level of mature fully glycosylated forms of κ-opioid receptors and facilitates trafficking of the κ-opioid receptor to the cell surface. GEC1 levels seem to be tightly regulated, as indicated by a toxic effect of over expressing GEC1.[74]

Using a targeted proteomic approach, Ribophorin I was identified as a chaperone for the μ-opioid receptor.[75] Overexpression of Ribophorin I was able to rescue the cell surface expression of mutant μ-opioid receptor. The chaperone activity of Ribophorin I is dependent on the glycosylation state of the μ-opioid receptor, as the chaperone is ineffective when overexpressed with glycosylation-deficient mutants. Similarly, the neuron-specific protein Norbin was shown to physically interact with the metabotropic glutamate receptor 5 and increase the cell surface localization of the receptor.[76]

Additional insights into the function of molecular chaperones come from studies performed characterizing knockout mice. Mice were created that lack the p11 protein.[77] P11 (also known as calpactin-I- or annexin-II-light chain) is a member of the S100 EF-hand calcium-dependent signaling-modulator family.[78] When p11 was depleted in mice, the animals exhibited a depression-like phenotype. Clinically, such depression-like phenotypes are commonly associated with abnormal serotonin signaling. In fact, the cell surface density and function of the serotonin 5-HT_{1B} receptor is decreased in p11 knockout mice. Additional studies demonstrated that the distribution of p11 mRNA in the brain resembles that of 5-HT_{1B} receptor mRNA. Also, p11 specifically interacts with 5-HT_{1B} and does not interact with other serotonin or dopamine receptors and p11 colocalizes with the receptor at the cell surface of transfected cells.[78] Interestingly, p11 levels are increased in the brain of mice treated with antidepressants and p11 levels are reduced in depressed patients.

VII. The Melanocortin-2 Receptor and MRAP

The melanocortin family of receptors is a group of five structurally related GPCRs that play diverse physiological roles in mammals.[79] All the melanocortin receptors cause an increase in cAMP levels when stimulated by agonists. However, the receptors differ in their affinity for physiological agonist. Melanocortin-2 is unique because it is the only melanocortin receptor that is selectively regulated by adrenocorticotropin hormone. Melanocortin-2 is also unique in the way that it requires an accessory protein for proper trafficking of the receptor to the cell surface, while the other related melanocortin receptors do not require this chaperone.[80] This accessory protein, the melanocortin-2 receptor accessory protein (MRAP) is required not only for the melanocortin-2 receptor to traffic from the ER to the plasma membrane, but is also required for the receptor to undergo proper glycosylation, binding of adrenocorticotropin hormone and stimulation of adenylyl cyclase.[80–82]

As is the case with odorant receptors, the functional expression of the melanocortin-2 receptor in heterologous systems proved difficult and could be achieved only in cells of adrenocortical origin, highlighting the fact that the melanocortin-2 receptor must be expressed in the presence of a cell-type-specific chaperone protein in order for proper biosynthetic trafficking and localization. Further investigation identified MRAP as the required chaperone protein. MRAP is a small single membrane-spanning protein with a conserved amino terminus. When MRAP was coprecipitated with melanocortin-2 it was discovered that MRAP forms antiparallel homomers.[82] Mutational analysis performed on MRAP revealed that the amino-terminal and transmembrane region, and not the carboxyl-terminal domain, is required for MRAP's function as a molecular chaperone.

In humans, inactivating mutations in the melanocortin-2 receptor lead to glucocorticoid deficiency resulting from adrenocorticotropin hormone resistance. The importance of MRAP's function is highlighted by the fact that mutations in MRAP can also lead to these disease states. Familial glucocorticoid deficiency is a genetic disease in which the adrenal cortex is resistant to adrenocorticotropin hormone. Inactivating mutations have been identified in the gene that encodes MRAP in patients that suffer from familial glucocorticoid deficiency. This example illustrates, particularly well, the physiopathological relevance of the interaction between a GPCR and a specific chaperone protein.

MRAP's specificity of action was demonstrated in the study that compared the effects of MRAP activity on the dimerization and trafficking of melanocortin-2 and melanocortin-5 receptors.[83] Melanocortin-5 shares 67% homology with the melanocortin-2 receptor at the amino acid level and functions in regulating exocrine gland secretion and behavioral responses following pheromone secretion.[84,85] While the melanocortin-2 receptor is activated by adrenocorticotropin hormone, melanocortin-5 responds to melanocyte-stimulating hormone. Unlike melanocortin-2, melanocortin-5 shows cell surface localization when expressed exogenously.[81]

As expected, coexpression of melanocortin-2 with MRAP revealed that MRAP facilitated the plasma membrane localization of melanocortin-2. In contrast, while melanocortin-5 receptors traffic to the plasma membrane in the absence of MRAP, coexpression of MRAP with melanocortin-5 leads to intracellular retention of the receptors.[83] MRAP was shown to form stable complexes with both melanocortin receptors as determined by co-immunoprecipitation experiments. In addition, coexpression of MRAP had no effect of melanocortin-2's ability to form homomers, but inhibited the formation of melanocortin-5 homomers. One interpretation of these results is that MRAP disrupts a dimerizaton step that is critical for the maturation and trafficking of the melanocortin-5 receptor to the cell surface. In the current model of MRAP's function, in cells expressing both melanocortin-2 and melanocortin-5 receptors, MRAP will have a dual role in which increased activity of MRAP will favor responsiveness to adrenocorticotropin hormone and decreased responsiveness to melanocyte-stimulating hormone, while decreased MRAP activity will have the opposite effect. This is one way in which a selective chaperone protein can enhance the signaling capabilities of a system comprising of structurally related receptors.

VIII. Opioid Receptors and RTP4

As discussed earlier, the three subtypes of opioid receptors, δ, κ, and μ have the ability to form both homomers and heteromers.[86] In addition, there are varied pharmacological profiles for different homomer and heteromer receptor pairs. The clinical importance of heteromerization of opioid receptors is highlighted by the fact that despite μ-opioid receptors mediating, most of the pain-relieving effects of morphine, antagonism of δ-opioid receptors leads to a reduction in the tolerance that develops after chronic morphine administration.[87] This observation suggests that there are potential links between opioid receptor dimerization and drug tolerance. Additionally, the analgesic potency of morphine has been linked to the presence of μ–δ-opioid receptor heteromers.[88] In the current model for the development of drug tolerance, factors that modulate the ratio of μ-and δ-opioid receptors that exist as homomers and receptors and which exist as μ–δ receptor heteromers serve as critical factors in determining the net effect of receptor stimulation.

Using the μ–δ-opioid receptor heteromer at a model system, it was reported that coexpression of the differentially tagged μ- and δ-opioid receptors leads to increased retention of both receptors in intracellular compartments.[89] Further investigation showed that the μ–δ receptor heteromers were accumulating in the Golgi complex and were subsequently targeted to the degradation pathway. In order to identify chaperones that could function at the Golgi and assist in folding and trafficking of these receptor heteromers, these opioid receptors were coexpressed with members of the RTP family of chaperones. As noted earlier, RTP proteins are known to act at the Golgi complex where they assist in the cell surface targeting of odorant and taste receptors.[66] Coexpression of RTP4 leads to an increase in the cell surface localization of both μ- and δ-opioid receptors.

RTP4 physically interacts with both opioid receptors and decreases the ubiquitination of the receptor heteromers. Subsequently, RTP4 was shown to be involved in the proper folding of the μ–δ-opioid receptor heteromers as it facilitates the trafficking of the heteromers from the Golgi complex to the cell surface. Furthermore, RTP4 was shown to be associated with the μ–δ-opioid receptor heteromer at the cell surface and could play a role in modulating heteromer-specific pharmacology. It has been proposed that the retention of μ–δ receptor heteromers in the Golgi may be the result of a mechanism in which dimerization of the receptors leads to the masking of motifs needed for recognition of the receptors by accessory proteins. Interaction of RTP4 at this motif could prevent the retention of the heteromers by serving as an adaptor protein for other factors involved in trafficking of the opioid receptors to the plasma membrane.

Similar to the role of RTP4 in decreasing the ubiquitination of opioid receptor heteromers, the ubiquitin-specific protease Usp4 has been shown to regulate the cell surface localization of the adenosine A_{2a} receptor.[90] Usp4 was shown to bind to the C-terminus of the adenosine receptor and lead to decrease in the ubiquitination of the receptor, leading to enhanced cell surface expression of the A_{2a} receptor. Additionally, a related ubiquitin-specific protease, Usp14, has been implicated in the deubiquitination of the chemokine receptor CXCR4.[91]

IX. Pharmacological Chaperones

As previously discussed, some receptor proteins do not achieve proper cell surface trafficking when expressed in heterologous systems. This is usually due to the absence of proper endogenous chaperones which lead to the intracellular retention of these receptors by the cell's quality-control system. One method to overcome this intracellular retention of receptor proteins is through the use of small molecules that act as chaperones and aid in the proper sorting and trafficking of receptors. Such molecules are referred to as pharmacological chaperones or "pharmacoperones." Pharmacological chaperones are small molecules that have the ability to enter cells, bind specifically to misfolded proteins and correct the folding of these proteins.[92,93] The result of the treatment with pharmacological chaperones is that these receptor proteins are able to escape retention by the cellular quality-control system and achieve proper cell surface localization. A proposed model for the action of pharmacological chaperones is illustrated in Fig. 1. In many cases, these pharmacological chaperones were initially identified as peptidomimetic antagonists selected from high-throughput screens.

In addition to having a practical utility in laboratory studies in which the functional expression of receptors in heterologous systems is required, pharmacological chaperones also have potential therapeutic value. As previously discussed, there are diseases that result from the inability of specific GPCRs to fold properly and traffic to the cell surface where they can function effectively (such as melanocortin-2 receptor mutations leading to familial glucocorticoid deficiency). There has been significant interest in identifying small molecules that can bind to disease-causing GPCR mutants that traffic poorly and by acting as pharmacological chaperones, enhance the surface expression and functionality of these receptors. Unlike other protein stabilizing compounds that may only work *in vitro* (such as polyols and other chemical chaperones) the goal is to develop pharmacological chaperones that can be used in a clinical setting to treat well-defined diseases involving misfolded/retained proteins.

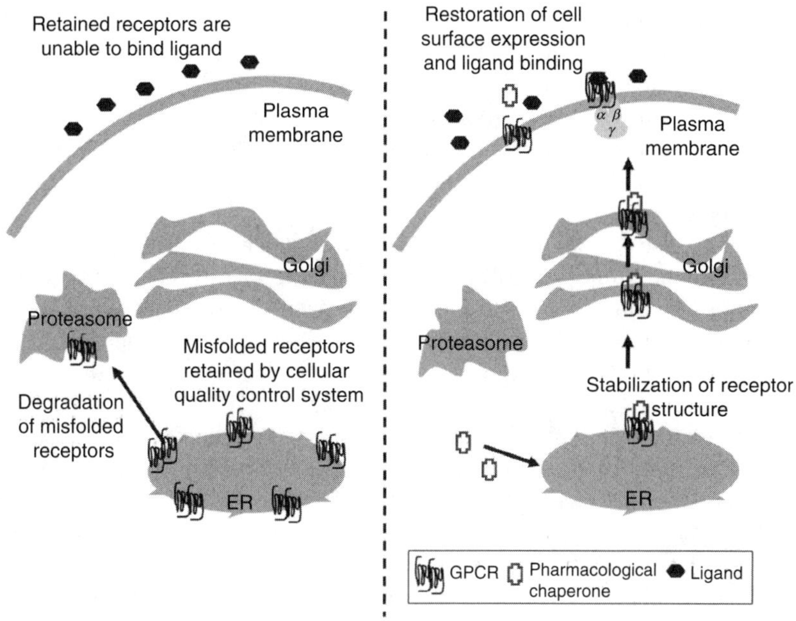

FIG. 1. Proposed model for the action of pharmacological chaperones. Left panel: Mutated receptors and wild-type receptors that are misfolded are retained in the ER by the cellular quality-control system and targeted for proteasomal degradation. Subsequently, these receptors are not able to bind ligand and participate in signaling. Right panel: Treatment of cells with cell permeable pharmacological chaperones leads to stabilization of the receptor structure and restoration of the cell surface expression of the receptors. The receptors are therefore able to bind ligands and function in signaling processes.

One example of the potential use of pharmacological chaperones for the treatment of a disease resulting from the ineffective sorting of a GPCR is the development of pharmacological chaperones for nephrogenic diabetes insipidus. Nephrogenic diabetes insipidus is a rare X-linked disease characterized by the loss of the antidiuretic response to the hormone arginine-vasopressin, resulting in an inability of the body to concentrate urine. Nephrogenic diabetes insipidus has been linked to a variety of mutations in vasopressin V_2 receptors. In fact, the majority of the V_2 receptor mutations are mutations that result in the V_2 receptors being retained in the ER and degraded.[94] Researchers have worked to identify a group of small molecules that are able to act as pharmacological chaperones for the V_2 receptor.

Treatment of cells with certain membrane-permeable V_2 receptor antagonists that act as pharmacological chaperones leads to the restoration of the cell surface expression of V_2 receptors mutants that were previously retained in the

ER.[95] These results are attributed to binding of the antagonists to the already misfolded V_2 receptors in the ER. Binding of these pharmacological chaperones results in stabilization of receptor structure and trafficking of V_2 receptors to the plasma membrane. Proof-of-concept evidence that such vasopressin receptor-targeted pharmacological chaperones can have beneficial therapeutic effects has been provided from clinical studies involving patients suffering from nephrogenic diabetes insipidus.[96]

Another disease that is caused from mutations in a specific GPCR resulting in intracellular retention and ineffective traveling of the receptor is retinitis pigmentosa. Retinitis pigmentosa is characterized by progressive photoreceptor degeneration and eventual retinal dysfunction. Retinitis pigmentosa has been linked to a number of mutations in proteins that serve as various components of the visual signaling pathway, including rhodopsin.[97] Mutations in the gene encoding rhodopsin are observed in one particular form of retinitis pigmentosa, autosomal dominant retinitis pigmentosa, in which there is a mutated rhodopsin receptor that is retained intracellularly and is unable to bind the ligand, 11-*cis*-retinal binding. The most common mutation of this type is P23H.[98] P23H rhodopsin mutants can be rescued by treatment of cells with an 11-*cis*-retinal analog, 11-*cis*-ring-retinal, resulting in restoration of receptor surface expression.[99]

Hypergonadotropic hypogonadism is a disease characterized by a decrease in gonadal function that manifests in delayed sexual development in patients who suffer from the disease. The clinical phenotypes associated with hypergonadotropic hypogonadism have been linked to mutations within the gonadotropin-releasing hormone receptor gene. Gonadotropin-releasing hormone is secreted by the hypothalamus and induces synthesis and release of follicle stimulating hormone and luteinizing hormone from the pituitary gland. Many of the mutations in the gonadotropin-releasing hormone receptor that have been linked to hypergonadotropic hypogonadism are mutations that result in intracellular retention of the receptor or degradation of the receptor.[100]

Studies have demonstrated that these mutant gonadotropin-releasing hormone receptors can be made to traffic to the cell surface and function effectively through the use of pharmacological chaperones.[101–103] When cells expressing gonadotropin-releasing hormone receptor mutants were treated with the IN3, a membrane-permeable gonadotropin-releasing hormone receptor antagonist the receptors were able to be rescued from intracellular retention, traffic to the plasma membrane and bind ligands.[101,102] Such examples highlight the potential utility of small molecule pharmacological chaperones in correcting receptor misfolding. Similarly, treating cells with the opioid antagonist naloxone, the agonist etorphine, and other membrane-permeable hydrophobic ligands enhanced the cell surface expression of μ-opioid receptor mutants.[104] These receptor mutants are usually retained in the ER and binding of the ligands resulted in concentration-dependent trafficking of the mutant receptors to the cell surface, indicating that these ligands can act as pharmacological chaperones. These *in vitro* findings are supported by observations that long-term use of these ligands leads to improved stability, folding efficiency of newly synthesized receptors *in vivo*.[105]

This approach of using pharmacological chaperones has the potential to be applied to an array of human diseases that result from ineffective receptor signaling due to protein misfolding. Evidence for the potential use of pharmacological chaperones has been presented for cystic fibrosis, hypercholesterolemia, cataracts and neurodegenerative diseases such as Huntington's, Alzheimer's, and Parkinson's diseases.[106–113] Also, while these pharmacological chaperones are being tested for the treatment of mutant receptors that are misfolded, there is increasing evidence that some wild-type GPCRs are also commonly misrouted, presumably as a result of misfolding. This suggests that pharmacological chaperone intervention can potentially be expanded to more general posttranslational control mechanisms and provide another level of potential therapeutic intervention.[114]

X. Conclusions

With recent advances in techniques to study the pharmacology, signaling, and trafficking of transmembrane receptors we are left with a better understanding of how this tightly regulated network of receptors work. We now have a great appreciation of the role of processes such as receptor oligomerization, receptor desensitization and interaction with accessory proteins and cofactors. Advances in the field of molecular chaperone biology have provided insights into what can happen when receptors are misfolded, retained in intracellular compartments, and degraded. Modulating of protein folding and/or molecular chaperone activity *in vitro* has provided initial clues to how mutant or otherwise misfolded receptors can be rescued and made to function properly.

As demonstrated by the studies looking at nephrogenic diabetes insipidus, retinitis pigmentosa, and hypergonadotropic hypogonadism, pharmacological chaperones represent the most promising therapeutic approach for the treatment of diseases resulting from protein misfolding. As more receptor-specific molecular chaperones (such as RTP4 for opioid receptor heteromers and MRAP for melanocortin-2 receptors) are identified additional studies will hopefully provide a good framework in which we can develop treatments that specifically modulate the levels/activity of these molecular chaperones to combat the receptor folding or trafficking problems that lead to various disease states.

The discoveries discussed in this chapter raise many new questions and interesting challenges that will certainly be addressed in the near future. This will inevitably expand the potential for the use of pharmacological tools to target these interactions between GPCRs and molecular chaperones and subsequently manipulate the expression, localization, and function of these receptors in order to gain therapeutic benefits. As we develop new therapies, the specificity of these therapies can be increased by using conventional pharmacological approaches to modify the activity of receptors in conjunction with novel pharmacological approaches to control which particular receptors are expressed at the cell surface of specific cells.

Acknowledgments

The authors thank Drs. Raphael Rozenfeld and Ivone Gomes for critically reading the manuscript. These studies were supported by National Institutes of Health Grants DA008863 and DA019521 (to L.A.D.).

References

1. Ellis J. Proteins as molecular chaperones. *Nature* 1987;**328**:378–9.
2. Higy M, Junne T, Spiess M. Topogenesis of membrane proteins at the endoplasmic reticulum. *Biochemistry* 2004;**43**:12716–22.
3. Krebs MP, Noorwez SM, Malhotra R, Kaushal S. Quality control of integral membrane proteins. *Trends Biochem Sci* 2004;**29**:648–55.
4. Ellgaard L, Helenius A. Quality control in the endoplasmic reticulum. *Nat Rev Mol Cell Biol* 2003;**4**:181–91.
5. Hebert DN, Molinari M. In and out of the ER: protein folding, quality control, degradation, and related human diseases. *Physiol Rev* 2007;**87**:1377–408.
6. Tsai B, Ye Y, Rapoport TA. Retro-translocation of proteins from the endoplasmic reticulum into the cytosol. *Nat Rev Mol Cell Biol* 2002;**3**:246–55.
7. Parsell DA, Lindquist S. The function of heat-shock proteins in stress tolerance: degradation and reactivation of damaged proteins. *Annu Rev Genet* 1993;**27**:437–96.
8. Siffroi-Fernandez S, Giraud A, Lanet J, Franc JL. Association of the thyrotropin receptor with calnexin, calreticulin and BiP. Effects on the maturation of the receptor. *Eur J Biochem* 2002;**269**:4930–7.
9. Mizrachi D, Segaloff DL. Intracellularly located misfolded glycoprotein hormone receptors associate with different chaperone proteins than their cognate wild-type receptors. *Mol Endocrinol* 2004;**18**:1768–77.
10. Rozell TG, Davis DP, Chai Y, Segaloff DL. Association of gonadotropin receptor precursors with the protein folding chaperone calnexin. *Endocrinology* 1998;**139**:1588–93.
11. Barlowe C. Signals for COPII-dependent export from the ER: what's the ticket out? *Trends Cell Biol* 2003;**13**:295–300.
12. Aridor M, Weissman J, Bannykh S, Nuoffer C, Balch WE. Cargo selection by the COPII budding machinery during export from the ER. *J Cell Biol* 1998;**141**:61–70.

13. Dong C, Filipeanu CM, Duvernay MT, Wu G. Regulation of G protein-coupled receptor export trafficking. *Biochim Biophys Acta* 2007;**1768**:853–70.
14. Appenzeller-Herzog C, Hauri HP. The ER-Golgi intermediate compartment (ERGIC): in search of its identity and function. *J Cell Sci* 2006;**119**:2173–83.
15. Moyer BD, Allan BB, Balch WE. Rab1 interaction with a GM130 effector complex regulates COPII vesicle *cis*-Golgi tethering. *Traffic* 2001;**2**:268–76.
16. Tisdale EJ. Rab2 interacts directly with atypical protein kinase C (aPKC) iota/lambda and inhibits aPKCiota/lambda-dependent glyceraldehyde-3-phosphate dehydrogenase phosphorylation. *J Biol Chem* 2003;**278**:52524–30.
17. Wu G, Zhao G, He Y. Distinct pathways for the trafficking of angiotensin II and adrenergic receptors from the endoplasmic reticulum to the cell surface: Rab1-independent transport of a G protein-coupled receptor. *J Biol Chem* 2003;**278**:47062–9.
18. Dong C, Wu G. Regulation of anterograde transport of adrenergic and angiotensin II receptors by Rab2 and Rab6 GTPases. *Cell Signal* 2007;**19**:2388–99.
19. Trucco A, Polishchuk RS, Martella O, Di Pentima A, Fusella A, Di Giandomenico D, et al. Secretory traffic triggers the formation of tubular continuities across Golgi sub-compartments. *Nat Cell Biol* 2004;**6**:1071–81.
20. Arvan P, Zhao X, Ramos-Castaneda J, Chang A. Secretory pathway quality control operating in Golgi, plasmalemmal, and endosomal systems. *Traffic* 2002;**3**:771–80.
21. Shetty KM, Kurada P, O'Tousa JE. Rab6 regulation of rhodopsin transport in *Drosophila*. *J Biol Chem* 1998;**273**:20425–30.
22. Petaja-Repo UE, Hogue M, Laperriere A, Walker P, Bouvier M. Export from the endoplasmic reticulum represents the limiting step in the maturation and cell surface expression of the human delta opioid receptor. *J Biol Chem* 2000;**275**:13727–36.
23. Dong C, Wu G. Regulation of anterograde transport of alpha2-adrenergic receptors by the N termini at multiple intracellular compartments. *J Biol Chem* 2006;**281**:38543–54.
24. Hein L, Ishii K, Coughlin SR, Kobilka BK. Intracellular targeting and trafficking of thrombin receptors. A novel mechanism for resensitization of a G protein-coupled receptor. *J Biol Chem* 1994;**269**:27719–26.
25. Brismar H, Asghar M, Carey RM, Greengard P, Aperia A. Dopamine-induced recruitment of dopamine D1 receptors to the plasma membrane. *Proc Natl Acad Sci U S A* 1998;**95**:5573–8.
26. Holtback U, Brismar H, DiBona GF, Fu M, Greengard P, Aperia A. Receptor recruitment: a mechanism for interactions between G protein-coupled receptors. *Proc Natl Acad Sci USA* 1999;**96**:7271–5.
27. Kim KA, von Zastrow M. Neurotrophin-regulated sorting of opioid receptors in the biosynthetic pathway of neurosecretory cells. *J Neurosci* 2003;**23**:2075–85.
28. Cahill CM, McClellan KA, Morinville A, Hoffert C, Hubatsch D, O'Donnell D, et al. Immunohistochemical distribution of delta opioid receptors in the rat central nervous system: evidence for somatodendritic labeling and antigen-specific cellular compartmentalization. *J Comp Neurol* 2001;**440**:65–84.
29. Cahill CM, Morinville A, Lee MC, Vincent JP, Collier B, Beaudet A. Prolonged morphine treatment targets delta opioid receptors to neuronal plasma membranes and enhances delta-mediated antinociception. *J Neurosci* 2001;**21**:7598–607.
30. Lucido AL, Morinville A, Gendron L, Stroh T, Beaudet A. Prolonged morphine treatment selectively increases membrane recruitment of delta-opioid receptors in mouse basal ganglia. *J Mol Neurosci* 2005;**25**:207–14.
31. Revankar CM, Cimino DF, Sklar LA, Arterburn JB, Prossnitz ER. A transmembrane intracellular estrogen receptor mediates rapid cell signaling. *Science* 2005;**307**:1625–30.
32. Rozenfeld R, Devi LA. Regulation of CB1 cannabinoid receptor trafficking by the adaptor protein AP-3. *FASEB J* 2008;**22**:2311–22.

33. Maggio R, Vogel Z, Wess J. Coexpression studies with mutant muscarinic/adrenergic receptors provide evidence for intermolecular "cross-talk" between G-protein-linked receptors. *Proc Natl Acad Sci USA* 1993;**90**:3103–7.
34. Bouvier M. Oligomerization of G-protein-coupled transmitter receptors. *Nat Rev Neurosci* 2001;**2**:274–86.
35. Couve A, Filippov AK, Connolly CN, Bettler B, Brown DA, Moss SJ. Intracellular retention of recombinant GABAB receptors. *J Biol Chem* 1998;**273**:26361–7.
36. Jones KA, Borowsky B, Tamm JA, Craig DA, Durkin MM, Dai M, et al. GABA(B) receptors function as a heteromeric assembly of the subunits GABA(B)R1 and GABA(B)R2. *Nature* 1998;**396**:674–9.
37. White JH, Wise A, Main MJ, Green A, Fraser NJ, Disney GH, et al. Heterodimerization is required for the formation of a functional GABA(B) receptor. *Nature* 1998;**396**:679–82.
38. Kuner R, Kohr G, Grunewald S, Eisenhardt G, Bach A, Kornau HC. Role of heteromer formation in GABAB receptor function. *Science* 1999;**283**:74–7.
39. Ng GY, Clark J, Coulombe N, Ethier N, Hebert TE, Sullivan R, et al. Identification of a GABAB receptor subunit, gb2, required for functional GABAB receptor activity. *J Biol Chem* 1999;**274**:7607–10.
40. Kaupmann K, Malitschek B, Schuler V, Heid J, Froestl W, Beck P, et al. GABA(B)-receptor subtypes assemble into functional heteromeric complexes. *Nature* 1998;**396**:683–7.
41. Margeta-Mitrovic M, Jan YN, Jan LY. A trafficking checkpoint controls GABA(B) receptor heterodimerization. *Neuron* 2000;**27**:97–106.
42. Margeta-Mitrovic M, Jan YN, Jan LY. Function of GB1 and GB2 subunits in G protein coupling of GABA(B) receptors. *Proc Natl Acad Sci USA* 2001;**98**:14649–54.
43. Jordan BA, Devi LA. G-protein-coupled receptor heterodimerization modulates receptor function. *Nature* 1999;**399**:697–700.
44. Cvejic S, Devi LA. Dimerization of the delta opioid receptor: implication for a role in receptor internalization. *J Biol Chem* 1997;**272**:26959–64.
45. Gomes I, Jordan BA, Gupta A, Trapaidze N, Nagy V, Devi LA. Heterodimerization of mu and delta opioid receptors: A role in opiate synergy. *J Neurosci* 2000;**20**:RC110.
46. Jordan BA, Trapaidze N, Gomes I, Nivarthi R, Devi LA. Oligomerization of opioid receptors with beta 2-adrenergic receptors: a role in trafficking and mitogen-activated protein kinase activation. *Proc Natl Acad Sci USA* 2001;**98**:343–8.
47. Limbird LE, Lefkowitz RJ. Negative cooperativity among beta-adrenergic receptors in frog erythrocyte membranes. *J Biol Chem* 1976;**251**:5007–14.
48. Hebert TE, Moffett S, Morello JP, Loisel TP, Bichet DG, Barret C, et al. A peptide derived from a beta2-adrenergic receptor transmembrane domain inhibits both receptor dimerization and activation. *J Biol Chem* 1996;**271**:16384–92.
49. Xu Y, Piston DW, Johnson CH. A bioluminescence resonance energy transfer (BRET) system: application to interacting circadian clock proteins. *Proc Natl Acad Sci USA* 1999;**96**:151–6.
50. Angers S, Salahpour A, Joly E, Hilairet S, Chelsky D, Dennis M, et al. Detection of beta 2-adrenergic receptor dimerization in living cells using bioluminescence resonance energy transfer (BRET). *Proc Natl Acad Sci USA* 2000;**97**:3684–9.
51. Gething MJ, Sambrook J. Protein folding in the cell. *Nature* 1992;**355**:33–45.
52. Bulenger S, Marullo S, Bouvier M. Emerging role of homo- and heterodimerization in G-protein-coupled receptor biosynthesis and maturation. *Trends Pharmacol Sci* 2005;**26**:131–7.
53. Herrick-Davis K, Weaver BA, Grinde E, Mazurkiewicz JE. Serotonin 5-HT2C receptor homodimer biogenesis in the endoplasmic reticulum: real-time visualization with confocal fluorescence resonance energy transfer. *J Biol Chem* 2006;**281**:27109–16.
54. Lee SP, O'Dowd BF, Ng GY, Varghese G, Akil H, Mansour A, et al. Inhibition of cell surface expression by mutant receptors demonstrates that D2 dopamine receptors exist as oligomers in the cell. *Mol Pharmacol* 2000;**58**:120–8.

55. Baker EK, Colley NJ, Zuker CS. The cyclophilin homolog NinaA functions as a chaperone, forming a stable complex in vivo with its protein target rhodopsin. *EMBO J* 1994;**13**:4886–95.
56. Ferreira PA, Nakayama TA, Pak WL, Travis GH. Cyclophilin-related protein RanBP2 acts as chaperone for red/green opsin. *Nature* 1996;**383**:637–40.
57. Shieh BH, Stamnes MA, Seavello S, Harris GL, Zuker CS. The ninaA gene required for visual transduction in Drosophila encodes a homologue of cyclosporin A-binding protein. *Nature* 1989;**338**:67–70.
58. Colley NJ, Baker EK, Stamnes MA, Zuker CS. The cyclophilin homolog ninaA is required in the secretory pathway. *Cell* 1991;**67**:255–63.
59. Chapple JP, Cheetham ME. The chaperone environment at the cytoplasmic face of the endoplasmic reticulum can modulate rhodopsin processing and inclusion formation. *J Biol Chem* 2003;**278**:19087–94.
60. Westhoff B, Chapple JP, van der Spuy J, Hohfeld J, Cheetham ME. HSJ1 is a neuronal shuttling factor for the sorting of chaperone clients to the proteasome. *Curr Biol* 2005;**15**:1058–64.
61. Herrmann JM, Malkus P, Schekman R. Out of the ER-outfitters, escorts and guides. *Trends Cell Biol* 1999;**9**:5–7.
62. Parameswaran N, Spielman WS. RAMPs: the past, present and future. *Trends Biochem Sci* 2006;**31**:631–8.
63. McLatchie LM, Fraser NJ, Main MJ, Wise A, Brown J, Thompson N, et al. RAMPs regulate the transport and ligand specificity of the calcitonin-receptor-like receptor. *Nature* 1998;**393**:333–9.
64. Hilairet S, Foord SM, Marshall FH, Bouvier M. Protein–protein interaction and not glycosylation determines the binding selectivity of heterodimers between the calcitonin receptor-like receptor and the receptor activity-modifying proteins. *J Biol Chem* 2001;**276**:29575–81.
65. Dwyer ND, Troemel ER, Sengupta P, Bargmann CI. Odorant receptor localization to olfactory cilia is mediated by ODR-4, a novel membrane-associated protein. *Cell* 1998;**93**:455–66.
66. Saito H, Kubota M, Roberts RW, Chi Q, Matsunami H. RTP family members induce functional expression of mammalian odorant receptors. *Cell* 2004;**119**:679–91.
67. Loconto J, Papes F, Chang E, Stowers L, Jones EP, Takada T, et al. Functional expression of murine V2R pheromone receptors involves selective association with the M10 and M1 families of MHC class Ib molecules. *Cell* 2003;**112**:607–18.
68. Olson R, Dulac C, Bjorkman PJ. MHC homologs in the nervous system–they have not lost their groove. *Curr Opin Neurobiol* 2006;**16**:351–7.
69. Bermak JC, Li M, Bullock C, Zhou QY. Regulation of transport of the dopamine D1 receptor by a new membrane-associated ER protein. *Nat Cell Biol* 2001;**3**:492–8.
70. Leclerc PC, Auger-Messier M, Lanctot PM, Escher E, Leduc R, Guillemette G. A polyaromatic caveolin-binding-like motif in the cytoplasmic tail of the type 1 receptor for angiotensin II plays an important role in receptor trafficking and signaling. *Endocrinology* 2002;**143**:4702–10.
71. Dupre DJ, Robitaille M, Richer M, Ethier N, Mamarbachi AM, Hebert TE. Dopamine receptor-interacting protein 78 acts as a molecular chaperone for Ggamma subunits before assembly with Gbeta. *J Biol Chem* 2007;**282**:13703–15.
72. Wruck CJ, Funke-Kaiser H, Pufe T, Kusserow H, Menk M, Schefe JH, et al. Regulation of transport of the angiotensin AT2 receptor by a novel membrane-associated Golgi protein. *Arterioscler Thromb Vasc Biol* 2005;**25**:57–64.
73. Parent A, Laroche G, Hamelin E, Parent JL. RACK1 regulates the cell surface expression of the G protein-coupled receptor for thromboxane A(2). *Traffic* 2008;**9**:394–407.
74. Chen C, Li JG, Chen Y, Huang P, Wang Y, Liu-Chen LY. GEC1 interacts with the kappa opioid receptor and enhances expression of the receptor. *J Biol Chem* 2006;**281**:7983–93.

75. Ge X, Loh HH, Law PY. mu-Opioid receptor cell surface expression is regulated by its direct interaction with ribophorin I. *Mol Pharmacol* 2009;**75**:1307–16.
76. Wang H, Westin L, Nong Y, Birnbaum S, Bendor J, Brismar H, et al. Norbin is an endogenous regulator of metabotropic glutamate receptor 5 signaling. *Science* 2009;**326**:1554–7.
77. Svenningsson P, Chergui K, Rachleff I, Flajolet M, Zhang X, El Yacoubi M, et al. Alterations in 5-HT1B receptor function by p11 in depression-like states. *Science* 2006;**311**:77–80.
78. Donato R. Functional roles of S100 proteins, calcium-binding proteins of the EF-hand type. *Biochim Biophys Acta* 1999;**1450**:191–231.
79. Abdel-Malek ZA. Melanocortin receptors: their functions and regulation by physiological agonists and antagonists. *Cell Mol Life Sci* 2001;**58**:434–41.
80. Metherell LA, Chapple JP, Cooray S, David A, Becker C, Ruschendorf F, et al. Mutations in MRAP, encoding a new interacting partner of the ACTH receptor, cause familial glucocorticoid deficiency type 2. *Nat Genet* 2005;**37**:166–70.
81. Sebag JA, Hinkle PM. Regions of melanocortin 2 (MC2) receptor accessory protein necessary for dual topology and MC2 receptor trafficking and signaling. *J Biol Chem* 2009;**284**:610–8.
82. Sebag JA, Hinkle PM. Melanocortin-2 receptor accessory protein MRAP forms antiparallel homodimers. *Proc Natl Acad Sci USA* 2007;**104**:20244–9.
83. Sebag JA, Hinkle PM. Opposite effects of the melanocortin-2 (MC2) receptor accessory protein MRAP on MC2 and MC5 receptor dimerization and trafficking. *J Biol Chem* 2009;**284**:22641–8.
84. Chen W, Kelly MA, Opitz-Araya X, Thomas RE, Low MJ, Cone RD. Exocrine gland dysfunction in MC5-R-deficient mice: evidence for coordinated regulation of exocrine gland function by melanocortin peptides. *Cell* 1997;**91**:789–98.
85. Morgan C, Cone RD. Melanocortin-5 receptor deficiency in mice blocks a novel pathway influencing pheromone-induced aggression. *Behav Genet* 2006;**36**:291–300.
86. Gomes I, Filipovska J, Jordan BA, Devi LA. Oligomerization of opioid receptors. *Methods* 2002;**27**:358–65.
87. Kieffer BL, Gaveriaux-Ruff C. Exploring the opioid system by gene knockout. *Prog Neurobiol* 2002;**66**:285–306.
88. Gomes I, Gupta A, Filipovska J, Szeto HH, Pintar JE, Devi LA. A role for heterodimerization of mu and delta opiate receptors in enhancing morphine analgesia. *Proc Natl Acad Sci USA* 2004;**101**:5135–9.
89. Decaillot FM, Rozenfeld R, Gupta A, Devi LA. Cell surface targeting of mu-delta opioid receptor heterodimers by RTP4. *Proc Natl Acad Sci USA* 2008;**105**:16045–50.
90. Milojevic T, Reiterer V, Stefan E, Korkhov VM, Dorostkar MM, Ducza E, et al. The ubiquitin-specific protease Usp4 regulates the cell surface level of the A2A receptor. *Mol Pharmacol* 2006;**69**:1083–94.
91. Mines MA, Goodwin JS, Limbird LE, Cui FF, Fan GH. Deubiquitination of CXCR4 by USP14 is critical for both CXCL12-induced CXCR4 degradation and chemotaxis but not ERK activation. *J Biol Chem* 2009;**284**:5742–52.
92. Sitia R, Braakman I. Quality control in the endoplasmic reticulum protein factory. *Nature* 2003;**426**:891–4.
93. Schroder M, Kaufman RJ. ER stress and the unfolded protein response. *Mutat Res* 2005;**569**:29–63.
94. Bernier V, Lagace M, Lonergan M, Arthus MF, Bichet DG, Bouvier M. Functional rescue of the constitutively internalized V2 vasopressin receptor mutant R137H by the pharmacological chaperone action of SR49059. *Mol Endocrinol* 2004;**18**:2074–84.
95. Bernier V, Lagace M, Bichet DG, Bouvier M. Pharmacological chaperones: potential treatment for conformational diseases. *Trends Endocrinol Metab* 2004;**15**:222–8.

96. Bernier V, Morello JP, Zarruk A, Debrand N, Salahpour A, Lonergan M, et al. Pharmacologic chaperones as a potential treatment for X-linked nephrogenic diabetes insipidus. *J Am Soc Nephrol* 2006;**17**:232–43.
97. Dejneka NS, Bennett J. Gene therapy and retinitis pigmentosa: advances and future challenges. *Bioessays* 2001;**23**:662–8.
98. Dryja TP, McGee TL, Hahn LB, Cowley GS, Olsson JE, Reichel E, et al. Mutations within the rhodopsin gene in patients with autosomal dominant retinitis pigmentosa. *N Engl J Med* 1990;**323**:1302–7.
99. Noorwez SM, Kuksa V, Imanishi Y, Zhu L, Filipek S, Palczewski K, et al. Pharmacological chaperone-mediated in vivo folding and stabilization of the P23H-opsin mutant associated with autosomal dominant retinitis pigmentosa. *J Biol Chem* 2003;**278**:14442–50.
100. Maya-Nunez G, Janovick JA, Ulloa-Aguirre A, Soderlund D, Conn PM, Mendez JP. Molecular basis of hypogonadotropic hypogonadism: restoration of mutant (E(90)K) GnRH receptor function by a deletion at a distant site. *J Clin Endocrinol Metab* 2002;**87**:2144–9.
101. Leanos-Miranda A, Janovick JA, Conn PM. Receptor-misrouting: an unexpectedly prevalent and rescuable etiology in gonadotropin-releasing hormone receptor-mediated hypogonadotropic hypogonadism. *J Clin Endocrinol Metab* 2002;**87**:4825–8.
102. Janovick JA, Maya-Nunez G, Conn PM. Rescue of hypogonadotropic hypogonadism-causing and manufactured GnRH receptor mutants by a specific protein-folding template: misrouted proteins as a novel disease etiology and therapeutic target. *J Clin Endocrinol Metab* 2002;**87**:3255–62.
103. Janovick JA, Goulet M, Bush E, Greer J, Wettlaufer DG, Conn PM. Structure–activity relations of successful pharmacologic chaperones for rescue of naturally occurring and manufactured mutants of the gonadotropin-releasing hormone receptor. *J Pharmacol Exp Ther* 2003;**305**:608–14.
104. Chaipatikul V, Erickson-Herbrandson LJ, Loh HH, Law PY. Rescuing the traffic-deficient mutants of rat mu-opioid receptors with hydrophobic ligands. *Mol Pharmacol* 2003;**64**:32–41.
105. Chen Y, Liu-Chen LY. Chaperone-like effects of cell-permeant ligands on opioid receptors. *Front Biosci* 2009;**14**:634–43.
106. Dormer RL, Derand R, McNeilly CM, Mettey Y, Bulteau-Pignoux L, Metaye T, et al. Correction of delF508-CFTR activity with benzo(*c*)quinolizinium compounds through facilitation of its processing in cystic fibrosis airway cells. *J Cell Sci* 2001;**114**:4073–81.
107. Benedek GB, Pande J, Thurston GM, Clark JI. Theoretical and experimental basis for the inhibition of cataract. *Prog Retin Eye Res* 1999;**18**:391–402.
108. Peng Y, Li C, Chen L, Sebti S, Chen J. Rescue of mutant p53 transcription function by ellipticine. *Oncogene* 2003;**22**:4478–87.
109. Heiser V, Scherzinger E, Boeddrich A, Nordhoff E, Lurz R, Schugardt N, et al. Inhibition of huntingtin fibrillogenesis by specific antibodies and small molecules: implications for Huntington's disease therapy. *Proc Natl Acad Sci USA* 2000;**97**:6739–44.
110. Soto C, Kascsak RJ, Saborio GP, Aucouturier P, Wisniewski T, Prelli F, et al. Reversion of prion protein conformational changes by synthetic beta-sheet breaker peptides. *Lancet* 2000;**355**:192–7.
111. Permanne B, Adessi C, Saborio GP, Fraga S, Frossard MJ, Van Dorpe J, et al. Reduction of amyloid load and cerebral damage in a transgenic mouse model of Alzheimer's disease by treatment with a beta-sheet breaker peptide. *FASEB J* 2002;**16**:860–2.
112. Forloni G, Terreni L, Bertani I, Fogliarino S, Invernizzi R, Assini A, et al. Protein misfolding in Alzheimer's and Parkinson's disease: genetics and molecular mechanisms. *Neurobiol Aging* 2002;**23**:957–76.
113. Muchowski PJ, Wacker JL. Modulation of neurodegeneration by molecular chaperones. *Nat Rev Neurosci* 2005;**6**:11–22.
114. Ulloa-Aguirre A, Janovick JA, Miranda AL, Conn PM. G-protein-coupled receptor trafficking: understanding the chemical basis of health and disease. *ACS Chem Biol* 2006;**1**:631–8.

The T1R2/T1R3 Sweet Receptor and TRPM5 Ion Channel: Taste Targets with Therapeutic Potential

Dennis Sprous and Kyle R. Palmer

Redpoint Bio Corporation, Ewing, New Jersey, USA

Taste signaling is a critical determinant of ingestive behaviors and thereby linked to obesity and related metabolic dysfunctions. Recent evidence of taste signaling pathways in the gut suggests the link to be more direct, raising the possibility that taste receptor systems could be regarded as therapeutic targets. T1R2/T1R3, the G protein coupled receptor that mediates sweet taste, and the TRPM5 ion channel have been the focus of discovery programs seeking novel compounds that could be useful in modifying taste. We review in this chapter the hypothesis of gastrointestinal taste signaling and discuss the potential for

Progress in Molecular Biology
and Translational Science, Vol. 91
DOI: 10.1016/S1877-1173(10)91006-5

T1R2/T1R3 and TRPM5 as targets of therapeutic intervention in obesity and diabetes. Critical to the development of a drug discovery program is the creation of libraries that enhance the likelihood of identifying novel compounds that modulate the target of interest. We advocate a computer-based chemoinformatic approach for assembling natural and synthetic compound libraries as well as for supporting optimization of structure activity relationships. Strategies for discovering modulators of T1R2/T1R3 and TRPM5 using methods of chemoinformatics are presented herein.

I. Ingestive Behaviors as Therapeutic Endpoints

The majority of an organism's biology is devoted to consumption and metabolism of food. As research has progressed on the multifarious and essential questions about food consumption and associated biology, a broad-ranging variety of disciplines and methodologies has grown together under the field of *ingestive behaviors*, the aim of which is to understand "food and fluid intake and its associated biological, psychological and social processes."[1] Setting aside inquiry into the social processes, the purely biological aspects of ingestive behavior can be abstracted to three principle areas of investigation—food seeking and consumption, nutrient uptake or digestion, and nutrient metabolism. Drug discovery technologies are amenable to all three.

The oral cavity contains the first distinctive features involved in ingestion. The tongue is the sensory organ that generates the signals that determine ingestion or rejection. Not only is the tongue a chemosensory organ but it is also sensitive to temperature, pH, tactile stimulation, and acts as an appendage to manipulate objects in the oral cavity. All of these sensations coalesce to form a mental conceptualization of substances that have entered the oral cavity. This is the subject matter of sensory perceptual psychology and is particularly important for the food and flavor ingredients industry, which, arguably, is interested not so much in increased absolute consumption but in consumption relative to other similar items—in other words, choice or preference.

Food consumption, on the other hand, is a chief concern for the healthcare industry. Overconsumption of food has been recognized as causally related to poor health and degenerative diseases, and intensive efforts have been aimed at therapeutic intervention in a variety of dysfunctions related to ingestive behaviors.[2] The discovery of drugs sufficiently effective at treating these dysfunctions has not been an easy task. As research progresses, it has become more apparent that a complex relationship exists among the processes of taste signaling, consumption, nutrient uptake and metabolism, and the challenges facing drug discovery in this area are formidable. Never the less, as knowledge of the biology underlying ingestive behaviors accumulates, new targets emerge, and new perspective on some familiar targets opens the door to additional possibilities for treatment.

Two targets in particular that have been historically regarded as taste signaling proteins, the T1R2/T1R3 sweet taste receptor and the transient receptor potential-melastatin 5 (TRPM5) ion channel, are gaining renewed attention for the broader roles they may have in ingestive behaviors and for their potential as the objects of programs designed to discover novel modulators of their activity.

II. Taste Signaling Proteins as Targets for the Drug Discovery Approach

A. Discovery of Taste Modulators

The food and flavor ingredients industry has been adopting the drug discovery approach for the purpose of rapidly identifying novel compounds that modify taste. With the elucidation of the receptor proteins and biochemical pathways that underlie the signals that generate taste sensations, the methods of molecular biology and pharmacology are gaining prevalence in the science of taste.[3] In this context, the interaction of a “tastant” with its taste receptor can be evaluated in the same way as any receptor-ligand pairing. Recombinant cell lines expressing G protein coupled receptors (GPCRs) that mediate sweet and umami (savory) tastes have been established and are in current use with high-throughput screening technologies in a systematic effort to discover novel agonists and other modulators with properties that provide advantage over existing tastants.[4,5] The advent of this approach holds the promise that with an understanding of ligand–receptor interactions, a broad range of taste modulators will be discovered that can achieve precise control over the signals generated at the peripheral sensory apparatus and thereby over the perceptual responses.

Currently the greatest focus in this arena is on the possibility of discovering allosteric modulators that can favorably alter the functional relationship between taste receptors and their cognate ligands. Of particular importance is the search for a positive allosteric modulator (PAM) of the T1R2/T1R3 sweet receptor that would be used in conjunction with carbohydrate sweeteners, such as sucrose, fructose, and glucose, to provide lower calories without changing any aspect of the desired sugar taste. Introduction of such molecules into food or beverage formulations would not only address the demands of consumers who desire fewer calories without compromising taste, but very likely would impart direct benefit on public health.

The interaction of sugars (or any sweet tasting ligand) with the T1R2/T1R3 sweet receptor sets into motion a biochemical chain of events that impacts on the activity of the TRPM5 ion channel, which is critical for further propagation of the taste signal.[6] In addition to sweet taste, the essential tastes of umami and bitterness also depend on the activity of TRPM5 for transduction of sensory signaling. It is thought that by controlling TRPM5 activity the whole taste signaling pathway can be manipulated.

B. Taste Signaling Proteins as Therapeutic Targets

Taste signaling proteins are now known to exist in tissues beyond the tongue. Recent discoveries of taste signaling proteins expressed throughout the gastrointestinal tract have suggested new and intriguing possibilities that novel ligands for sweet receptors and TRPM5 could hold potential as therapeutics. Although in early stages, research directed at attempting to understand the functions of these proteins in the gut has implicated sweet receptors and TRPM5 in biological activities more broadly associated with nutrient processing and coordination of metabolism. If true, then new strategies could emerge for confronting one of the most significant problems of current public health concern.

III. Obesity and Diabetes—The Result of Dysfunctional Ingestive Behavior

In recent years, questions over the control of food intake and its relation to health and disease have intensified with the recognition that we are, and have been, in the midst of obesity and diabetes epidemics. The Center for Disease Control has posted a map on its website[7] depicting an alarming trend over the years from 1986 to 2008 of dramatic increases, state by state, of percentages of obese adults (defined as having body mass index (BMI) of greater than 30). By 2008, only one state (Colorado) reported fewer than 20% of adults with BMI of 30 or greater. Thirty two states reported greater than 25% obese adults, and in six of these, more than 30% of adults were obese. Not surprisingly, a map of the prevalence of Type II Diabetes in adults is strikingly similar.[8]

Prior to 1990, obesity was a relatively rare condition in the United States. Something happened over the ensuing decades to alter this picture. The question over the source of the epidemic has been a vexing one, generating much speculation about causes. Underlying most of the explanations is the implicit assumption that body fat composition is a direct function of the balance between calories ingested and calories burned. Are we more sedentary than we used to be or are we consuming more calories? Certainly, the calories burned from 1 h of vigorous exercise can be completely compensated by one bowl of ice cream ingested in a few minutes. The calorie burn side of the equation merits as a plausible explanation for why Americans, with less daily physical activity, are becoming obese.

On the other hand, since the late 1970s, daily per capita caloric intake has risen steadily by at least 200 calories.[9] Meal sizes have increased[10] and meal content has shifted to calorie-dense carbohydrate and fat sources.[11] Evidence corroborating the increases in caloric intake is abundant, but poorly understood are the causes for the changes in diet that underlie the obesity problem. The question of what has happened to change our diets is complex, and most likely involves many factors, both environmental as well as biological. Never the less, there is a tendency to search for and assign blame to a few causes or even a single culpable agent. In search of the easiest explanation, the finger is usually pointed at something that is consumed, at the expense of attention to environmental contributors, which are more difficult to characterize and control.

A. Calorie-Seeking Organisms

Food seeking and consumption are governed by rules laid down by evolution over generations of organisms interacting with nutrient sources in their habitats.[12] Simply put, items in the environment that contain nutrient value are sought and ingested and those with the highest nutrient value are most consumed. Organisms are guided to nutrient sources in their habitats through chemoreception, both olfactory and gustatory. For the most part, it is the gustatory chemoreception that determines the fate of something in the oral cavity. Consequent ingestive responses can be characterized as *appetitive*, that is, those which lead to consumption, or *aversive*, which result in rejection. Potential food items can therefore be regarded as appetitive or aversive based on the animal's reaction to the presence of the substance in the oral cavity. Appetitive responses occur with substances rich in macronutrients—proteins and amino acids, lipids, and carbohydrates. Substances having little or no nutritional significance, or that are potentially harmful, elicit aversive reactions.

Aversiveness and appetitiveness are qualities that exist on a continuum that can be quantified.[3,13] A clear relationship can be gleaned from laboratory models of foraging that indicate an optimization in the balance between calorie expenditure in food seeking with calorie intake.[14] Without resorting to sophisticated methods of quantification, however, a casual reflection of the current obesity situation can lead reasonably to a conclusion that calorie density is a chief determinant of appetitiveness, perhaps even at the expense of other essential nutrients. Foods that are calorie-dense are being consumed in greater quantities,[15] and the ease at which these types of foods can now be procured in most human habitats has upset the evolutionary balance in favor of excessive caloric consumption.[12,16] Calories not burned are calories stored, hence obesity.

B. Sweet-Seeking Organisms

Sugars used to sweeten foods and beverages are highly appetitive macronutrients and have become much more prevalent in the diets of industrialized nations over the last few decades.[11,17] Sugars certainly have been present in the diet long before obesity reached epidemic status. Easily metabolized, simple carbohydrate sugars are a major source of calories and their contribution to caloric intake has been increasing. Several studies have indicated that increased consumption of sugars, particularly in sweetened beverages,[11] can account for a large percentage of the rising incidence of obesity.[18,19]

Some specific changes that have occurred in sugar usage and consumption correlate with the development of the obesity epidemic and have led to much speculation of the underlying causality. Considerable attention and vilification has been directed to high fructose corn syrup (HFCS), a blend of fructose and glucose. For HFCS use in soft drinks, the proportion of fructose to glucose in the blend usually is 55:45, respectively.[20] The rapid increase in consumption of HFCS, concurrently replacing the prevalence of sucrose in the American diet,[21] often has been cited as a potential cause of the obesity epidemic.[22] Although the appearance and rise of HFCS consumption somewhat mirrors the growth of the obesity epidemic, its culpability is unlikely. The disaccharide sucrose has long been a mainstay in the American diet, yet is hydrolyzed in the stomach to yield equal proportions of glucose and fructose. Therefore the proportions of the glucose and fructose absorbed by the body following ingestion of sucrose are almost the same as when HFCS is consumed.[20]

Consumption of nonnutritive, or "high intensity," sweeteners as alternatives to caloric sugars also has raised concerns, since their use similarly has increased in parallel with incidences of obesity and diabetes.[23] The correlation between the per capita rates of consumption and prevalence of disease has spawned conjecture of an insidious relationship between sweet taste sensation and ingestive behaviors. Generally, signals (sensory or otherwise) initiated by nonnutritive sweeteners have been hypothesized to prepare the body to expect ingestion of sugar, creating behavioral or physiological dysfunctions when no sweet-associated calories appear. Waves of studies investigating the potential of nonnutritive sweeteners to intensify postingestive hunger have failed to yield a conclusive link to increases in calorie intake.[24]

However, recent studies have reported that the ability of rats to maintain a proper balance in caloric intake was impaired when the nonnutritive sweeteners saccharin and acesulfame-K were included in their diets, resulting in a paradoxical weight gain.[25,26] The investigators suggest that sweet taste sensations normally serve as Pavlovian cues to initiate physiologic responses appropriate for an incoming calorie bolus. Experiences with nonnutritive sweeteners were hypothesized to reduce the predictive value of sweet taste, losing their physiologic meaningfulness. A situation would then be created where compensatory mechanisms of ingestive behavior and metabolism normally set into motion by sweet taste are reduced when carbohydrates are present in the gut.[25,26]

Others still have proposed that sweet receptor ligands, whether nutritive or noncaloric, could directly stimulate receptor-mediated signal transduction cascades in the gut.[27] Evidence has accumulated over the last decade indicating the presence of taste GPCRs and other taste signaling proteins throughout the gastrointestinal system, leading many to surmise that nutrient sensory mechanisms were operating there in a manner analogous to those that exist in the tongue.[28,29] If entire signaling chains could be determined in the gastrointestinal tract, then nutrient sensing for some physiological purpose would be clearly implied. Given the significance of glucose to metabolism, much attention thus has focused on the search for a complete, functional sweet receptor signaling apparatus. The existence of a sweet receptor signaling pathway in the gut that presumably evolved to detect glucose could erroneously actuate nutrient uptake mechanisms and metabolic processes when stimulated by nonnutritive sweet ligands, resulting in dysfunctional ingestive responses.

Before discussing this important hypothesis further, we will present some details of current knowledge regarding two of the key elements of sweet taste signaling, the T1R2/T1R3 sweet taste receptor and the TRPM5 ion channel. Both of these targets are "druggable" in the sense that they are credible targets for a discovery program based on approaches developed in the pharmaceutical industry for identifying novel therapeutics.[30] Indeed, classic compound screening campaigns have been conducted resulting in novel ligands that enhance the sweetness intensity of sucrose[4] as well as enhancers and inhibitors of TRPM5 activity.[31] A full understanding of the established relationship between these two proteins in taste signaling is crucial to the direction of strategies designed to discover new molecules to modulate their activity. Although momentum in the current scientific discourse fosters expectations that the functional association of sweet receptors with TRPM5 is essentially the same in the gut as it is in the tongue, the anticipated relationship ultimately may not hold up under exhaustive scrutiny. Considering the differences in the cellular physiology of taste cells and gut secretory cells suggests a different type of functionality for TRPM5 in particular, and offers a strategic direction for a discovery program with a therapeutic objective for intervention in metabolic disease.

IV. The T1R2/T1R3 Sweet Receptor

The receptor that mediates the sweet taste sensation is a heterodimer composed of monomer units from the T1R family of GCPRs. Three members of this receptor family are currently known and designated T1R1, T1R2, and T1R3; all are class C GPCRS. The T1R2 and T1R3 units combine to make a functional sweet receptor.[32] Typical of class C GPCRs, the N terminus of the T1R2/T1R3 sweet receptor is relatively large, containing a venus flytrap (VFT) domain structurally similar to that found in metabotropic glutamate receptors (mGluR) and gamma-amino butyric acid type B (GABAB) receptors. A cystiene-rich domain connects the N terminus to the 7TM portion of the molecule.[33]

Perhaps the most striking feature of the T1R2/T1R3 receptor is its capacity to accommodate extensive structural diversity among agonist ligands. Agonists for T1R2/T1R3 include mono- and polysaccharide carbohydrate sugars, alcohols, glycosidic terpines, amino acids and proteins, lead salts, and a variety of small molecules.[33,34] Equally remarkable is the broad range of affinities exhibited by sweet ligands that none the less retain selectivity with respect to integrity of the sweet taste signal. Apparent affinities for sweet ligands range from submicromolar for guanidinyl compounds[35] to well into the millimolar for carbohydrates.[32] Presumably multiple binding sites must exist on T1R2/T1R3 to accommodate such a variety of ligands.

Domain-swapping experiments have determined that the N terminus of the T1R2 subunit is critical for binding carbohydrate sugars, and for humans, the sweet proteins monellin and thaumatin and aspartame.[36–38] Spectroscopic studies of solubilized N termini from murine T1R2 and T1R3 suggest that T1R3 might also contribute to form the binding site for sucrose, glucose, and the sucrose analog sucralose, in the VTF domain of the complete heterodimer.[39] Domain-swapping also has indicated a binding site for the small molecule sweetener cyclamate as well as the sweet taste inhibitor lactisole located in the transmembrane region of T1R3.[38] Evidence for a third binding site, for the sweet protein brazzein, in the cysteine-rich region between residues 536 and 545 of the human T1R3 N terminus has been obtained by selective mutagenesis.[37] However, others have argued that this region is critical to the functional integrity of the receptor and might not represent a specific binding site.[40]

Computational modeling predicated on the structural similarities of the N termini of T1R2/T1R3 and the mGluR1 receptor suggest ways in which binding sites for a variety of ligands can be created from a complex interaction between T1R2 and T1R3 N termini.[33,41] The N termini of T1R2 and T1R3 are thought to occur in dynamic equilibrium between "open" and "closed" conformations. The model predicts that lower molecular weight sweet ligands, such as aspartame, are bound in a closed cavity of the T1R2 VTF that intimately interacts with T1R3 in an active "open-closed" state. Separate clefts within the

VTF of T1R3 host larger protein and nonproteic sweet ligands. The synergism in sweetness intensity that occurs when two sweeteners from different structural classes are mixed together in human taste tests[42] is a result of the intercommunication between the binding sites as predicted by the model.[33,41]

The T1R2/T1R3 receptor is selectively expressed in type II taste cells, the cells within the taste bud that are the source of signals for the essentials bitter, sweet, and umami.[43] Transcripts for T1R2 and T1R3 also have been detected throughout the gastrointestinal tract, though expression appears to be relatively low. For example, low levels of mRNA have been reported to occur in both proximal and distal gastrointestinal tract of rodents[44–46] and humans.[47] The minimal mRNA detection might be due to highly localized expression within specialized cells of the gut. Use of antibodies selective for T1R subunits has indicated a punctate expression pattern thought to visualize the presence of receptors in specialized enteroendocrine cells.[48,49] However, there is some controversy over whether the observed expression pattern represents T1R2/T1R3 heterodimers; while some reports show overlap of T1R2 and T1R3 expression in murine intestinal cells,[48,49] others have found no immunoreactive signal for T1R2 anywhere in the gut.[50]

V. The TRPM5 Ion Channel

Stimulation of T1R2/T1R3 receptors results in the activation of the inositol phosphate signal transduction cascade, causing the release of calcium from intracellular stores in taste cells. Intracellular calcium is the ligand that gates the opening of the TRPM5 channel, an obligatory downstream component of GPCR-mediated taste signal transduction.[3]

TRPM5 is a member of a large family of ion channels, many of which are involved in sensory function[51] and also are under investigation for their roles in human diseases.[52] All of the TRP channels are thought to have six transmembrane spanning domains, with the ion pore being formed by a loop connecting the fifth and sixth. In contrast to other TRP channels, "M" class TRP channels do not have ankyrin repeats. Instead, they are characterized by a large domain of approximately 700 residues, with in the N terminal intracellular domain, currently of unknown function.[53,54] Also characteristic of TRPM channels is a region of coiled-coil structure in the C terminus that is likely to enable multimerization.[55] A functional TRP channel is formed from an assembly of four monomer units. The evidence to date indicates that TRPM5 channels are homotetrameric.[53–55]

TRPM5 is a monovalent cation channel with a conductance on the order of 25 pS and characterized by an outwardly rectifying steady-state current–voltage relationship.[56] Following activation by intracellular calcium, TRPM5 desensitizes with an inactivation–deactivation time constant of approximately 3–6 s.[56,57] A major contributor to the desensitization of TRPM5 is the rundown of inositol lipids in the membrane due to the activity of phospholipase C.[58] It has been suggested that these functional properties ideally suit TRM5 for coupling the activity of calcium-linked GPCRs to changes in membrane potential,[54] and indeed, its critical role in taste signaling is due to its function as a trigger for the tastant-stimulated depolarization of taste cells. The initial calcium-gated sodium conductance of TRPM5 alters local membrane potential, which is sensed by voltage-dependent sodium channels, resulting ultimately in taste cell depolarization.

The expression pattern of TRPM5 is limited and highly localized to specialized cell types. As expected from its role in taste signaling, TRPM5 is co-expressed with T1R2/T1R3 other receptors and signaling proteins for taste in type II taste cells.[59,60] Selective expression of TRPM5 also has been observed in the gastrointestinal tract, where its presence in enteroendocrine cells[49] and pancreas[56,61] suggest a connection to processes in nutrient uptake and metabolism. Consistent with these findings are further results showing the presence of TRPM5 in enteroendocrine cell lines, such as murine SCT-1[62] and NCI-H716 cells,[49] and pancreatic beta-cell lines, such as MIN-6 and INS-1, as well as isolated human islets.[56] The localization of TRPM5 to the pancreas naturally leads to the speculation of its involvement in regulation of glucose metabolism, possibly in some way connected to insulin secretion. It further has been suggested that TRPM5 is likely to act as an initiator of membrane depolarization in pancreatic beta cells, similar to its role taste cells. However, consideration of the differences in the operations of taste cells and secretory cells might imply a different function for TRPM5.

VI. Morphological and Functional Characteristics of Type II Taste Cells

The taste bud is the sensory organelle on the dorsal surface of the tongue that contains the highly differentiated and specialized taste cells that act in coordination to generate and propagate signals in response to chemical agents that contact the tongue.[43] Morphologically, taste cells are polarized, having a basal end from which humoral intercellular signaling factors are released and an apical end that is differentiated into a microvillus structure. Taste cells are identified as type I, II, or III, by protein markers[60] as well as physiological functions that are characteristic of each cell type.[62]

It is the type II taste cells that express the GPCR taste receptors that mediate the signals for sweet, bitter, and umami tastant ligands.[6,63] Type II taste cells are further distinguished by the expression of phospholipase Cβ2 (PLCβ2), IP3R3 (a channel activated by inositol 1,4,5-trisphosphate), and TRPM5, all of which are required for propagation of the taste signals.[60] Voltage-dependent calcium channels are not expressed in type II taste cells.[64] However, voltage-gated sodium channels SCN3A and SCN9A have been identified as being the major mediators of sodium conductances in type II taste cells.[65]

Tastant molecules, acting as agonist ligands, bind to and stabilize the active conformation of GPCRs. The active receptor interacts with the heterotrimeric G protein gustducin (as well as other G_i-class G proteins; Ref. 66), setting into action classic second messenger signal transduction through phosphatydal inositol lipid hydrolysis (reviewed in Ref. 3). Beta–gamma subunits, dissociated from activated G protein, stimulate PLCβ2, which in turn hydrolyses phosphatidyl inositol 4,5-bisphosphate in the plasma membrane. The resulting inositol 1,4,5-trisphosphate binds to and opens IP3R3 to release calcium from the endoplasmic reticulum. Calcium released into the cytoplasm interacts with TRPM5, causing it to open and conduct sodium from the extracellular space across the plasma membrane. The conductance for TRPM5 is relatively small, but the resulting change in local membrane potential is sensed by voltage-gated sodium channels SCN3A and SCN9A.[65] TRPM5 thus serves as the trigger for taste cell depolarization.[67] High intracellular calcium concurrent with a depolarized membrane potential causes the opening of panexin-1 hemichannels, allowing cytosolic ATP to flow out of the cell where it acts as a parcrine intercellular signaling factor[68] to propagate the taste signal on to other intermediary taste cells.

VII. Morphological and Functional Characteristics of Gut Chemosensory Cells

The nutrient-sensing cells of the gut are specialized secretory epithelial cells.[28,29] A variety of different types of chemosensory cells are expressed throughout the gut but are relatively isolated in expression pattern, positioned well as sentinals of nutrient processing as food moves through the GI system. Among the candidate chemosensory cell types, enteroendocrine cells and brush cells have received much recent attention due to their expression of taste-signaling proteins.

Two general morphologic categories of enteroendocrine cells are described as "open cell" or "closed cells." It is the "open cell" type that is regarded as most similar to type II taste cells and therefore as nutrient-sensing chemoreceptor cells. Similar to type II taste cells, enteroendocrine cells are polarized with a basal and a microvillus apical end. However, unlike taste cells, humoral signaling factors are packaged into secretory vesicles in the basal end of enteroendocrine cells, and upon stimulation, are secreted by exocytosis into the extracellular space.[28,29]

Enteroendocrine cell types can be further distinguished by the factors they secrete and by their location within the alimentary tract. For example, L-cells are found in the highest density in the distal ileum and colon and are identified by their secretion of the incretin glucagon-like petide-1 (GLP-1) and peptide YY 3-36 (PYY). Enteroendocrine K-cells are found mostly in the duodenum and jejunum and characteristically secrete the incretin glucose-dependent insulinotropic peptide (GIP).[69]

Less is known about brush cells, which are distinct from enteroendocrine cells.[70] Solitary brush cells are found in the luminal epithelial layer throughout the gastrointestinal tract. Brush cells are identified by their highly differentiated "tufted" microvillus and by the markers cytokeratin 18, tubulin, and villin.[46,70,71] Though the presence of secretory granules has not been clearly established, brush cells release factors upon stimulation and also appear to make direct synaptic contact with mesenteric neurons.[70,71] Brush cells identified in mouse duodenum also might be involved in inflammation, as they have been found to express the entire biosynthetic pathway for leukotriene C4 as well as other inflammatory markers.[70]

The stimulus-secretion coupling of gut secretory cells is more complex than that of taste cells. Secretion of these cells can be accomplished through any of several pathways, all of which share a common endpoint of increased intracellular calcium. Whether from extracellular or intracellular sources, elevation of cytosolic calcium is the final causal step that effects fusion of secretory vesicles with the plasma membrane.[69,72,73]

The classic metabolism-dependent pathway that has been well established for pancreatic beta cells appears also to be present in enteroendocrine cells.[68,72–76] In both beta cells and enterocytes, glucose is taken up by transporters and then metabolized to yield ATP, which in turn binds to and causes the closure of ATP-sensitive potassium (K_{ATP}) channels. Blocking the outward flow of potassium ions through K_{ATP} channels upsets the equilibrium of resting membrane potential bringing about depolarization. Voltage-gated calcium channels are then opened to conduct extracellular calcium across the plasma membrane and into the cytosol.

While the uptake of glucose in pancreatic beta cells is chiefly accomplished by the facilitative glucose transporter 2 (GLUT2; Ref. 77,78), uptake by enterocytes additionally occurs by means of the sodium-dependent glucose co-transporter 1 (SGLT1; Ref. 48,75,79). Since sodium ions are transported across the plasma membrane along with glucose, the actions of SGLT1 are electrogenic and result in membrane depolarization. Metabolism of the glucose does not appear to be necessary, since α-methylglucopyranoside, a nonmetabolizable analog of glucose, serves as a substrate for SGLT1 and can elicit secretion of GLP-1 from primary cultures of murine intestinal cells and the murine L-cell line GLUTag.[73,79]

Agonist activation of GPCRs represents yet another mechanism of stimulus-secretion coupling in enteroendocrine cells. Some GPCRs of enteroendocrine cells associated with secretory responses are linked to the elevation of intracellular calcium through their interactions with G_q proteins, thereby actuating the inositol phosphate signal transduction cascade. For example, stimulation of the L-cell receptor GPR120 by free fatty acids results in the mobilization of calcium from intracellular stores and secretion of GLP-1.[80]

Other secretagogue receptors, such as GPR119, are G_s-linked and therefore upon agonist occupancy increase production of intracellular cAMP.[81,82] Phosphorylation of a variety of intracellular targets involved in the secretory response consequently occurs as a result of the activity of protein kinase A. For example, K_{ATP} channels are inactivated upon phosphorylation, whereas the activity of voltage-gated calcium channels as well as ryanodine and IP3 receptors of the endoplasmic reticulum, are enhanced to lead to further increases in intracellular calcium from both extracellular and intracellular sources.[83]

Finally, evidence for the expression in enterocytes of GPCRs for tastant ligands has mounted in recent years. The presence of these receptors and accompanying signaling proteins would suggest that enteroendocrine cells could act as gut chemosensory cells that respond to nutrients in a manner analogous, or perhaps even identical, to taste signaling in the tongue.

VIII. Taste Signaling in the Gut?

Several lines of evidence point to the existence of taste signaling proteins expressed throughout the gut. However, conflicting data from different laboratories have created an unclear picture of where in the gut expression of specific proteins might occur, whether they are expressed to functional significance, and whether the different signaling proteins coexist within the same cells to form a complete signal transduction system. Although some studies provide strong evidence of intact functional taste signaling systems, others still provide contradictory evidence.

In support of the gut taste signaling hypothesis, T1R2, T1R3, the alpha subunit of the G protein gustducin, PLCβ2, and TRPM5 have been shown by immunofluorescence to co-localize in rat jejunum[45] and in L cells of the human duodenum.[49] Several laboratories report that rodent and human enterocytes express alpha gustducin,[48,74,84,85] and genetically engineered mice deficient of alpha gustducin or T1R3 exhibited abnormal glucose metabolism and an impaired incretin response to ingested glucose.[49,86] Gavage of glucose directly into the duodenum resulted in elevated circulating GLP-1 in wild type, but not in gustducin−/− or T1R3−/− mice.[49,86] In experiments using a human L-cell line, NCI-H716, the release of incretins also was stimulated by nonnutritive sweeteners, such as sucralose and acesulphame-K, and could be blocked by the sweet taste inhibitor lactisole, strongly suggesting that a sweet receptor could be the glucose sensor of enteroendocrine cells in the intestines.[49]

Furthermore, mice maintained on diets that included high glucose, or the artificial sweeteners sucralose or acesulfame-K, were shown to upregulate the illeal sodium-dependent glucose transporter 1 (SGLT1). Aspartame, which does not elicit a sweet taste in mice, similarly had no effect on the expression of SGLT1. The sweetener-mediated upregulation of SGLT1 was dependent on the presence of T1R3 and gustducin, since mice lacking functional expression of these genes showed no effect of sweeteners on SGLT1.[49] This series of reports provides strong evidence that the T1R2/T1R3 sweet receptor is the luminal glucose sensor that upon activation by sweet ligands actuates a taste signal transduction cascade. If this proposal holds true then so-called "high intensity" sweeteners would have metabolic effects in spite of the fact they contribute few if any calories.

In contrast, other means of detecting proteins of taste signaling pathways have yielded a different picture of protein expression patterns in the gut. The expression of green fluorescent protein (GFP) can be placed under the control of promotors for a second protein. Cells expressing the protein of interest then can be visualized and collected by virtue of the fluorescence from GFP. Using this method, GFP-labeled TRPM5-expressing cells of the mouse small intestine were determined to be predominantly brush cells.[46,50,71] Although many components of a taste-like signaling pathway were detected by microarray or quantitative PCR in these cells, T1R2 was not.[46] Moreover, T1R2 was not detected anywhere in the mouse small intestine.[50] If this is true, then an intact sweet receptor does not exist in the gut, and therefore the glucose-sensing mechanism cannot be equivalent to taste signaling in the tongue. Nevertheless, some mechanism must be in place to sense glucose in the gut.

Using a similar approach, enteroendocrine L cells were identified by coupling the fluorescent protein Venus to the expression of proglucagon, the transcript precursor of GLP-1.[75] Quantitative PCR of Venus-expressing colonic L cells were found to have low, but detectable, mRNA levels of T1R2, T1R3, and alpha gustducin. In contrast, expression of SGLT-1, K_{ATP} channels, and glucokinase, all of which could comprise a glucose-sensing pathway, was enhanced in L cells. These investigators further examined the functional relationship between sweeteners and GLP-1 release from primary cultures of cells from mouse small intestine and colon. The cells released GLP-1 in response to submillimolar concentrations of glucose and the nonmetabolizable SGLT-1 substrate alpha. However, no GLP-1 response was elicited by sucralose at concentrations below 20 mM.[75] The concentrations at which glucose and sucralose were reported to be active in these assays are not inconsistent with the established pharmacology of the T1R2/T1R3 sweet receptor.

Perhaps the greatest objection to the hypothesis of physiologically relevant taste signaling systems in the gut is that no evidence has surfaced that artificial sweeteners have any impact on metabolism *in vivo*. Nonnutritive sweeteners have been used safely for decades.[87] Nonnutritive sweeteners such as steviol glycosides[88] and sucralose[89] are without effect in normals and diabetics with respect to glucose homeostasis. Moreover, sucralose has been shown to have no impact on incretin responses in both rats[90] and human subjects.[91]

IX. Type II Taste Cells Are Not Secretory Cells

Nutrient sensing cells of the tongue and those of the gut bare some shared resemblance not only the expression profiles of some signaling proteins but also in overt morphology. The focus on the similarities has led to the implicit supposition that the physiological processes of these different cell types are in essence the same. But the emphasis on the similarities between taste cells of the tongue and nutrient-sensing cells in the GI tract might be an overgeneralization and potentially misleading. The question of function should be carefully reflected upon now since current understanding of the activities of T1R2/R3 and TRPM5 in the gut is sparse, and programs for discovery of new therapeutic ligands for these targets could be launched in wrong directions. In this context, we shall compare the characteristics of taste cells with gastrointestinal nutrient-sensing cells, which might then suggest alternative functions of identical proteins expressed in different cell types or locations in the alimentary canal.

Type II taste cells are not structured for exocytotic secretion in the same way as secretory cells. The cell stimulus (tastant activation of a GPCR) is coupled to a biological endpoint that essentially can be described as a "leak" of a cytosolic paracrine factor (ATP). GPCR and TRPM5 are configured in series in this "stimulus-leak coupling." Type II taste cells must rely on type III cells as intercommunicating cells that pass on the taste signal exocytotically to sensory neurons.[43] The functions of both T1R2/R3 and TRPM5 in the type II taste cell are reasonably well understood. It would be very simple, and therefore tempting, to expect that these signaling proteins serve identical functions both in type II taste cells and secretory cells of the gut.

However, as described above, the coupling of stimulus to secretion is laid down in multiple pathways of gut secretory cells, whether enteroendocrine or pancreatic cells. Secretion that occurs by means of GPCR activation is one of several options for effecting secretion. Enteroendocrine cells appear to have a glucose sensing mechanism in place that is essentially the same as that of pancreatic beta cells—the activity of glucose transporters is quite sufficient to explain the secretory response to carbohydrate sweeteners.[76] Never the less, biology is replete with examples of redundancy, and the evidence that nonnutritive sweeteners can impact on intestinal secretory cells requires consideration.

It is clear that TRPM5 is expressed in intestinal cells that are important in nutrient processing and metabolism. Again, however, the signal transduction pathways of these cells are configured in ways that are more complex than is the case for taste cells. It should not necessarily be the case then, that TRPM5 conductances amount to the same trigger function for depolarization in secretory cells as in taste cells. Unfortunately, no direct evidence presently is available to indicate what position in stimulus-secretion coupling could be occupied by TRPM5. However, some plausible guesses can be made based on current knowledge of the physiology of taste cells and secretory cells. We will consider two hypothetical mechanistic scenarios and what the consequences might be for each with respect to the identification of useful TRPM5 modulators.

A. Trigger for Depolarization

In this case, TRPM5 would serve the same function in both secretory and taste cells. This mechanism would be expected if the relationship of TRPM5 to GPCRs of taste cells generalizes to other chemosensory cell types and GPCRs. It would be reasonable to assume that this well-understood mechanism might also operate to control secretory responses of enteroendocrine cells, pancreatic beta cells, or brush cells. If the existence of functional T1R2/T1R3 in enteroendocrine cells is confirmed, then TRPM5 should be the critical component of the signaling cascade as it is in type II taste cells. Other secretogogue GPCRs (such as GPR120) not related to taste, but that can be activated by fatty and bile acids, also would be expected to rely on TRPM5 to operate as the calcium sensing-trigger for membrane depolarization.

B. Affecting Calcium Entry or Homeostasis

The way in which cells of the gut respond to luminal glucose is the main concern. As discussed earlier, glucose sensing by enteroendocrine cells or pancreatic beta cells is adequately explained by the actions of transporters, such as SGLT1 or GLUT2, and the uptake of glucose. Changes in membrane potential are achieved by closure of K_{ATP} channels resulting from metabolism of glucose, and by the electrogenic effects of sodium-glucose cotransport of SGLT1. Although glucose-induced secretion could be amplified through concurrent stimulation of GPCRs, the critical step in stimulus-secretion coupling is calcium entry through voltage-gated calcium channels that are activated by membrane depolarization. The steps to membrane depolarization in response to glucose then do not require the actions of a trigger like TRPM5 (see Fig. 1).

Yet, the evidence that TRPM5 is present in gut secretory cells is significant. What function could TRPM5 channels serve in these cells if different from their established role as a trigger for depolarization? An alternative function for TRPM5 is suggested by the actions of a closely related channel thought to be

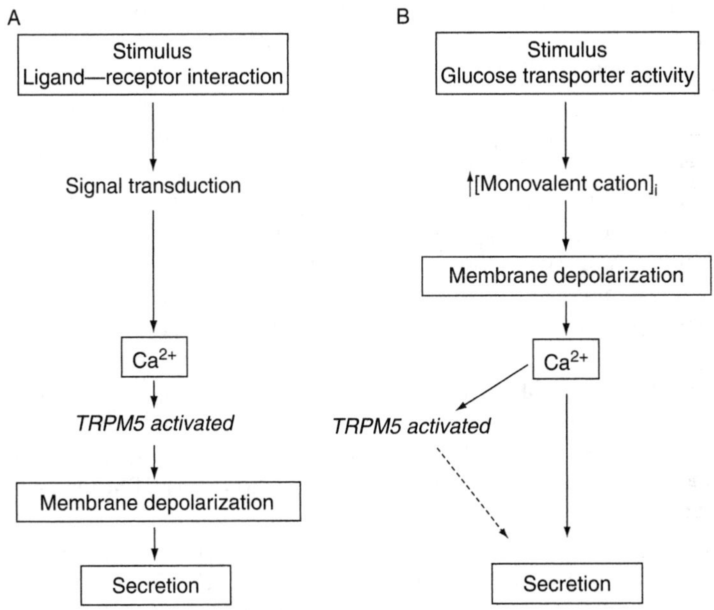

FIG. 1. Schematic comparison of the signaling pathways of taste cells and gut secretory cells. TRPM5 is activated by calcium that is released from intracellular stores or that enters the cell across the plasma membrane. The position of TRPM5 relative to membrane depolarization and calcium mobilization in the signal transduction cascades suggests its role in secretion. (A) TRPM5 is activated by mobilization of intracellular calcium. This scenario occurs in type II taste cells and in secretory cells that express GPCRs linked to secretion through classic second messenger cascades. TRPM5 conductance serves as a trigger for membrane depolarization. (B) TRPM5 is activated by calcium influx through voltage-gated calcium channels. Membrane depolarization occurs before TRPM5 activation; therefore, TRPM5 cannot serve as a trigger. This scenario would occur in enterocytes or pancreatic beta cells when glucose uptake is electrogenic, either by co-transport of Na^+ (SGLT-1) or as a result of metabolism and closure of K_{ATP}. TRPM5 activity is more likely to influence events that occur after secretion has been initiated.

involved in release of cytokines from T cells. TRPM4 also is a calcium-activated nonselective cation channel, but with a much wider expression pattern than for TRPM5.[52,92] TRPM4 is expressed in T-lymphocytes and lymphocytic cells lines, such as Jurkat T cells[93] and is thought to be involved in regulating the intracellular calcium oscillations that underlie release of the cytokine IL-2. The model for this function proposes that TRPM4 sodium currents, activated by high intracellular calcium following receptor stimulation, reduce the driving force for further calcium entry across the plasma membrane. Thus, after intracellular calcium has reached a peak, the influx of more positive charge in the form of sodium ions dampens continued calcium influx during the sustained phase of intracellular calcium mobilization.[93,94] In support of this model are results showing that receptor-stimulated release of IL-2 was enhanced in Jurkat T cells that expressed a dominant-negative form of TRPM4. IL-2 release also was increased by transfecting cells with siRNA designed to specifically inhibit the activity of TRPM4.[93]

It should be noted that the regulation of intracellular calcium in T cells[95] is understood to be reliant upon somewhat disparate mechanisms than are under operation in pancreatic beta cells[74,78] or enteroendocrine cells.[83] Sustaining the elevated levels of intracellular calcium nevertheless is important for the secretory responses of both cell types. By analogy with TRPM4, as intracellular calcium rises in gut secretory cells, the consequent activation of sodium current through TRPM5 could influence calcium entry, or calcium hemeostasis, after intracellular calcium has reached its peak. Again, it is important to emphasize that the key event in triggering secretion is the influx of calcium through voltage-gated calcium channels. Activation of TRPM5, which depends on calcium, must occur subsequent to the initial calcium influx and therefore is not likely to be involved with *initiation* of secretion. In this scenario, calcium influx has already been accomplished without the aid of TRPM5. Therefore, if TRPM5 is involved in secretion, then its impact will be felt on the mechanisms that sustain secretion.

Very recent evidence for a role for TRPM5 in secretion of insulin from pancreatic beta cells has come from studies of TRPM5−/− mice.[96] These results showed that secretion of insulin was impaired when functional TRPM5 channels were not expressed. The effect apparently is due to TRPM5 action on the calcium oscillations that arise after depolarization has initiated calcium entry. The authors believe that TRPM5 plays a role in depolarization during rapid oscillations, helping to raise intracellular calcium as it increases after the trough in the periodic pattern. Rapid oscillations were absent from the calcium signals of TRPM5−/− pancreatic cells, whereas slow oscillating cells were observed in equal percentages in wild-type and knockout mice.[96]

Some important indication that TRPM5 is involved in the physiology of intestinal secretory cells has recently been provided by other studies using TRPM5−/− mice. When a variety of nutrients was perfused directly into an *ex vivo* duodenal preparation, β-endorphin was secreted into the lumen of duodenum from wild-type, but not from TRPM5−/− mice. The response appears to be a TRPM5-dependent response to the hypertonicity sensed by brush cells of the duodenum.[71] Since β-endorphin is a regulator of intestinal motility,[97] the results show that TRPM5 is likely to have significant involvement in nutrient processing.

X. Discovery of Useful Ligands for Sweet Receptors and TRPM5

Having discussed at length the potential of T1R2/T1R3 and TRPM5 as therapeutic targets, we now will turn our attention to the proposal of programs to discover new ligands that modulate their activities. A discovery program must be founded on assays sufficiently robust and rapid to enable primary screens, hit validation, and thorough pharmacologic characterization. Suitable assays have been developed for both T1R2/T1R3 and TRPM5.

A. Assays for T1R2/T1R3

Functional human sweet receptors have been expressed in HEK293 cells by cotransfection of the T1R2 and T1R3 monomers.[4,32] Activation of the natively expressed receptor in taste cells results in calcium responses believed to be mediated by the interaction between PLCβ2 and beta–gamma subunits released from gustducin. To enhance calcium signaling in recombinant cells lines, the promiscuous G protein Gα_{16} often is additionally cotransfected. Reconstituting the signaling pathway for T1R2/T1R3 in HEK293s accordingly facilitates the use of calcium-sensitive fluorescent dyes and real-time measurement of responses in high-throughput plate readers, such as FLIPR. Full dose–response functions for a variety of sweet receptor ligands have been generated using this type of assay,[32,38] as well as high-throughput primary screens for the discovery of PAMs.[4] A similar assay using the murine T1R2/T1R3 also was developed and has been instrumental in locating binding sites on the receptor.[34,37]

B. Assays for TRPM5

We previously have reported the development of a high-throughput assay for human and mouse TRPM5 that was successfully used to identify modulators from a random library of small molecules.[31] The screen relied on a FLIPR-based assay that used recombinant HEK-293 cells engineered to express human TRPM5 that were loaded with a membrane potential-sensitive fluorescent dye. Endogenously expressed P2Y receptors were activated by the addition of ATP to HEK293 cells plated in 384-well plates to generate calcium responses. The recombinantly expressed TRPM5 was consequently activated by raised intracellular calcium concentrations resulting in membrane depolarization that was detected as an increase in fluorescence output. The assay was robust, with Z' values exceeding 0.5 for over 90% of the plates screened. A relatively large number of both enhancers and inhibitors of TRPM5 were obtained.

Since TRPM5 was indirectly activated by stimulating a GPCR, and its activity indirectly measured by fluorescence output, two phases of validation were implemented to ensure that the effects of library compounds were on target. In phase 1 validation (P1V), the possibility that the activity of hits from the library were due to disturbances in the integrity of the cell membrane or due to affecting the signal transduction cascade upstream of TRPM5 was evaluated. P1V consisted of three assays—two membrane potential and one calcium assay—evaluating hits at 30, 10, and 3 μg/ml. The first membrane potential assay was conducted using the original condition of ATP stimulation so that some indication of dose-responsiveness to compound effects on TRPM5 activity could be observed. In the second membrane potential assay, changes in membrane potential were achieved by addition of 20 mM KCl to identify compounds that produced a general dampening of the plasma membrane's ability to conduct ions. Additionally at this stage, compounds that effected changes in basal fluorescence were removed from further consideration, eliminating false positives that were due to autofluorescence or absorbance. The third assay of P1V measured the effects of compounds on calcium signaling. Only those hits that had no impact on the calcium responses were progressed on to phase 2 validation (P2V).

The purpose of P2V was to evaluate selectivity of the hits in full dose–response function of activity against TRPM5 in the HEK293 cells and also when expressed in CHO cells, and also against other related TRP channels, such as murine TRPM5 and human TRPM4, TRPV1, and TRPA1. Generally, compounds that produced well-behaved dose–response functions on human TRPM5 were approximately 3–10-fold less potent on murine TRPM5. We found very few examples of TRPM5-active compounds that also influenced the activity of other TRP channels tested in P2V.

XI. Creating Focused Libraries through Chemoinformatics

With robust high-throughput assays for both sweet receptors and TRPM5 and sets of useful ligands that demonstrate intrinsic efficacy and reasonable selectivity, the next step in the design of an efficient discovery program is the generation of compound libraries. We favor the use of a chemoinformatics approach that relies on the structures within training sets of compounds having known functional properties.

With some more validation yet to be done on sweet receptors for their potential as therapeutic targets related to metabolism, a discovery program designed for ligands presently should be aimed at new sweeteners. Given the increased focus of attention to the connection between caloric sweeteners and obesity, one could even regard noncaloric sweeteners that meet the demand for a satisfactory taste profile as therapeutic agents useful for combating obesity. We propose that programs should be aimed at meeting public demand for discovery of new natural product sweeteners.[98]

Screening for natural products has been problematic in the past, but new technological advancements have improved the situation.[99] Also, some new technologies that enable rapid assessment and discovery of sweet compounds using *in vivo* taste models are currently under development and have recently been reported.[100] But the use of a chemoinformatics approach to create sweet libraries should increase the likelihood of discovering new natural product sweeteners and reduce reliance on and avoid some of the problems associated with cell-based high-throughput screening.

XII. Computational and Chemoinformatics Methods: GPCRs, Ion Channels, Taste and Therapeutics

At RedPoint Bio, we have assembled a 177 member database of known GPCR binding compounds from two literature sources.[101,102] These compounds are known to be active on the sweet receptor, T1R2/T1R3. The dataset is the union of 57 artificial compounds presented across several papers from Walters et al.[101] and 120 natural products reviewed by Kim and Kinghorn.[102] The molecules span a wide range of structures and are best discussed statistically as functions of key metrics (Table I) in comparison to other key databases (Figs. 2–8). In the next section, we will discuss techniques which can be applied to work with a database of this sort from the standpoint of computational chemistry and chemoinformatics. In the second section, we will apply appropriate techniques to this dataset. Specifically, we will discuss this dataset from the standpoint of molecular statistics. The artificial and natural compounds will be compared to one another as groups and to pharmaceuticals and flavorants at a descriptor level. Thereafter, we will outline how this dataset can be used to develop models to search for novel compounds to create a T1R2/T1R3 focus screening library. Note, we are not claiming to devise a model that will allow one to pull a picomolar sweetener with no detectable side taste from a library of botanical compounds. We are devising models to develop libraries from off-the-shelf catalog compounds which will be enriched with compounds able to bind T1R2/T1R3. Binding, in turn, can manifest a signal for sweet, inhibit sweet or possibly enhance the signal of sweet from another sweetener.

TABLE I
DESCRIPTORS USED FOR ANALYSIS OF THE PRESTWICK, GRAS, SWEET NATURALS AND SWEET ARTIFICIAL COMPOUND DATABASES (SEE FIGS. 1–7)

Description	MOE tag	Shorthand used in chapter	Descriptor in QSAR model
Molecular weight	MW	MW	Y
Heavy atom count	a_heavy	Heavy atoms	Y
Aromatic atom count	a_aro	Aromatics	Y
Halogen atom count		Halogens	Y
Nitrogen atom count	a_nN	Nitrogens	Y
Acceptor atom count	a_acc	Acceptr	Y
Donor atom count	a_don	dnr	Y
Ester count (SMILE: O=CO)		Esters	Y
Ether count (SMILE: COC excluding O=CO)		Ethers	Y
OH count ([OH])		Hydroxyls	Y
Presence of one of the listed SMILE cores:		Bg	
C14CCC(C(CCCC3)C3CC2)C2(CC4)C1			
C14CCC(C(CCCC3)C3CC2)C2(C=C4)C1			
C14CCC(C(CCCC3)C3CC2)C2C1(CCC4)			
C1CCC(C(CCCC3)C3=CC2)C2C1			
C1CC=C(C(CCCC3)C3CC2)C2C1			
C1CC(C(CCCC3)C3CC2)C2C1			
OCC(C)CCCC(C)=CCCC(C)=CO OCC(C) = CCCXC(C)=CCCC(C)=CO			

A. Methodologies

Working with GPCR and ion channels from the standpoint of computational chemistry and chemoinformatics is by definition a ligand-based design exercise. This is dictated by the brutal reality of the lack of reliable structural information for the targets. GPCRs and ion channels are membrane bound and ill-represented in the RCSB Protein Data Bank (the public repository for protein crystal structures). There are ample structure *fragments* for GPCR and ion channels but complete proteins which actually give details adequate for molecular design are limited. GPCR's are represented in the RCSB Protein Data Bank by only the mammalian bovine rhodopsin structure (1U19 current lowest resolution). Recently, MacKinnon and coworkers presented a mammalian K^+ channel (2A79). While homology models based on these reference

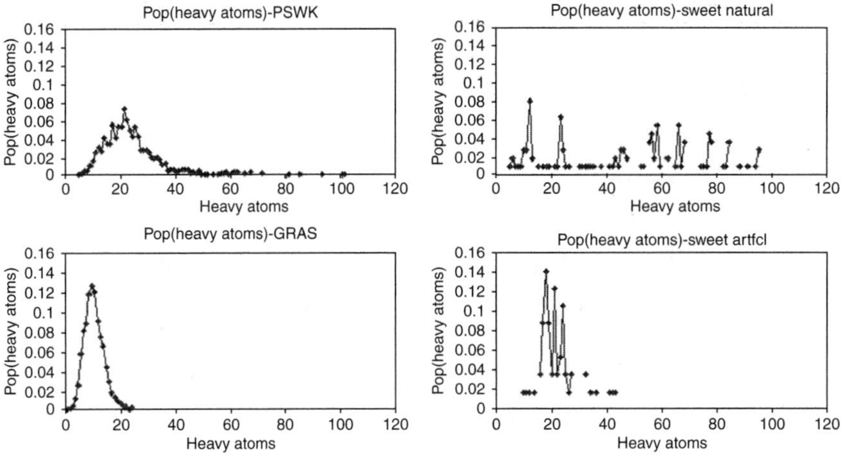

FIG. 2. Population as a function of heavy atom count for the Prestwick, GRAS, Sweet Naturals and Sweet Artificial datasets (top to bottom).

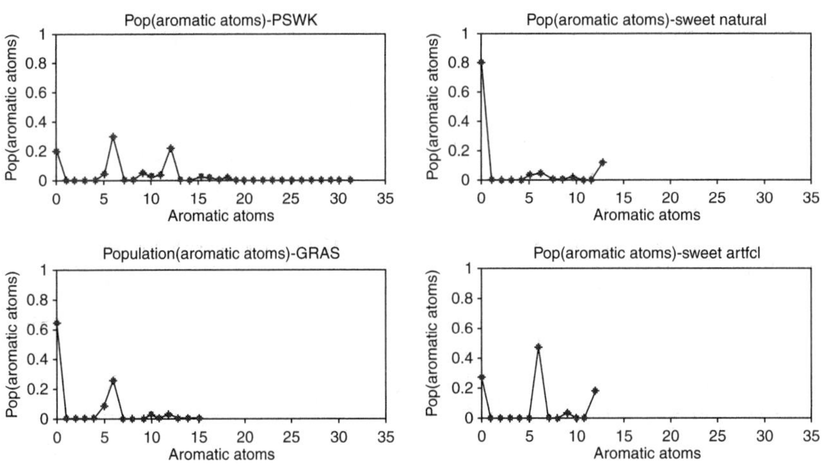

FIG. 3. Population as a function of aromatic atom count for the Prestwick, GRAS, Sweet Naturals and Sweet Artificial datasets (top to bottom).

structures are common, it is dubious whether these are useful for lead optimization due to the strong possibility of a frame shift in the alignment used to build the model and thereafter the problematic issue of the nature of the

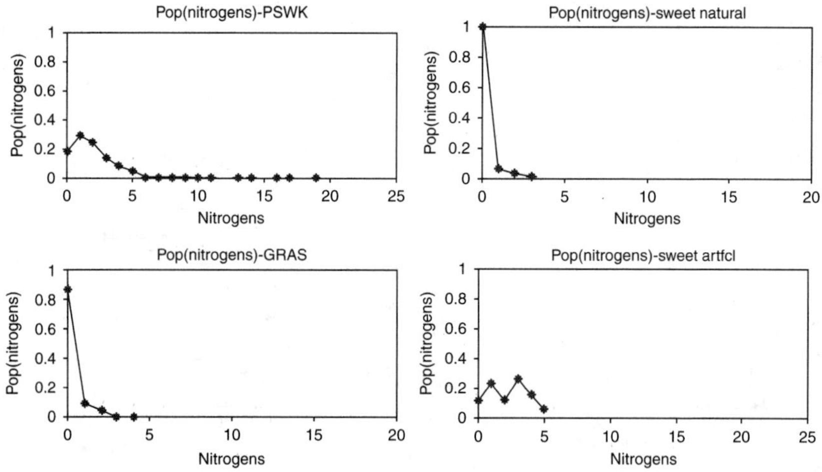

FIG. 4. Population as a function of nitrogen atom count for the Prestwick, GRAS, Sweet Naturals and Sweet Artificial datasets (top to bottom).

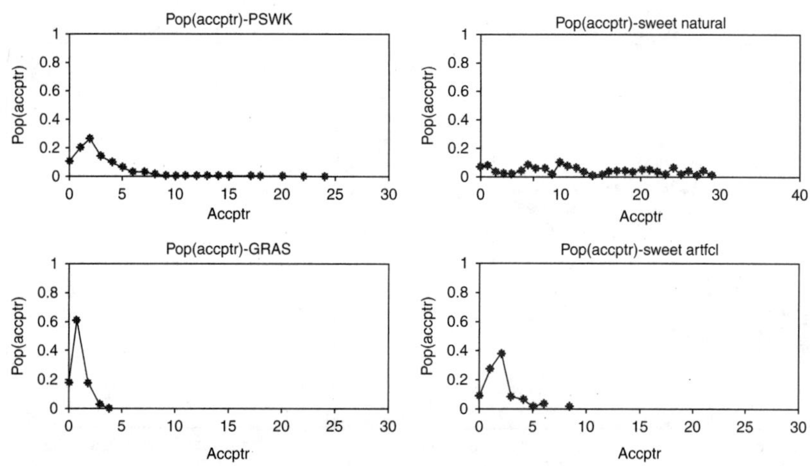

FIG. 5. Population as a function of hydrogen bond acceptor count for the Prestwick, GRAS, Sweet Naturals and Sweet Artificial datasets (top to bottom).

binding mode between the protein and ligand. This lack of structural data has effectively pushed GPCR computational chemistry into a ligand-design-based paradigm. Examples of ligand-based design strategies include QSAR, pharmacophore modeling and similarity based approaches. Developing picomolar

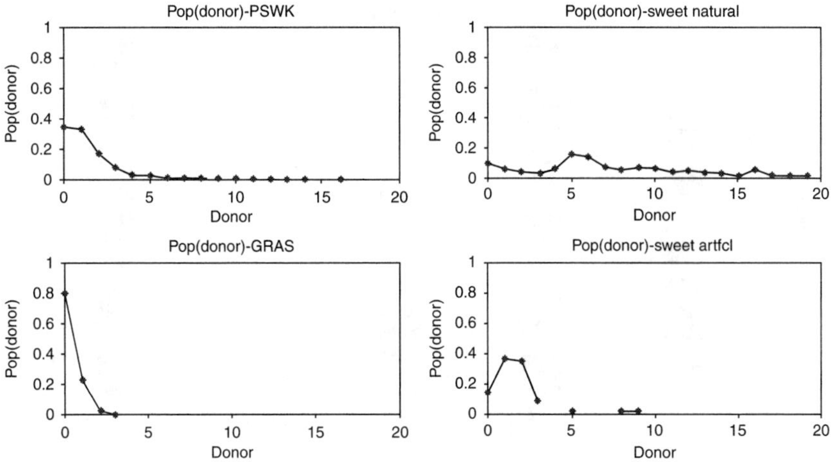

FIG. 6. Population as a function of hydrogen bond donor count for the Prestwick, GRAS, Sweet Naturals and Sweet Artificial datasets (top to bottom).

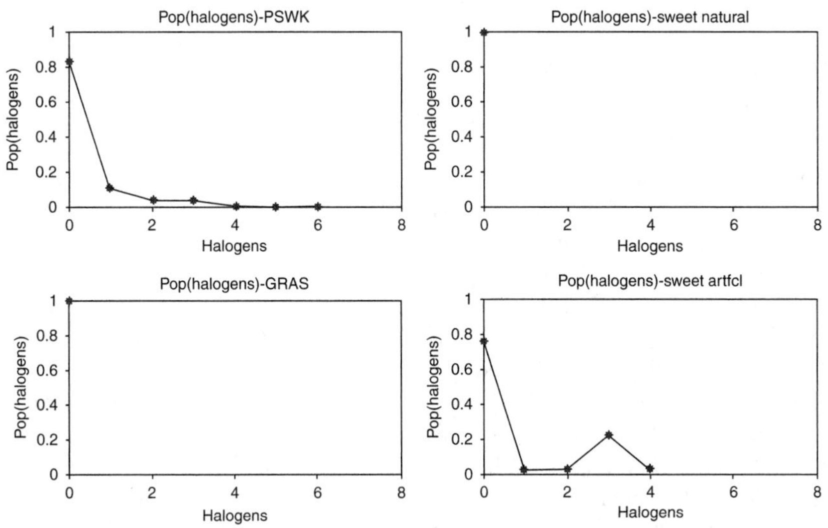

FIG. 7. Population as a function of halogen atom count for the Prestwick, GRAS, Sweet Naturals and Sweet Artificial datasets (top to bottom).

compounds with no side effects in a single step is not the goal of the use of these techniques. Broadly speaking, the goal is to reduce cost for the whole research endeavor.

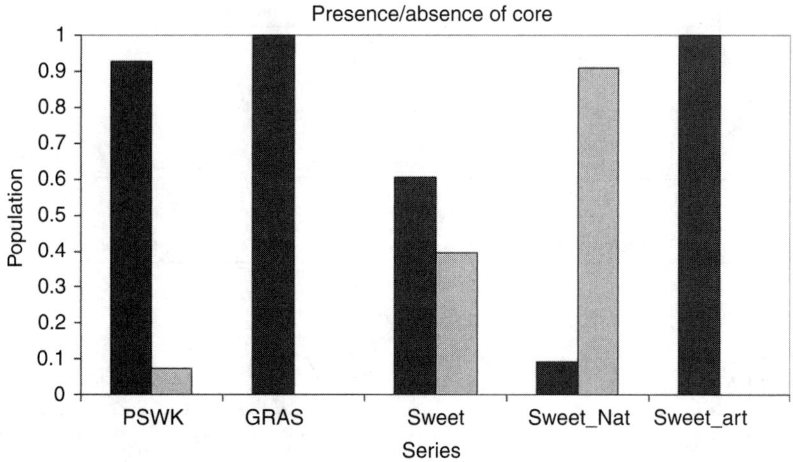

FIG. 8. Population as a function of the Presence (gray) or Absence (black) for the various datasets. Core is one of the substructures defined in Table I.

1. Ligand-Based Design Techniques

All ligand-based design techniques are based on metrics (usually called descriptors) calculated from the structure of the compound. To calculate these metrics, compounds need to be stored at minimum as a 2D representation such as a SMILE[103] and possibly as a 3D representation. Once metrics are calculated, techniques such as similarity analysis/searching, QSAR, and/or pharmacophore analysis can be applied.

Descriptors are properties calculated from the compound structure. There are a bewildering number of available descriptors and no two randomly named descriptors should be considered independent until checked. There are *3224* descriptors available from the program Dragon[104] *alone* and this may not be the upper bounds for the actual available set of descriptors. MolConnZ[105] can calculate more than a thousand. Other programs available calculate hundreds or so descriptors including MOE,[106] SYBYL,[107] ADAPT,[108] and Codessa.[109] Given the number of available descriptors, the editorial demand that future papers submitted to *J. Chem. Inf. Model.*[110] not introduce new descriptors unless there is a clear superiority over the existing descriptors seems prudent.

A simple start for classifying descriptors is by the requirements for structure representation. All descriptor calculations require at minimum a molecular formula (molecular weight and per element atom counts could be determined from a formula alone). More traditionally, even the simplest descriptors are described as being 2D and represented as a connection table:

a list of atoms and connecting bonds. Others require additional atomic coordinates (3Di—internal) and grid methods require coordinates of aligned molecules (3D).

A key disadvantage of 3D descriptors is that coordinates must be generated properly. While CPSA,[111] VolSurf,[112] Chekasov's inductive descriptors[113] and Almond/GRIND[114] do not require molecular alignment there is frequently a conformational dependency which requires conformational sampling, making the claim of alignment independence somewhat hollow. Grid-based methods such as CoMFA[115] and CoMSIA[116] require aligned databases. The process for alignment or conformational searching can be CPU and individually demanding. This makes training over more than hundred of compounds problematic for 3D QSAR methods. Sometimes, the requirements of alignment and conformational searching can make application to searching/evaluation over a larger number of compounds problematic as well.

2D methods, by contrast, are dependent on only the connection table, making the calculation of these descriptors rapid. A nonexhaustive list of examples of classes of 2D descriptors include atom and bond counts, molecular weight, fragment-based calculations done for items like logS, logP[117] and so forth, Merz-Dixon "1D" descriptors,[118] molecular connectivity indices,[119] molecular shape,[120] E-state indices,[121] radial descriptors under various names[122,123] and TPSA.[124] Despite the low CPU demand of 2D descriptors, they frequently perform as well or superior to 3D descriptors. A simple example is the well known correlation between PSA and TPSA.[124] Further, head to head comparisons of 2D versus 3D QSAR on the same datasets can perform equally well.[125] Descriptors can be used with various techniques including similarity, QSAR and pharmacophores. They can also be analyzed directly to understand underlying phenomena.

Similarity analysis can hypothetically use just about any descriptor or collection of descriptors. However, it is most commonly used in conjunction with MACCS fingerprints.[126] Molecular fingerprints are a means of representing molecules as sets where each set element is a count of a specific functional group. These functional groups are overlapping and redundant: by example, there are counts for halogens and for F, Cl, Br and I as elements. Once the fingerprints are generated, comparisons between molecules can be done rapidly. A key measure is the Tanimoto similarity coefficient:

$$T(A, B) = 1 - \{[|\mathrm{A} \cup \mathrm{B}| - |\mathrm{A} \cap \mathrm{B}|]/[|\mathrm{A} \cup \mathrm{B}|]\} \qquad (1)$$

where $\mathrm{A} \cup \mathrm{B}$ is the union for sets A and B and $\mathrm{A} \cap \mathrm{B}$ is the intersection of sets A and B where A and B are the fingerprints from molecules *A* and *B*. In the case where A and B are identical with respect to their list of elements, then the intersection and union are the same and T(*A*, *B*) is 1. In the case where there are no common elements, the intersection is zero then the equation becomes:

$$\begin{aligned} \mathrm{T}(A,B) &= 1 - \{[|\mathrm{A} \cup \mathrm{B}| - 0]/[|\mathrm{A} \cup \mathrm{B}|]\} \\ &= 1 - \{[|\mathrm{A} \cup \mathrm{B}|]/[|\mathrm{A} \cup \mathrm{B}|]\} \\ &= 1 - 1 \\ &= 0 \end{aligned} \tag{2}$$

Whenever the intersection is nonzero, then the term $\{[|\mathrm{A} \cup \mathrm{B}| - |\mathrm{A} \cap \mathrm{B}|]/[|\mathrm{A} \cup \mathrm{B}|]\}$ is a fraction less than unity and T(*A*, *B*) is between 0 and 1.0. A good deal of empirical experience is available for working with similarity based on MACCS fingerprints. Early on, it was seen for a limited example that if *A* was active and T(*A*, *B*) was greater than 0.85, then B has an 80% chance of being active.[127] More complete analysis has led to the estimate that *B* has a 30% chance of being active under these conditions.[128] Despite the fact that 30–80% is a large range, this is magnitudes greater than a randomly picked compound's chance to be active. After fingerprints are calculated for a database, similarity searching relative to a reference compound is a quick means to assemble compounds that are (or alternatively not) similar to the reference.

QSAR (quantitative structure activity relationships) are based on determining a function that relates activity to structures, or more specifically, descriptors calculated from the structures. Typically, functions are of the following form:

$$A(\mathrm{X}) = F(\mathrm{X}) = w_1 * d_1(\mathrm{X}) + w_2 d_2(\mathrm{X}) + \cdots + w_n d_n(\mathrm{X}) \tag{3}$$

where X is the structure of compound X, *A*(X) is the *calculated activity* of compound X as a function of structure, w_i is the solved *i*th weight associated with the *i*th descriptor d_i(X) calculated relative to the structure X. QSAR requires a database of compounds with numeric activities. After descriptors are calculated based on the structures, the weights can be solved using various techniques. In the case where there are more activities (dependent variables) than descriptors (independent variables), traditional MLR (multiple linear regression) can be applied. In the case where there are more descriptors than activities various other methods including PLS (partial least squares), genetic algorithms, neural networks or SVM (support vector machine) need to be used as an alternative. In the case where one is working with larger datasets (hundred and greater), MLR seems to be less favored than alternatives. A last alternate QSAR technique is LaBute's binary QSAR.[129] This QSAR method works off of products of probabilities rather than a sum, and returns the chance that a given compound is in one set or another. By example, there can be a database of compounds composed of two sets: some are potent GPCR compounds while others are marketed drugs which do not act on GPCR proteins. Binary QSAR develops a model based on a series of descriptors and returns the chance that a given compound belongs to the reference set.

Regardless of the type of QSAR employed, the model created needs to be validated. Generally, it is accepted that leave-one-out cross-validation (LOO-cross-validation) and validation schemes based only on the training set used to produce the model are not adequate to validate a model.[110,130–135] By editorial policy, they are inadequate for publication in *J. Chem. Inf. Model.*[110] LOO-cross-validation is especially inappropriate when applied to large redundant datasets, since the chance of any compound having a sibling with essentially identical properties is unity.[130] Hence, any QSAR needs to be tested for predictivity against compounds not employed to develop the QSAR. A variety of means are available for this that largely revolve around created multiple models in the same manner trained over a subset with a reserve test set available to establish predictivity.[130–135] These methods have been reviewed elsewhere.[136]

The pharmacophore concept was defined in 1909 as "a molecular framework that carries (phoros) the essential features responsible for a drug's (=pharmacon's) biological activity" [Paul Ehlich, 1909. *Dtsch. Chem. Ges.* 42: 17]. Modern concept is firmly 3D in nature, practice using computer-based modeling is widespread and has been comprehensively reviewed.[137,138] Packaged software is available from all the common computational chemistry software houses.[106,107,139,140] Regardless, of the program's author, the fundamental process is the same within each implementation. The process of posing a pharmacophore starts with at minimum an active compound in a conformation which is assumed to be the active conformation. More normally, an alignment of multiple active compounds is performed. Thereafter, for each molecule, features are determined such as donors, acceptors, aromatic and hydrophobic groups. Typically, a feature is used in the pharmacophore hypothesis if it is present—within 3D distance tolerances—in some fraction of the active molecules. This fraction is a user controlled value. During the process of searching with the hypothesis, the user may require some specific features essential for a compound to be a match and may require a minimal count of feature overall that can be met for a compound to be a match. Searching is usually done over databases of 3D conformations, but flexible calculation of conformations is not unknown. The great advantage of this method is that it can frequently find active compounds that are chemically remote from the original compounds used to build the hypothesis. Frequently, aligning multiple flexible molecules can limit the conformations available—there may not be a large number of means to place critical donors, acceptors, aromatic and hydrophobic groups atop one another. However, there is no guarantee of this situation and sometimes a common alignment is difficult to determine, mandating the use of multiple pharmacophore hypotheses or simply dictating that one alignment be chosen arbitrarily.

2. Databases

Ligand-based design requires two distinct databases. The first database is the training database used to develop a model. In the case of internal lead optimization, this database will be hits from in-house assays and/or data provided by literature references. In the case of the design of a target family focus library (e.g., a GPCR-amine ligand library or an ion-channel ligand focus library), the training set can come from one or a combination of the following: (a) the MDDR,[141] (b) a proprietary database such as that offered by certain vendors[142,143] that provides curated data drawn from literature for GPCR or ion channels, (c) a carefully assembled dataset made from review articles, and (d) internal databases of large biopharmaceutical or pharmaceutical firms. Armed with this first dataset for training, a model can be created and applied to the second database discussed below.

The second database, frequently undervaled, is composed of structures of compounds that are obtainable and can be screened as "electronic hits." Several companies or organizations[144–146] provide prepackaged databases for compounds for electronic screening. The typical size of electronic catalogs of this type is approximately 5 million compounds. At RedpointBio, we assembled our own from scratch concentrating on 10 reliable suppliers[147–156] and developed a unique set of 3 million compounds. Databases are stored electronically in file formats specific to different modeling packages, though the SDF format serves as a common interchange format. When needed, 3D coordinates must be generated for use and stored. These databases are viable for electronic screens but can be prohibitively expensive to purchase in their whole, and many of the compounds may not be desirable.[157] Therefore, actual assembled screening collections often will be subsets from which target family focus libraries may be assembled.

3. Screening Library Design

A key means by which chemoinformatics has improved research costs is by improving screening collections. This has been done over stages spanning roughly 10 years and is partly a backlash against problems found in libraries assembled from massive combinatorial chemistry programs. Herein, we will review the progression of changes in screening library conceptual design through drug likeness, lead likeness, GRAS-likeness and target focus library design. A greater emphasis will be placed on target focus library design, and it will serve as a template for work with a library of known T1R2/T1R3 binding compounds.

4. Drug Likeness and Lead Likeness

The first stage in design of ligands against any target involves a screen against the target using an appropriate assay. In the early 1990s, there was an assumption that if compound libraries were simply made large enough, high-throughput screening would be able to deliver drug candidates directly.[158] By the late 1990s and early 2000s, it was clear that it was not practical to "just make everything." Large compound libraries that were, practical to make with combinatorial chemistry, tended toward structural redundancy and to consist of hydrophobic, poorly soluble, higher molecular weight compounds,[159–162] resulting in frequent false positives from promiscuous binders.[163]

The above situation led to a more careful screening library design. We have already mentioned that the issue of similar compounds having similar activities has been measured relative to the similarity as defined by molecular fingerprints: if A is active and $T(A, B) \geq 0.85$ then B has a 30–80% chance of being active.[127,128] This observation has the profound implication that the library needs a balance between diversity and redundancy. If the database is designed so that no two molecules are $T(A, B) \geq 0.85$, there is a 70% chance of missing an active compounds series. This situation leads logically to the desire to have five or six compounds—but not likely more—forming a cluster that are $T(A, B) \geq 0.85$ to one another for large screening libraries 10 thousand molecules or more.

High-throughput screening is capable of volume. A FLIPR-based system with staff on the order of six people can process more than a hundred thousand compounds within a month. In this case, the desire is for a large compound library having a suitable property profile as well as a balance of diversity and redundancy (discussed in the previous paragraph). By property profile, we are referring to a usual need for a large population of druglike[159,160] or, more desirably, lead-like[161,162] compounds. The desirability for lead-likeness comes from the fact that if a compound is smaller and more soluble, there is more room for a synthetic chemist to try modifications which will make the compound larger and potentially less soluble but more potent. In the event that one starts from a drug-like compound from screening, the room for modification is less, since addition of functional groups will push the compound outside of normal oral bioavailability ranges. Drug-likeness has been recognizable by QSAR programs[164,165] for more than a decade and the process for developing drug-likeness QSAR models is routine. We will discuss the rough idea of how this is done when we discuss target focus library design; the methods employed there are linear descendants of the drug-likeness methods and differ in starting with a training dataset specific for the target library rather than a training dataset of marketed drugs. Lipinski and coworkers analyzed[160] a large database of pharmaceuticals across several metrics and developed the Rule-of-5 for oral bioavailability:

$$\begin{aligned}&\text{donors} \leq 5\\&\text{acceptors} \leq 10\\&\text{MW} \leq 500\\&\text{logP} \leq 5\end{aligned}$$

This rule can be summarized as a Goldilocks type rule: not too polar (donors and acceptors), not too large (MW) and not too apolar (logP).

Lead-likeness is more defined by logic since assembly of a lead library is more difficult—that is, one can say with certainty that a compound is a drug but saying that a compound is or is not a lead is a matter of opinion which makes assembly of a large database problematic. The lead-likeness concept was crystallized by Oprea and coworkers in response to problems seen during the lead optimization process[159,161]: if beginning an optimization process with a drug-like molecule, any addition of a functional group leads to molecules too big and bulky for practical use. A study by Oprea and coworkers in 2001[161] compared 96 lead/drug pairs. This comprised 62 leads that are not marketed as drugs and 75 drugs. Molecular weight, calculated molar refractivity, ring count, rotatable bond count, donors, acceptors, clogP, logD and Daylight-fingerprint drug score were the applied descriptors. The reversal from drug to lead led to smaller, lighter compounds that were less complex, less hydrophobic and less drug-like. Review and revision of these concepts led to the formulation of the lead-like filter rule[162]:

$$\begin{aligned}&\text{MW} \leq 460,\\&-4 \leq \text{clogP} \leq 4.5,\\&\text{donors} \leq 5\\&\text{acceptors} \leq 9\end{aligned}$$

All of the above assumes a project geared for a pharmaceutical development program. The general features of the above molecules are amenable to a taste target. However, with a taste target there are different considerations. The first consideration is that the targets are in the oral cavity and systemic oral bioavailability is logically neither needed nor desired. From a toxicology and safety consideration, a compound altering taste which then passes through the gastrointestinal tract with zero absorption would be a good situation. This generally permits the consideration of molecules which violate the Rule-of-5 above.

A second consideration of taste compounds is one of commercial cultural considerations: food corporations and customers are more comfortable with the compounds if they look like other food ingredients. Alternatively, compounds which come from natural sources are especially preferred and are considered "label friendly." Most of the compounds used as flavorants are GRAS (generally recognized as safe).[166,167] This is a self regulatory mechanism permitted by the 1958 Food Additive Amendment to the Federal Food, Drug and Cosmetic Act. All compounds in common use as flavorants in 1958 are considered GRAS. This amendment was necessary since the law poised to take effect at the time would have required just about all flavorants in common use at that time to undergo toxicological testing, which would have taken such ingredients as vanilla and cinnamon off the market for approximately two years. In addition to the grandfather clause, any individual, corporation, or group can convene an expert panel qualified by training and experience to determine food safety and ascertain if a new compound is or is not GRAS. A significant fraction of GRAS compounds were designated by the Flavor and Extract Manufacturers Association (FEMA) Expert Panel which has operated since the early 1960s.[166,167] This particular list of compounds is easily obtainable as a dataset,[168] and was used in an analysis and to define a QSAR[169] which can discriminate between GRAS compounds and pharmaceuticals.

Sprous and Salemme[169] are clear that this is not especially original work but rather is application of methods from the drug-likeness, lead-likeness and target focus library design literature genres to the problem of asking "Are GRAS compounds distinct and are they especially distinct as compared to pharmaceuticals?" To approach this problem, the authors assembled a dataset of GRAS compounds based on the FlavorBase dataset[168] to define GRAS space with a 1882 member set. Pharmaceutical space was defined by the Prestwick dataset,[170] this is a dataset of off patent pharmaceuticals sold as a screening library and comprises 1120 compounds. Thereafter, descriptors were calculated over both datasets. Much of the paper centers around presentation of plots of population of molecular weight, flexibility, acceptors, donors, acid atoms, basic atoms, halogens, aromatics atom counts, nitrogen and oxygen atom counts which show exactly how pharmaceuticals and GRAS compounds differ (similar plots will be presented with analysis of our T1R2/T1R3 dataset as it compares to GRAS and pharmaceutical compounds). GRAS compounds can be summarized as smaller, more flexible, less aromatic, essentially absent any halogens, oxygen rich and nitrogen poor as compared to pharmaceuticals.

To create a QSAR to predict what is GRAS versus what is pharmaceutical, the authors employed LaBute's binary QSAR methods.[129] The resulting QSAR can accurately predict GRAS as 94% and pharmaceuticals as 92%. More significantly, if the dataset was randomly partitioned into a 50/50 training and test set, the accuracy for both GRAS and pharmaceutical was 93%. The authors would say this is a signature of the GRAS and Prestwick datasets being strongly distinct over several variables rather than anything to do with the inherent strength of the methodology.

The binary QSAR can easily be applied to datasets on the order of tens of thousands of compounds. It cannot be applied to hundreds of thousands of compounds. This led the authors to develop a GRAS-like filter rule based on analysis of the data in the paper[169] and develop a screening library of approximately 12,500 compounds.[171] A compound was considered GRAS-like and slated for purchase if it met the following criteria:

MW $\leq$ 250
aromatic atom count $\leq$ 6
halogen atom count = 0
nitrogen atom count $\leq$ 2
require an ether, ester or alcohol.

This library was used in a screening program for a taste target. An unintended consequence of the above criteria is that the restrictions frequently returned hits from within compounds marketed as being for fragment-based design,[172] which roughly follow the following "Rule-of-3" selection rules developed by scientists at Astex: MW < 300, donors $\leq$ 3, acceptors $\leq$ 3 and clogP $\leq$ 3. Note the limits on MW and donors or nitrogen atoms are similar in the two independently derived rules. This model will be referenced later and the molecular statistics of GRAS compared to artificial and natural sweeteners.

5. Target Focus Library Development

The previous several paragraphs have covered traits desirable for a large screening library for initial HTS run. In some cases, there may be a motive for starting from a more modest size library due to assay cost. Alternatively, the target may be well understood to be a member of a target family where a specific target family focus library is relevant. Flavor type and related targets covered in this chapter are either GPCRs or ion-channels. There is a well developed literature around the development of GPCR ligand-like libraries.[173] For ion-channels, there is less literature available, though ion-channel libraries are marketed by a variety of vendors. It is important to note that GPCR focus libraries are more difficult to develop than kinase focus libraries. The classification scheme[174] for GPCR is not especially more complex than for kinases.[175] However, where kinases naturally have ATP as a substrate, GPCR substrates range among ions and molecules of highly variable size and shape. It is not logical for GPCR ligands to form as coherent a group as is the case for kinases.

One of the earliest GPCR ligand-like selection QSAR was reported by Ford and coworkers.[176] It is essentially a reapplication of techniques and concepts developed previously to select for drug-likeness. The first critical step was to naturally develop datasets over which to train the QSAR. GPCR ligands are diverse, leading to the need to focus on ligands which target two subgroups of the so-called Class A rhodopsin-like receptors.[174] GPCR amine ligands (1368 compounds) and GPCR peptide ligands (1702 compounds) were modeled. These datasets were extracted from the MDDR database.[141] For both the GPCR datasets, a random set of compounds from the MDDR was created as to represent pharmaceutical chemical space. BCUT metrics[177–179] were calculated across all compounds and neural networks trained to classify membership for set choices between GPCR-amine versus pharmaceutical and between GPCR-peptide versus pharmaceutical. The GPCR amine model had an accuracy of 81.47% while the GPCR peptide model had an accuracy of 88.14. This chapter is an early demonstration that GPCR ligands are in a distinct chemical space as compared to drug-like chemical libraries.

Balakin et al.[180] developed another neural network-based method for selecting GPCR-ligand-like from non-GPCR-ligand-like compounds. Key differences in this study versus that of Ford and coworkers[176,] is that no partitioning of the GPCR ligand set by class was done and more interpretable descriptors were employed. Training was over 5736 known GPCR ligands and 7505 non-GPCR binding compounds. Molecular weight, clogP, donor count, acceptor count, rotatable bond count, logS and fraction of dose absorbed were used as the descriptors. Several parameters showed a distinct difference in distribution between GPCR ligands and non-GPCR ligands. The overall trend was for GPCR ligands to be less polar and more rigid with specific shifts being fewer donors, fewer acceptors, few rotatable bonds and higher lipophilicity (seen in logS and logP). Discrimination between GPCR and non-GPCR ligands was reported as excellent. This particular study is the basis of the methodology used to develop compounds offered from ChemBridge[148] as a GPCR focus library.

Rolland et al.[181] developed a global QSAR for 1939 compounds systematically screened against a 40 GPCR panel. The dependent variable modeled was the average pIC50 per compound across the 40 GPCRs. Inactive compounds were defined by having a percent inhibition less than 30% at 10 uM and set to pIC50 = 3.5 as a default. In-house specialty descriptors were used as the base for three complimentary approaches: linear regression, predictive neighborhood behavior and synergy models which are unions of the two approaches. Models were trained over 1551 compounds as to retain a 388 compound dataset as a validation reserve. The split was done in a manner to "ensure an equivalent relative distribution of actives and inactives throughout both sets." The best synergy model had an $r^2 \sim 0.67$–0.74 and r.m.s. = 0.2–0.24 where the first number is for the training set and the second for the validation set. This best synergy model was used to prioritize compounds from virtual combinatorial libraries for synthesis and led to the synthesis of 240 compounds which showed 14.3% hit rate at a threshold of 70% inhibition at 10 μM.

Givehchi and Schneider[182] developed a method able to recognize GPCR ligands with 71% accuracy. One key result of their work was the observation that a consensus result from several models trained over different descriptors was superior to a model trained over the same multiple descriptors. The models were based on perception, multilayer neural networks, and radial basis function networks. Overall accuracy was improved by joining multiple models together which were of different type and topology.

In addition to QSAR models developed for selecting GPCR ligand-likeness, there has been systematic structural analysis done to identify privileged substructures. Specifically, analysis[183] has shown that the 2-aryl-indole framework is associated frequently with potent ligands to the biogenic amine and peptide binding GPCR subfamilies.

In contrast to GPCRs ligand focus library selection, much less has been done with ion-channels with respect to focus libraries. Indeed, a PubMed search over "Ion Channel-focused Library" returns only one hit.[184] As with GPCRs, screening compound houses offer ion channel focus libraries with little publication or documentation of how they were created. In the end, ion-channel focus libraries can be assembled in the same manner as GPCR libraries. The process would start with developing a dataset to train over, which can be obtained from the same possible sources as used for GPCRs: proprietary data provided by subscription such as Aureus,[142] literature review and in-house data. Thereafter, any of the methods used for developing GPCR ligand target libraries[173,176,180–182] could, in principle, be applied. There are, however, two specific considerations that need to be addressed with respect to ion channel ligand focus library design: classification and avoidance of hERG liabilities.

Exactly what type of classification is needed to organize ion-channel focus libraries into coherent training sets for developing models is not easy to know without actually doing the science. Partitioning the datasets according to specific gene family subgroups (e.g., all TRP ligands in one group) is one possibility. Another possible means of attack would be to consider the range of molecular structures by analysis of population as functions of simple descriptors such as heavy atom, aromatic atom, aromatic nitrogen, nitrogen and oxygen atom counts. If there are obvious patterns, irrespective of any correlation with specific gene subgroups of ion channel proteins, it may be advantageous to consider developing multiple models specific for those ligands as a group.

It is an unfortunate possibility that design of an ion channel focus compound library may inadvertently design a library with the ability to interact with a critical ion channel: that encoded by the hERG related gene. Compounds which interact with the specific K^+ion channel coded by the hERG related gene are a critical, literally lethal, concern in pharmaceutical research. Indeed, terfenadine, astemizole, and grepafloxin were withdrawn from market when hERG liabilities were determined in patients and other therapeutics have had use restricted after determination of hERG liabilities—obviously unintended and expensive experiments which are to be avoided.[185–187] The expense of the patch clamp technique,[188,189] the most reliable assay for hERG liability, has stimulated interest in models to screen virtually by using computational methods to filter out compounds with hERG liabilities from the earliest stages prior to synthesis or purchase. There are large repositories of private data for hERG liabilities, the models presented by Hansen[190] used > 500 and Johnson[191] used more than 1000 proprietary compounds. Those wishing or needing to rely on public data can obtain data from the MDDR,[141] proprietary subscription databases[142] and publish reviews. The recent paper by Polak and coworkers[192] is of special quality due to the final > 700 collated compounds and careful comparison of effects of cell lines and/or other variables in the patch clamp assay. The example of previous practitioners is available[190–196] as is the insight that a basic amine associated with multiple aromatic or hydrophobic groups will frequently lead to a hERG liability. Given this, building a model for hERG liability is a reasonably defined process and should be a check of compounds intended for an ion channel ligand focus library.

The above is applicable to ion channel libraries assembled for either pharmaceutical or taste targets. However, taste target projects may desire compounds which are more in the comfort zone of food corporation interest. To satisfy this desire, the GRAS model previously[169,171] alluded to and/or especially the filter could be applied as a restriction. While this is not a guarantee for nontoxic compounds, the requirements for a GRAS like compound are so divergent from what is needed for a hERG ligand that it would help steer the compounds away from this liability.

B. Application Example

1. Development of a Sweetener Database and Use to Determine T1R2/T1R3 Ligands

All of the 177 compounds in our sweet compound database are measurably sweet. Most of the compounds from the Kim and Kinghorn[102] paper are used by local populations as traditional herbs while the artificials from the Walters et al.[101] have generally not been used outside the laboratory. Do all of these compounds bind the same site on the T1R2/T1R3 protein? Inspection of the structures, and molecular statistics (Figs. 2–8) offers some suggestions: do 5–7 heavy atom amino acids with simply perceivable sweetness bind in the same manner and place as 94–96 heavy atom telosmosides? Do the compounds with no aromatic atoms bind in the same manner and place as the compounds with 12 or even 48 aromatic atoms? However, we do not need to guess, the clear evidence that they do not is provided by synergestic measurements by Schiffman.[197]

Though this dataset is diverse, it is still redundant and there are subgroups present. Specifically, if the dataset is clustered in such a manner as that all compounds equal or more similar than 0.85 Tanimoto correlation coefficient (TCC) with MACCS fingerprints,[126] then there are 101 clusters which fall out by consequence of the redundancy in the such groups as the albiziasaponins and telomosides. A significance of this is that it illustrates a principle[127,128] that compounds equal or more similar than 0.85 TCC have a good chance of having the same activities. In this case, the compounds illustrate that if *A* is sweet and *B* is similar to *A*, then *B* has a good chance of being sweet as well. *B* also has the chance of binding (as *A* does) without producing a sweet sensation. In this case, though *B* has no taste, *B* could be inhibitory to *A* or possibly synergestic to other compounds which are noncompetitive against A. A concrete example of this is the sweetness inhibitor ziziphin (See Fig. 9 for structure), which falls into a cluster with the albiziasaponins (see Fig. 10)—this was found by doing a search with a saponin as a reference. This situation leads one to ask does ziziphin bind at the same place as the saponins but somehow alter T1R2/T1R3 in some manner that its ability to detect even small molecule nutritive sweeteners is altered? We do not have concrete data to answer this.

The aforementioned model to recognize GRAS and pharmaceuticals[169] was applied to the database of sweeteners. Sixteen compounds were recognized as GRAS by the model and all the remaining were classified as pharmaceutical. Three compounds from the artificial list were classified as GRAS. These compounds were the oxygen rich naringin/neohesperidin dihydrochalcone and the small aromatic atom poor compound cyclamate. The remainder of the compounds classified as GRAS were compounds from the Kim and Kinghorn[102] review. All other 54 artificial sweeteners—saccharin, sucralose, aspartame, sucronic acid and so forth—clearly were classified as pharmaceuticals by the model. Careful inspection of Figs. 2–8 provide a good deal of insight into why this classification occurred.

Figure 2 presents population as a function of heavy atom count. Prestwick has a broader and higher reaching distribution than GRAS. Where the Prestwick dataset is platykurtotic vs. the GRAS, the natural sweeteners are polykurtotic and reach to even larger sizes than GRAS ever does. The natural sweeteners reach beyond the customary boundary for GRAS compounds seen by analysis. The artificial sweeteners are polykurtotic as well, but are closer to the range seen for the Preswick set.

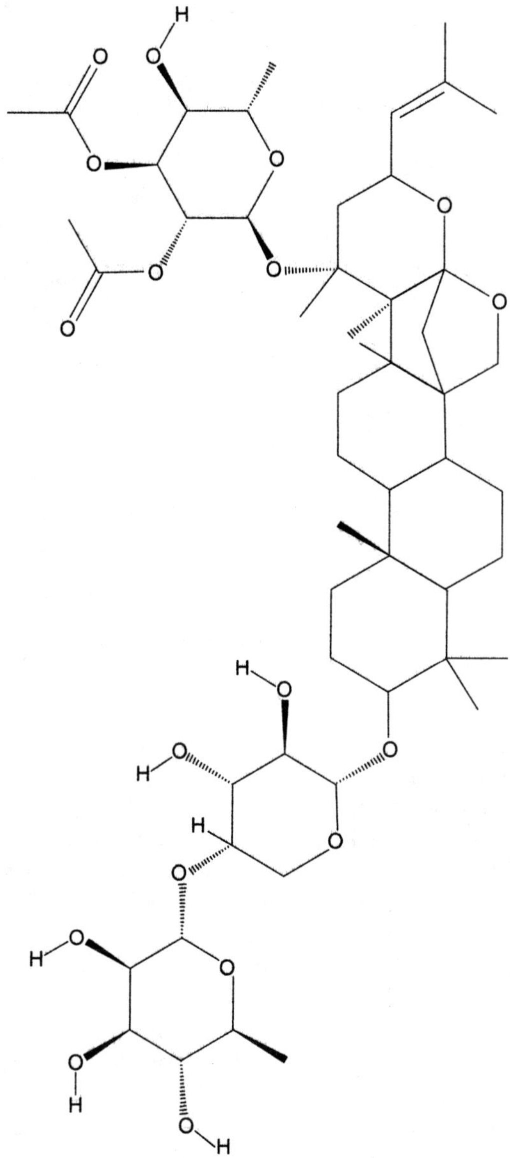

FIG. 9. Structure of sweet taste inhibitor ziziphin.

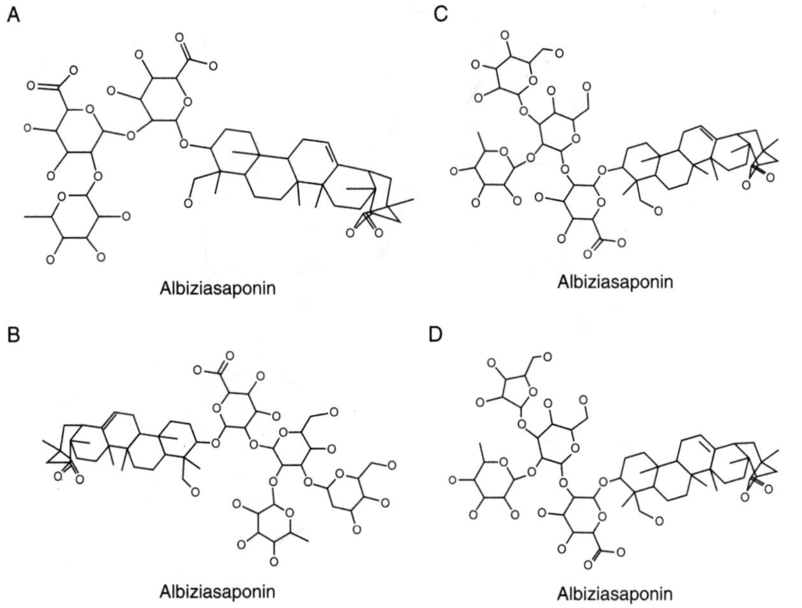

FIG. 10. Structures of the albiziasaponins.

The difference between GRAS and pharmaceuticals with respect to aromatic atom count is the difference between very rarely having fused and two aromatic rings for GRAS compounds and frequently having this situation for pharmaceuticals (Fig. 3). The sweet naturals have a different distribution than either GRAS or pharmaceutical. It is heavy to compounds with no aromatic atoms but does have a noted population with two aromatic rings. This population at 12 aromatic atoms is in a region where pharmaceuticals have some population and where GRAS compounds have less. Hence, it favors the QSAR assigning against GRAS for the natural sweet compounds in general when they have this feature. The sweet artificials have a distribution that differs only in detail to the pharmaceuticals in terms of exact relative heights of the peaks at 0, 5–6 and 12 aromatic atoms. Compounds with these features are easily accepted by the QSAR to be pharmaceutical-like.

The distribution patterns for nitrogens are simple (Fig. 4). The sweet naturals clearly follow the pattern of the GRAS compounds and the sweet artificials, though bimodal, clearly follow the pattern of the Prestwick database. Nitrogen is frequently associated with base which has a bitter flavor. In the case of the artificial sweeteners, nitrogen is embedded in complex structures which may be part of the off taste of artificial sweeteners but are also essential for the high potency.

Acceptors are presented in Fig. 5. The sweet naturals are noted by a broad distribution of acceptor counts and donor counts. This appearance is distinct from either the Prestwick or GRAS datasets. These datasets differ in that GRAS is more leptokurtotic. The sweet artificials are somewhat intermediate between GRAS and Prestwick. Most of what was written above for acceptors can be written for the donors. A key difference though in the distribution is that GRAS compounds favor no donor at all so stating a degree of leptokurtosis would be somewhat misleading. As was mentioned before, base (and the highly related donor count seen in Fig. 6) are associated with bitter taste, and hence the shift to high populations at low counts is expected for donors. The sweet artificials have a distribution close to that of the Prestwick dataset.

Halogens are not especially common for Prestwick compounds (Fig. 7). However, they are effectively unknown for GRAS compounds. The sweet naturals are perfectly matched in having no halogens versus GRAS while the artificial sweeteners have a halogen count very similar overall to pharmaceuticals. Specifically, both are approximately 80% with no halogens while the remainder is distributed in distinct means. Specifically, several artificial sweeteners have CF3 as a group which produces a characteristic peak where the Prestwick dataset simply slowly drops from slightly less than 10% population with a single halogen to no population with any halogens after 5.

In Table I, eight redundant cores from the natural sweeteners are listed as SMILES strings. These were queried and counted. In Fig. 8, black is the population for compounds without the cores and gray is the population for compounds with the cores. A small percentage of Prestwick compounds have these cores where the GRAS dataset lacks any representation for these cores. Almost all the natural product sweeteners have one of these cores where none of the artificial sweeteners have these cores.

Careful inspection of Figs. 2–8 supports the behavior that the QSAR marks the artificials as pharmaceuticals. The set's distributions are within the ranges seen for the pharmaceutical Prestwick dataset. Figures 2–8, however, suggest that the natural sweeteners are beyond the acceptable ranges for either the pharmaceutical dataset or the GRAS dataset. Simply put, the natural sweeteners are typically too large to be considered GRAS, though their halogen and N profiles and tendency to oxygen richness are similar. This is not to say these could not be determined GRAS-natural sweeteners do not look statistically like that which was GRAS in 2007 when the model was made.

Knowing beforehand the diversity in the sweetener dataset (Figs. 2–8), it would be unwise to attempt to develop a rule or QSAR over the whole dataset. Applying the information provided by Schiffman[197] is a possible option to develop a means of grouping the compounds into sets which logically have common binding sites. This is, however, tedious and leads to questions of what is truly synergestic by acting through independent sites versus merely additive.

Given this situation, we would choose in this case to perform searches using similarity and MACCS fingerprints.[126] Again, the objective in this search is to create a library rich in T1R2/T1R3 binding ligands. They may be sweet, inhibitory to sweet (as example ziziphin illustrates) or possibly synergistic without inherent sweetness.

Given the pharmaceutical-likeness appearance of most of the artificials, searching in the usual screening house catalogs[147–156] would be a good strategy relative to these 57 compounds. Exact setting used with the similarity search would vary dependent on the expense of the final assay of interest. Application of searches with Tanimoto set to 0.85 across the three million compound catalog assembled in-house returns ~1200 hits. Some of the hits are known to be sweet including derivatives of such compounds as P4000 and naringin/neohesperidin dihydrochalcone. Loosening this value will result in more hits but they will be progressively more divergent than the training set. An animal-model-based assay could handle possible several dozen to hundreds of compounds. A cell-based assay could handle several thousands. Regardless of the type of assay involved, the screen will need to be constructed to search for sweet compounds, sweet suppressing compounds and sweet enhancing compounds to achieve its true scientific and commercial potential.

The natural sweeteners present a problem for database searching. The previous 15 years of screening library design has been spent trying to eliminate compounds that are remotely similar to those due to poor potential oral bioavailability. The review performed by Baurin[157] is clear that compounds beyond MW 500 are rare in the commercial catalog space. However, there are several natural product catalogs[198–202] which do have compounds which look like the statistics of the natural product sweeteners though they are far from druglike.[203] Searches at Tanimoto 0.85 across these databases leads to a dataset of ~500 compounds. Beyond the issues of the assay, the cost of natural products must include a consideration of the purification process and these compounds are significantly more expensive than compounds from more traditional sources.[147–156] An advantage of these compounds though is that they have little possibility of oral bioavailability, which for a taste program would be a positive characteristic.

2. Development of a TRP Channel Ligand Database and Library

The same chemoinformatic process for developing GPCR libraries with select properties can be applied to a TRPM5 discovery program. With very few exceptions, the structures of ligands that have been discovered for modulation of TRPM5 are not yet publically available. Starting with compounds that we have identified from earlier screening efforts, we currently are using a chemoinformatic approach to discover novel and optimized TRPM5 ligands. Due to sensitivities concerning intellectual property over the structures of the discovered TRPM5 structures at this early stage, we cannot directly describe our strategy for TRPM5. However, the basic principles of our approach can be exemplified by using a related target, the TRPV1 ion channel.

Figure 11 lists a panel of 17 TRPV1 ligands.[204–208] Thirteen are the results of pharmaceutical research while the remaining six are natural products (capsacin analogs) responsible for the burning sensation associated with

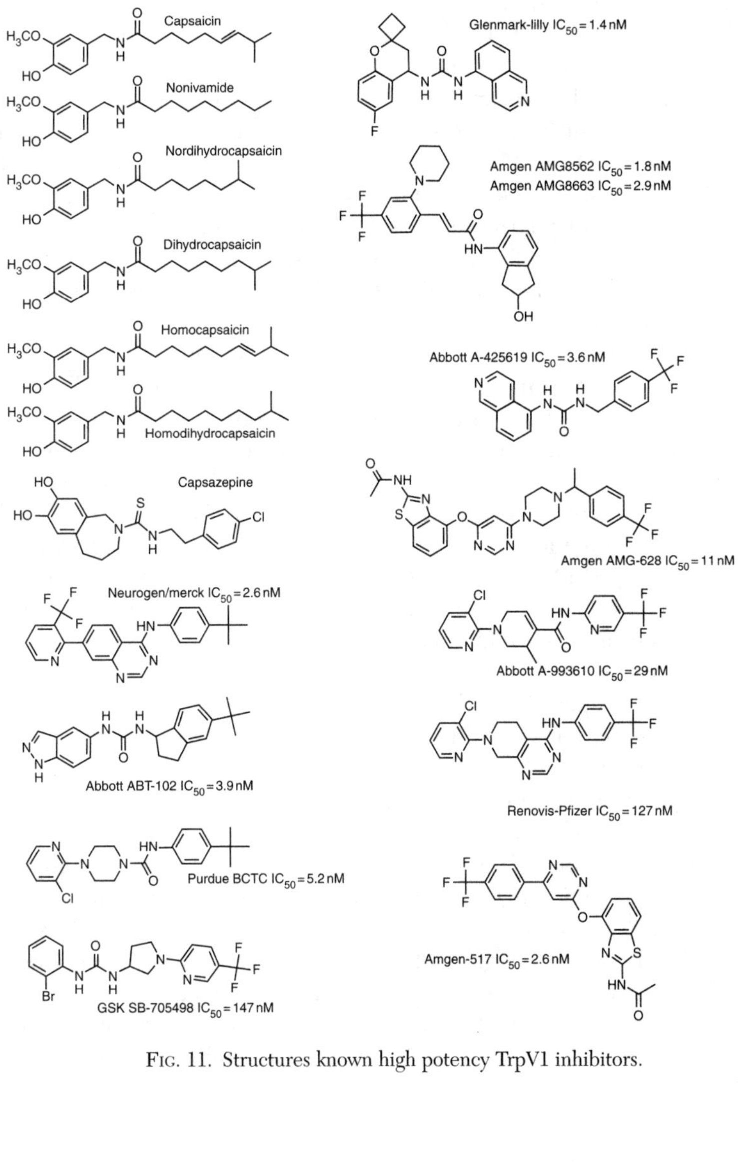

FIG. 11. Structures known high potency TrpV1 inhibitors.

the pepper plant (genus Capsicum). The natural products show obvious GRAS-like features with the presence of oxygen and the absence of nitrogen or halogens and low aromatic content where the pharmaceuticals are clearly not GRAS-like. Since this is a pain target, the issue of GRAS is merely a remark; the overriding issue here is the desire for traditional drug-like properties to support oral bioavailability. All of the pharmaceuticals are antagonists of capsacin and its "hot" analogs. When this dataset is clustered at a Tanimoto correlation coefficient of 0.85, all the natural products fall into one cluster; the isomers AMG-517 and AMG-8562 form a second cluster and every other compound is a singleton cluster to itself. Capsazipine, which is described as an analog to capsacin, does not cluster with capsacin. Given the high similarity of the natural products, we can assume that they act on the same binding site of the TRPV1 protein. The same logic holds that AMG-517 and AMG-8562 do as well. It is not a guarantee that capsacin, the natural product analogs and pharmaceutical research compounds all operate at the same site on TRPV1. Unlike the case with respect to the sweeteners, there is no detailed study of synergistic effects to clarify if we are looking at a single site or multiple sites.

The first operation we would consider for this dataset is to use it as the basis for a similarity search. Since we are searching using a dataset which is a mix of agonist and antagonist, we are certain to get both in the resultant hit list. Requiring a compound match at 0.85 Tanimoto correlation coefficient to the reference dataset, we search across our in-house three million compound catalog database and retrieve 527 electronic hits. These are obtainable at about $45 per 10 mg and could be ordered and screened for TRPV1 agonist and antagonist providing a rich SAR for further analysis. While antagonists have the obvious pharmaceutical value in the treatment of pain, knowing that an agonist and antagonist may be closely related structurally is naturally a useful SAR datum.

After the compounds had been assayed, the next desire would be to improve the compounds by optimizing the mix for assayed potency, properties which favor oral bioavailability and patent position. In this specific case, one could construct an alignment and a pharmacophore, as has been done elsewhere.[209,210] In the case of the work done by Kym et al.,[209] a rather diverse set of compounds was placed in an alignment with a high degree of compound overlap. Whether this alignment is correct or not is difficult to be certain. Cheung et al.[210] focuses on a single series of compounds where there is greater certainty for a common alignment. In both the referenced cases, the referenced authors are using these alignments and pharmacophores as post-hoc exercises to rationalize data. In our case, we would be constructing alignments and pharmacophores to be used as a query for determining new compounds for assay.

As an example, we align capsaicin, BCTC and A-993610 (Fig. 11A). Thereafter, a pharmacophore was created comprised of the features outlined in Table II and visualized in Fig. 11B. All features are present in all three molecules and all features are required for a match. The search was executed over a 3D dataset of ~1 million lead-like molecules provided as with the MOE release.[106] A total of > 10 thousand hits were retrieved from this exercise. This being an unmanageable number, we applied a restriction that VDW_AREA (2D approximate as calculated by MOE) be between 325 and 425 since capsaicin, BCTC and A-993610 have values of VDW_AREA of 353, 356 and 401 Å^2, respectively. This limit puts us at a more manageable set of 3 thousand hits. This could be paired down if desired by further restrictions such as, for example, by requiring a minimal common volume with the alignment in Fig. 12A. A random set of 12 from this list are shown in Fig. 12C (as part of the alignment and pharmacophore) and in Fig. 13. Note that though a donor–acceptor pairing is conserved and relative sequential positioning of hydrophobes and aromatics, the exact structures are chemically distinct.

The search methods outlined above have different, complimentary utility. Similarity searches return compounds which will be structurally very close to the reference compounds and allow a thorough investigation of the structure–activity landscape around a given reference whether it is obtained from the literature or from an in-house screen. A key advantage is that similarity searches do not require an alignment to proceed. A key disadvantage is that

TABLE II
RADIUS (IN Å) FOR A SUCCESSFUL MATCH BETWEEN THE PHARMACOPHORE MODEL AND A POTENTIAL ELECTRONIC HIT FOR THE VARIOUS PHARMACOPHORE POINTS

Radius	Type
0.59	Acceptor
0.70	Donor
0.70	Aromatic π cloud
0.90	Aromatic or hydrophobic group
1.00	Donor extension point
1.25	Hydrophobic group
2.10	Aromatic or hydrophobic group

the results will typically be structurally close to the references and in the case of compounds starting from patent literature, the chance of finding compounds exterior to the patent is small.

Pharmacophore searching, by contrast, allows one to break out of a given space by finding compounds which share features without necessarily sharing chemical structure to a high degree. However, a pharmacophore requires an alignment, which in turn requires molecules which are logically alignable. While the molecules shown in Fig. 11 can be aligned, it is not obvious if they should be aligned and doing so will produce several molecules with outlying structural features. However, several subsets can easily be assembled which can be aligned with relative ease and used as the basis for a search.

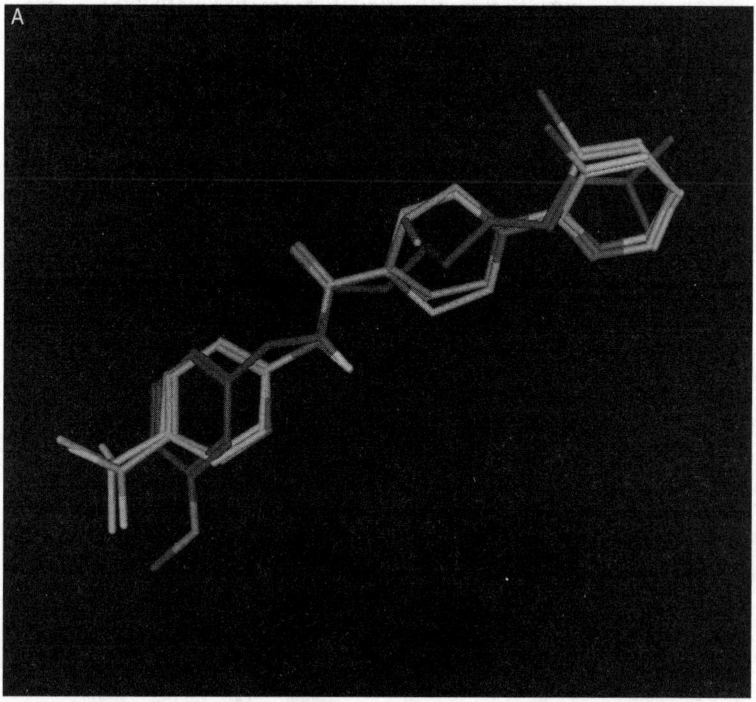

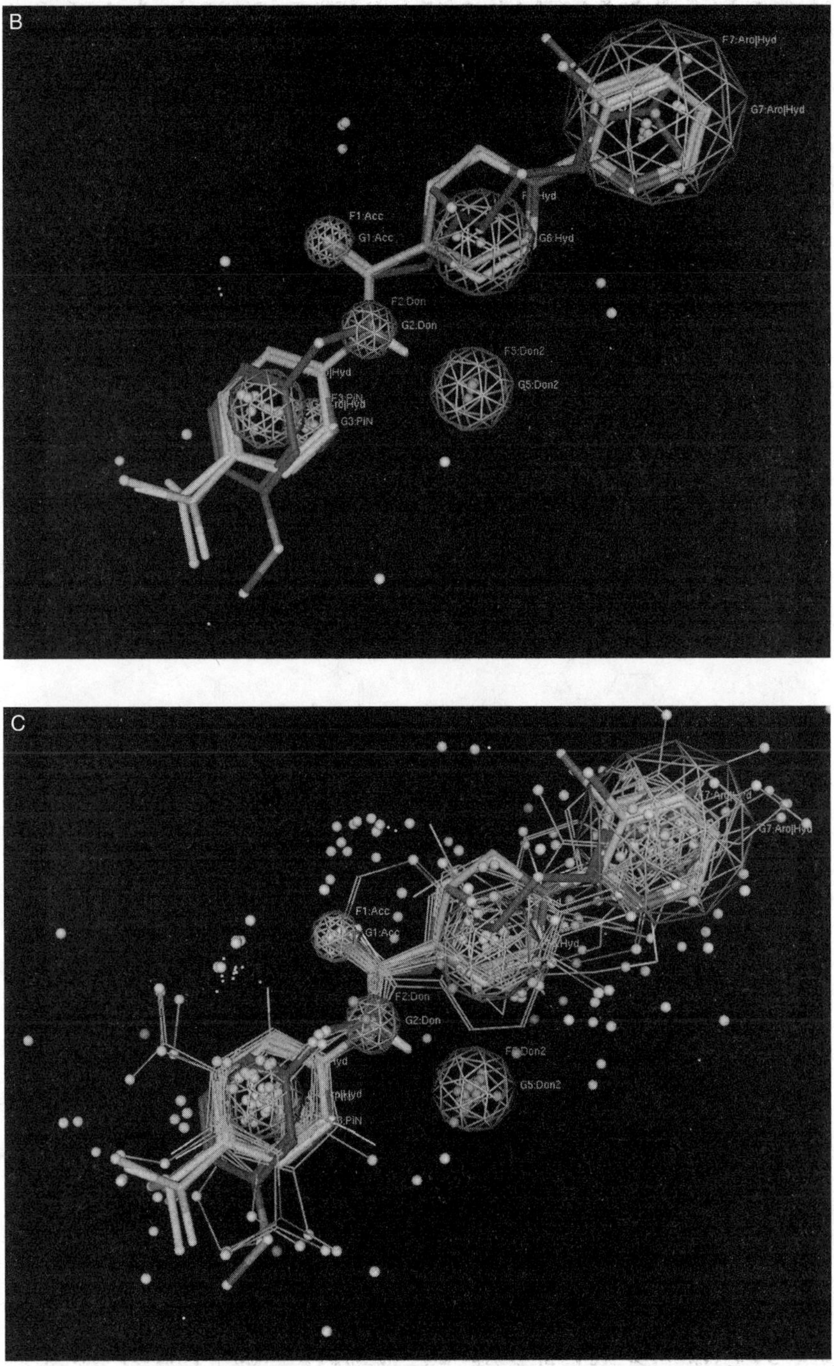

FIG. 12. (A) Alignment of capsaicin, BCTC and A-993610. (B) Pharmacophore around alignment, donor, acceptors and hydrophobe points are labeled on image (Don, hydrogen bond donor; Acc, hydrogen bond acceptor; Aro, aromatic group center; Hyd, hydrophophic group center; PiN, π cloud center. Presence of a "2"—Don2—signifies an extension point rather than a on the compound donor or acceptor). (C) Some of the better hits from a pharmacophore search performed with the model.

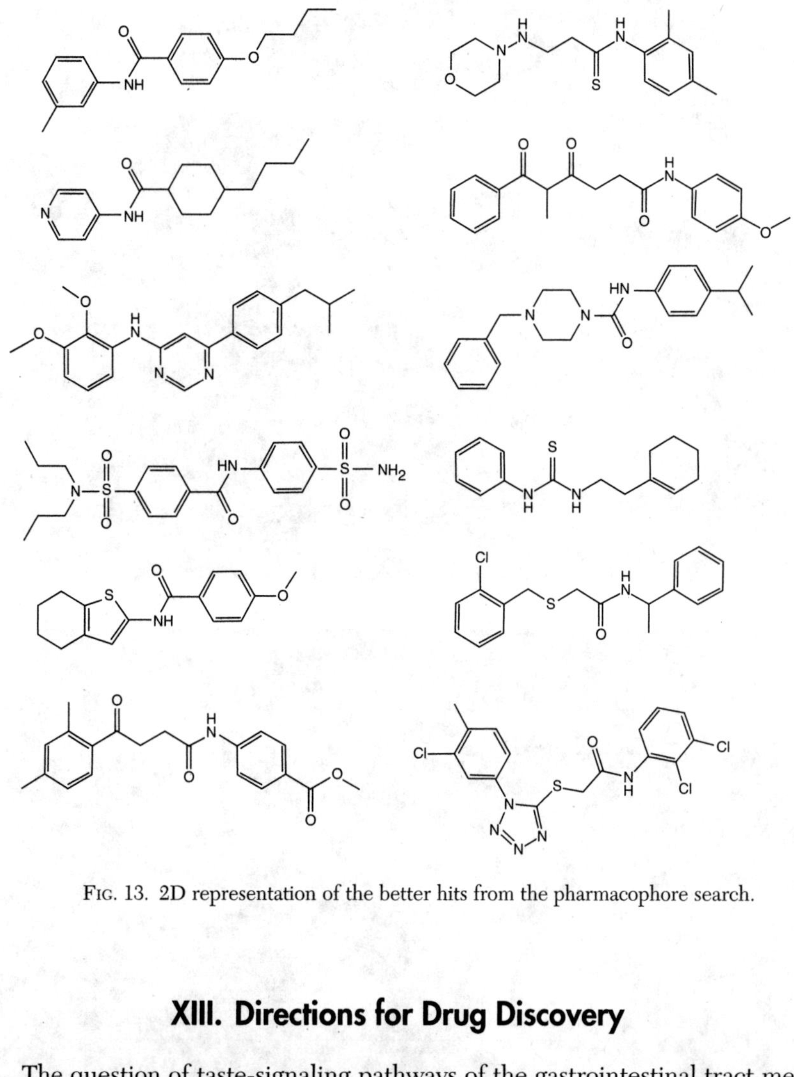

FIG. 13. 2D representation of the better hits from the pharmacophore search.

XIII. Directions for Drug Discovery

The question of taste-signaling pathways of the gastrointestinal tract mediating responses to ingested sugars or other sweeteners begs for further research. Further studies challenging the physiology of TRPM5−/− mice are eagerly awaited and will undoubtedly point the way to other important functions of TRPM5. Attempts to elucidate the roles for either sweet receptors or TRPM5 expressed in the gastrointestinal tract are in the earliest stages and have not yet provided information sufficient to reach consensus. However, momentum seems to be gaining that favors a role for TRPM5 in secretory processes of enteroendocrine cells or pancreatic beta cells, and the evidence is sufficient to at least consider strategies for drug discovery aimed at TRPM5 as a therapeutic target. The recent evidence available from knock-out mouse studies might lead to the conclusion that pharmacologic enhancers of TRPM5 activity should be sought for therapeutic purposes. However, constitutive knock-out mice have been without TRPM5 since inception and the animal has developed and matured in ways that could have compensated for the lost gene. Caution in predicting the desired pharmacologic activity in otherwise normal individuals is warranted. Although a knock-out phenotype can implicate a protein in a specific physiological function or its place in a disease state, the need for an agonist versus an inhibitor should not be determined until the ligands are tested in appropriate primary cell assays or *in vivo* models using wild type animals. Identification of high potency, selective ligands with good bioavailability certainly will aid in the validation of TRPM5 as a target for metabolic diseases or other dysfunctions related to ingestive behaviors.

Though evidence has been offered supporting the notion that sweet receptors of the gut might be involved in nutrient handling and metabolism, at this early stage the jury is still out. Ultimately the question will be answered by establishing pharmacologic profiles of secretion or other gastrointestinal activities that compare with the known pharmacology of the sweet receptor. Potent, selective T1R2/T1R3 antagonists will be helpful additions to the pharmacologic tool box. If the hypothesis of sweet receptor signaling in the gut is upheld, then T1R2/T1R3 potentially could assume a new identity as a therapeutic target for drug discovery programs.

References

1. Society for the Study of Ingestive Behavior. Society for the Study of Ingestive Behavior. http://www.ssib.org/web; 2010. Last accessed 12-Feb-2010.
2. Nair RP, Ren J. Pharmacotherapy of obesity—benefits, bias and hyperbole. *Curr Med Chem* 2009;**16**:1888–97.
3. Palmer RK. The pharmacology and signaling of bittr, sweet, and umami taste sensing. *Mol Interv* 2007;**7**:87–98.
4. Li X, Servant G. Functional characterization of the human sweet receptor: high-throughput screening assay development and structural function relation. In: Weerasinghe DK, Dubois G, editors. *Sweetness and sweeteners: biology, chemistry and psychophysics*. Washington, USA: Oxford University Press; 2008. p. 368–385, Vol. 979.
5. Zhang F, Klebansky B, Fine RM, Xu H, Pronin A, Liu H, et al. Molecular mechanism for the umami taste synergism. *Proc Natl Acad Sci USA* 2008;**105**:20930–4.
6. Zhang Y, Hoon M, Chandrashekar J, Mueller K, Cook B, Wu D, et al. Coding of sweet, bitter, and umami tastes, different receptor cells sharing similar signaling pathways. *Cell* 2003;**112**:293–301.
7. Center for Disease Control and Prevention. Obesity and overweight for professionals. http://www.cdc.gov/obesity/data/trends.html; 2010. Last accessed 12-Feb-2010.

8. Mokdad AH, Ford ES, Bowman BA, Dietz WH, Vinicor F, Bales VS, et al. Prevalence of obesity, diabetes, and obesity-related health risk factors, 2001. *JAMA* 2003;**289**:76–9.
9. Popkin BM, Nielsen SJ, Siega-Riz AM. Trends in energy intake in U.S. between 1977 and 1996: similar shifts seen across age groups. *Obes Res* 2002;**10**:370–8.
10. Steenhuis IH, Vermeer WM. Portion size: review and framework for interventions. *Int J Behav Nutr Phys Act* 2009;**6**:58.
11. Malik VS, Schulze MB, Hu FB. Intake of sugar-sweetened beverages and weight gain: a systematic review. *Am J Clin Nutr* 2006;**84**:274–88.
12. Bellisari A. Evolutionary origins of obesity. *Obes Rev* 2008;**9**:165–80.
13. Glendinning J, Spector A. A high-throughput screening procedure for identifying mice with aberrant taste and oromotor function. *Chem Senses* 2002;**27**:461–74.
14. Collier GH. Satiety: an ecological perspective. *Brain Res Bull* 1985;**14**:693–700.
15. Ello-Martin JA, Ledikwe JH, Rolls BJ. The influence of food portion size and energy density on energy intake: implications for weight management. *Am J Clin Nutr* 2005;**82**(1 Suppl.):S236–41.
16. Lieberman LS. Evolutionary and anthropological perspectives on optimal foraging in obesogenic environments. *Appetite* 2006;**47**(1):3–9.
17. Popkin BM, Nielsen SJ. The sweetening of the world's diet. *Obes Res* 2003;**11**:1325–32.
18. Brownell KD, Farley T, Willett WC, Popkin BM, Chaloupka FJ, Thompson JW, et al. The public health and economic benefits of taxing sugar-sweetened beverages. *N Engl J Med* 2009;**361**:1599–605.
19. Putnam J, Allshouse J, Kantor LS. U.S. per capita food supply trends: more calories, refined carbohydrates, and fats. *Economic Research Service, USDA, Food Rev* 2002;**25**(Issue 3):2–15.
20. White JS. Straight talk about high-fructose corn syrup: what it is and what it ain't. *Am J Clin Nutr* 2008;**88**:236S1716–21.
21. Duffy KJ, Popkin BM. High-fructose corn syrup: is this what's for dinner? *Am J Clin Nutr* 2008;**88**:S1722–32.
22. Bray GA. Fructose: should we worry? *Int J Obes* 2008;**32**:S127–31.
23. Sharon P, Fowler SP, Ken Williams K, Resendez RG, Kelly J, Hunt KJ, et al. Fueling the obesity epidemic? Artificially sweetened beverage use and long-term weight gain. *Obesity* 2008;**16**:1894–900.
24. Bellisle F, Drewnowski A. Intense sweeteners, energy intake and the control of body weight. *Eur J Clin Nutr* 2007;**61**:691–700.
25. Swithers SE, Davidson TL. A role for sweet taste: calorie predictive relations in energy regulation by rats. *Behav Neurosci* 2008;**122**:161–73.
26. Swithers SE, Baker CR, Davidson TL. General and persistent effects of high intensity sweeteners on body weight gain and caloric comensation in rats. *Behav Neurosci* 2009;**123**:772–80.
27. Egan JM, Margolskee RF. Taste cells of the gut and gastrointestinal chemosensation. *Mol Interv* 2008;**8**:78–81.
28. Rozengurt E, Sternini C. Taste receptor signaling in the mammalian gut. *Curr Opin Pharmacol* 2007;**7**:557–62.
29. Sternini C, Anselmi L, Rozengurt E. Enteroendocrine cells: a site of 'taste' in gastrointestinal chemosensing. *Curr Opin Endocrinol Diab Obes* 2008;**15**:73–8.
30. Grigoriadis DE, Hoare SRJ, Lechner SM, Slee DH, Williams JA. Drugability of extracellular targets: discovery of small molecule drugs targeting allosteric, functional, and subunit-selective sites on GPCRs and ion channels. *Neuropsychopharmacology* 2009;**34**:106–25.
31. Bryant RW, Atwal KS, Bakaj I, Buber MT, Carlucci S, Cerne R, et al. Development of transient receptor potential melanostatin 5 modulators for sweetness enhancement. In:

Weerasinghe G, Dubois GE, editors. *Sweetness and sweeteners: biology, chemistry and psychophysics*. Washington, USA: Oxford University Press; 2008. p. 386–399, Vol. 979.
32. Nelson G, Hoon MA, Chandrashekar J, Zhang Y, Ryba NJ, Zuker CS. Mammalian sweet taste receptors. *Cell* 2001;**106**:381–90.
33. Temussi PA. The sweet taste receptor: a single receptor with multiple sites and modes of interaction. *Adv Food Nutr Res* 2007;**53**:199–239.
34. Cui M, Jiang P, Maillet E, Max M, Margolskee RF, Osman R. The heterodimeric sweet taste receptor has multiple potential ligand binding sites. *Curr Pharm Des* 2006;**12**:4591–600.
35. Muller GW, Walters DE, DuBois GE. N, N'-disubstituted guanidine high-potency sweeteners. *J Med Chem* 1992;**35**:740–3.
36. Zhao G, Zhang Y, Hoon M, Chandrashekar J, Erlenbach I, Ryba N, et al. The receptors for mammalian sweet and umami taste. *Cell* 2003;**115**:255–66.
37. Jiang P, Ji Q, Liu Z, Snyder L, Bernard L, Margolskee R, et al. The cysteine-rich region of the T1R3 determines responses to intensely sweet proteins. *J Biol Chem* 2004;**279**:45068–75.
38. Xu H, Staszewski L, Tang H, Adler E, Zoller M, Li X. Different functional roles of T1R subunits in the heteromeric taste receptors. *Proc Natl Acad Sci USA* 2004;**101**:14258–63.
39. Nie Y, Vigues S, Hobbs J, Conn G, Munger S. Distinct contributions of T1R2 and T1R3 taste receptor taste receptor subunits to the detection of sweet stimuli. *Curr Biol* 2005;**15**:1948–52.
40. Temussi PA. Sweet, bitter and umami receptors. *Trends Biochem Sci* 2009;**6**:296–302.
41. Morini G, Bassoli A, Temussi PA. From small sweeteners to sweet proteins: anatomy of the binding sites of the human T1R2–T1R3 receptor. *J Med Chem* 2005;**48**:5520–9.
42. Schiffman SS, Booth BJ, Carr BT, Losee ML, Sattely-Miller EA, Graham BG. Investigation of synergism in a binary mixture of sweeteners. *Brain Res Bull* 1995;**38**:105–20.
43. Roper SD. Signal transduction and information processing in mammalian taste buds. *Pflugers Arch* 2007;**454**:759–76.
44. Dyer J, Salmon KSH, Zibrik L, Shirazi-Beechey SP. Expression of sweet taste receptors of the T1R family in the intestinal tract and enteroendocrine cells. *Biochem Soc Trans* 2005;**33**:302–5.
45. Mace OJ, Affleck J, Patel N, Kellett GL. Sweet taste receptors in rat small intestine stimulate glucose absorption through apical GLUT2. *J Physiol* 2007;**582**:379–92.
46. Bezençon C, Fürholz A, Raymond F, Mansourian R, Métairon S, Le Coutre J, et al. Murine intestinal cells expressing Trpm5 are mostly brush cells and express markers of neuronal and inflammatory cells. *J Comp Neurol* 2008;**509**:514–25.
47. Young RL, Sutherland K, Pezos N, Brierley SM, Horowitz M, Rayner CK, et al. Expression of taste molecules in the upper gastrointestinal tract in humans with and without type 2 diabetes. *Gut* 2009;**58**:337–46.
48. Margolskee RF, Dyer J, Kokrashvili Z, Salmon KSH, Ilegems E, Daly K, et al. T1R3 and gustducin in gut sense sugars to regulate expression of Na-glucose cotransporter 1. *Proc Natl Acad Sci USA* 2007;**104**:15075–80.
49. Jang HJ, Kokrashvili Z, Theodorakis MJ, Carlson OD, Kim BJ, Zhou J, et al. Gut-expressed gustducin and taste receptors regulate secretion of glucagon-like peptide-1. *Proc Natl Acad Sci USA* 2007;**104**:15069–74.
50. Bezencon C, le Coutre J, Damak S. Taste-signaling proteins are coexpressed in solitary intestinal epithelial cells. *Chem Senses* 2007;**32**:41–9.
51. Damann N, Voets T, Nilius B. TRPs in our senses. *Curr Biol* 2008;**18**:R880–9.
52. Nilius B, Owsianik G, Voets T, Peters JA. Transient receptor potential cation channels in disease. *Physiol Rev* 2007;**87**:165–217.
53. Gaudet R. TRP channels entering the structural era. *J Physiol* 2008;**586**:3565–75.
54. Fleig A, Trends PennerR. The TRPM ion channel subfamily: molecular, biophysical and functional features. *Pharmacol Sci* 2004;**25**:633–9.

55. Jiang L-H. Subunit interaction in channel assembly and functional regulation of transient receptor potential melastatin (TRPM) channels. *Biochem Soc Trans* 2007;**35**:86–8.
56. Prawitt D, Monteilh-Zoller M, Brixel L, Spangenberg C, Zabel B, Fleig A, et al. TRPM5 is a transient Ca2+-activated cation channel responding to rapid changes in $[Ca^{2+}]_i$. *Proc Natl Acad Sci USA* 2003;**100**:15166–71.
57. Hofman T, Chubanov V, Gudermann T, Montell C. TRPM5 is a voltage-modulated and Ca^{2+}-activated monovalent selective cation channel. *Curr Biol* 2003;**13**:1153–8.
58. Liu D, Liman E. Intracellular Ca2+ and the phospholipid PIP2 regulate the taste transduction ion channel TRPM5. *Proc Natl Acad Sci USA* 2003;**100**:15160–5.
59. Perez C, Huang L, Rong M, Kozak J, Preuss A, Zhang H, et al. Transient receptor potential channel in taste receptor cells. *Nat Neurosci* 2002;**5**:1169–76.
60. DeFazio R, Dvoryanchikov G, Maruyama Y, Kim J, Pereira E, Roper S, et al. Separate populations of receptor cells and presysnaptic cells in mouse taste buds. *J Neurosci* 2006;**26**:971–3980.
61. Fonfria E, Murdock P, Cusdin F, Benham C, Kelsell R, McNulty S. Tissue distribution profiles of the human TRPM cation channel family. *J Recept Signal Transduct Res* 2006;**26**:159–78.
62. Chen MC, Wu SV, Reeve Jr JR, Rozengurt E. Bitter stimuli induce Ca^{2+} signaling and CCK release in enteroendocrine STC-1 cells: role of L-type voltage-sensitive Ca^{2+} channels. *Am J Physiol Cell Physiol* 2006;**291**:C726–39.
63. Clapp TR, Medler KF, Damak S, Margolskee RF, Kinnamon SC. Mouse taste cells with G protein-coupled taste receptors lack voltage-gated calcium channels and SNAP-25. *BMC Biol* 2006;**4**:7–16.
64. Gao N, Lu M, Echeverri F, Laita B, Kalabat D, Williams ME, et al. Voltage-gated sodium channels in taste bud cells. *BMC Neurosci* 2009;**10**:20.
65. Romanov R, Kolesnikov S. Electrophysiologically identified subpopulations of taste bud cells. *Neurosci Lett* 2006;**395**:249–54.
66. Sainz E, Cavenagh MM, LopezJimenez ND, Gutierrez JC, Battey JF, Northup JK, et al. The G-protein coupling properties of the human sweet and amino acid taste receptors. *Dev Neurobiol* 2007;**67**:948–59.
67. Yarmolinsky DA, Zuker CS, Ryba NJ. Common sense about taste: from mammals to insects. *Cell* 2009;**139**:234–44.
68. Huang Y, Maruyama Y, Dvoryanchikov G, Pereira E, Chaudhari N, Roper S. The role of pannexin-1 hemichannels in ATP release and cell-cell communication in mouse taste buds. *Proc Natl Acad Sci USA* 2007;**104**:6436–41.
69. Tolhurst G, Reimann F, Gribble FM. Nutritional regulation of glucagon-like peptide-1 secretion. *J Physiol* 2009;**587**:27–32.
70. Sbarbati A, Osculati F. A new fate for old cells: brush cells and related elements. *J Anat* 2005;**206**:349–58.
71. Kokrashvili Z, Rodriguez D, Yevshayeva V, Zhou H, Margolskee RF. Release of endogenous opioids from duodenal enteroendocrine cells requires TrpM5. *Gastroenterology* 2009;**137**:598–606.
72. Kim W, Egan JM. The role of incretins in glucose homeostasis and diabetes treatment. *Pharmacol Rev* 2008;**60**:470–512.
73. Gauthier BR, Wollheim CB. Synaptotagmins bind calcium to release insulin. *Am J Physiol Endocrinol Metab* 2008;**295**:E1279–86.
74. Bertram R, Sherman A, Satin LS. Metabolic and electrical oscillations: partners in controlling pulsatile insulin secretion. *Am J Physiol Endocrinol Metab* 2007;**293**:E890–900.
75. Reimann F, Habib AM, Tolhurst G, Parker HE, Rogers GJ, Gribble FM. Glucose sensing in L cells: a primary cell study. *Cell Metab* 2008;**8**:532–9.

76. Reimann F, Gribble FM. Glucose-sensing in glucagon-like peptide-1—secreting cells. *Diabetes* 2002;**51**:2757–63.
77. Thorens B. GLUT2 in pancreatic and extra-pancreatic gluco-detection (review). *Mol Membr Biol* 2001;**18**:265–73.
78. Henquin JC. Regulation of insulin secretion: a matter of phase control and amplitude modulation. *Diabetologia* 2009;**52**:739–51.
79. Gribble FM, Williams L, Simpson AK, Reimann F. A novel glucose-sensing mechanism contributing to glucagon-like peptide-1 secretion from the GLUTag cell line. *Diabetes* 2003;**52**:1147–54.
80. Hirasawa A, Tsumaya K, Awaji T, Katsuma S, Adachi T, Yamada M, et al. Free fatty acids regulate gut incretin glucagon-like peptide-1 secretion through GPR120. *Nat Med* 2005;**11**:90–4.
81. Overton HA, Babbs AJ, Doel SM, Fyfe MC, Gardner LS, Griffin G, et al. Deorphanization of a G protein-coupled receptor for oleoylethanolamide and its use in the discovery of small-molecule hypophagic agents. *Cell Metab* 2006;**3**:167–75.
82. Overton HA, Fyfe MCT, Reynet C. GPR119, a novel G-protein-coupled receptor target for the treatment of type 2 diabetes and obesity. *Br J Pharmacol* 2008;**153**:S76–81.
83. Reimann F, Ward PS, Gribble FM. Signalling mechanisms underlying the release of glucagon-like peptide-1. *Diabetes* 2006;**55**(Suppl. 2):S78–85.
84. Rozengurt N, Wu SV, Chen MC, Huang C, Sternini C, Rozengurt E. Colocalization of the alpha-subunit of gustducin with PYY and GLP-1 in L cells of human colon. *Am J Physiol Gastrointest Liver Physiol* 2006;**291**:G792–802.
85. Sutherland K, Young RL, Cooper NJ, Horowitz M, Blackshaw LA. Phenotypic characterization of taste cells of the mouse small intestine. *Am J Physiol Gastrpomtest Liver Physiol* 2007;**292**:G1420–8.
86. Kokrashvili Z, Mosinger B, Margolskee RF. Taste signaling elements expressed in gut enteroendocrine cells regulate nutrient-responsive secretion of gut hormones. *Am J Clin Nutr* 2009;**90**:S822–5.
87. American Dietetic Association . Position of the American Dietetic Association: use of nutritive and nonnutritive sweeteners. *J Am Diet Assoc* 2004;**104**:255–75.
88. Grotz VL, Henry RR, McGill JB, Prince MJ, Shamoon H, Trout JR, et al. Lack of effect of sucralose on glucose homeostasis in subjects with type 2 diabetes. *J Am Diet Assoc* 2003;**103**:1607–12.
89. Barriocanal LA, Palacios M, Benitez G, Benitez S, Jimenez JT, Jimenez N, et al. Apparent lack of pharmacological effect of steviol glycosides used as sweeteners in humans. A pilot study of repeated exposures in some normotensive and hypotensive individuals and in type 1 and type 2 diabetics. *Regul Toxicol Pharmacol* 2008;**51**:37–41.
90. Fujita Y, Wideman RD, Speck M, Asadi A, King DS, Webber TD, et al. Incretin release from gut is acutely enhanced by sugar but not by sweeteners in vivo. *Am J Physiol Endocrinol Metab* 2009;**296**:E473–9.
91. Ma J, Bellon M, Wishart JM, Young R, Blackshaw LA, Jones KL, et al. Effect of the artificial sweetener, sucralose, on gastric emptying and incretin hormone release in healthy subjects. *Am J Physiol Gastrointest Liver Physiol* 2009;**296**:G735–9.
92. Vennekens R, Nilius B. Insights into TRPM4 function, regulation and a physiological role. *Handb Exp Pharmacol* 2007;**179**:269–85.
93. Launay P, Cheng H, Srivatsan S, Penner R, Fleig A, Kinet JP. TRPM4 regulates calcium oscillations after T-cell activation. *Science* 2004;**306**:1374–7.
94. Winslow MM, Crabtree GR. Decoding calcium signaling. *Science* 2005;**307**:56–7.
95. Oh-hora M, Rao A. Calcium signaling in lymphocytes. *Curr Opin Immunol* 2008;**20**:250–8.

96. Colsoul B, Schraenen A, Lemaire K, Quintens R, Van Lommel L, Segal A, et al. Loss of high-frequency glucose-induced Ca^{2+} oscillations in pancreatic islets correlates with impaired glucose tolerance in Trpm5-/- mice. *Proc Natl Acad Sci USA* 2010;**107**:5208–13.
97. Holzer P. Opioid receptors in the gastrointestinal tract. *Regul Pept* 2009;**155**:11–7.
98. Goyal SK, Samsher , Goyal RK. Stevia (*Stevia rebaudiana*) a bio-sweetener: a review. *Int J Food Sci Nutr* 2010;**61**:1–10.
99. Li J, Vederas JC. Drug discovery and natural products: end of an era or an endless frontier? *Science* 2009;**325**:161–5.
100. Long DJ, Brennan FX, Devantier HR, Bryant RW, Salemme FR, Palmer RK. *Development of a high throughput (HT) in vivo sweetness enhancing assay in rats. Program No. 64.7. 2008 neuroscience meeting planner.* Washington, DC: Society for Neuroscience; 2008 [online].
101. Walters DE, Orthoefer FT, DuBois GE. *Sweeteners: discovery, molecular design and chemoreception. ACS Symposia Series 450.*
102. Kim NK, Kinghorn AD. Highly sweet compounds of plant origin. *Arch Pharm Res* 2002;**25**:725–46.
103. Weininger D. SMILES. 1. Introduction and encoding rules. *J Chem Info Comp Sci* 1988;**28**:31–8.
104. Dragon. *Talete SRL.* Italy: Milan; 2009.
105. Molconn-z. Edusoft LLC: Ashland, VA, USA; 2009.
106. MOE. Chemical Computing Group: Montreal, Canada; 2009.
107. SYBYL. Tripos Inc.: St. Louis, MO, USA; 2009.
108. ADAPT. Pennsylvania State University: College Station, PA USA; 2009.
109. Codessa. Semichem Inc.: Shawnee Mission, KS, USA; 2009.
110. Jorgensen WL. QSAR/QSPR and proprietary data. *J Chem Inf Model* 2006;**46**:3.
111. Stanton D, Jurs P. Development and use of charged partial surface-area structural descriptors in computer-assisted quantitative structure-property relationship studies. *Anal Chem* 1990;**62**:2323–9.
112. Cruciani G, Crivori P, Carrupt P-A, Testa B. Molecular fields in quantitative structure permeation relationships: the VolSurf approach. *J Mol Struct* 2000;**503**:17–30.
113. Cherkasov A. Inductive descriptors. *Curr Comput Aided Drug Des* 2005;**2005**:21–42.
114. Almond. *Molecular discovery.* UK: Pinner; 2009.
115. Cramer III RD, Patterson DE, Bunce JD. Comparative molecular field analysis. *J Am Chem Soc* 1988;**110**:5959–67.
116. Klebe G, Abraham U, Mietzner T. Molecular similarity indexes in a comparative-analysis (CoMSIA) of drug molecules to correlate and predict their biological activity. *J Med Chem* 1994;**37**:4130–46.
117. Wildman SA, Crippen GM. Prediction of physiochemical parameters by atomic contributions. *J Chem Inf Comput Sci* 1999;**39**:868–73.
118. Dixon SL, Mertz Jr. KM. One-Dimensional molecular representations and similarity calculations: methodologies and validation. *J Med Chem* 2001;**44**:3795–809.
119. Hall LH, Kier LB. *Molecular connectivity in structure-activity analysis.* New York: Wiley; 1986.
120. Kier LB. Inclusion of symmetry as a shape attribute in kappa-index analysis. *Quant Struct Activ Relat* 1987;**6**:8–12.
121. Kier LB, Hall LH. *Molecular structure description. The electrotopological state.* New York: Academic Press; 1999.
122. Xing Li, Glen RC, Clark RD. Predictiong pK_a by molecular tree structured fingerprints and PLS. *J Chem Comput Sci* 2003;**43**:870–9.
123. Rodgers S, Glen RC, Bender A. Characterizing bitterness. *J Chem Inf Model* 2006;**46**:569–76.

124. Ertle P, Rohde B, Selzer P. Fast calculation of molecular polar surface area as a sum of fragment-based contributions and its approach to the prediction of drug transport properties. *J Med Chem* 2000;**43**:3714–7.
125. Zheng W, Tropsha A. Novel variable selection quantitative structure property relationship approach based on the k-nearestneighbor principle. *J Chem Inf Comput Sci* 2000;**40**:185–94.
126. Durant JL, Leland BA, Henry DR, Nourse JG. Reoptimization of MDL keys for use in drug discovery. *J Chem Inf Comput Sci* 2002;**42**:1273–80.
127. Brown RD, Martin YC. An evaluation of structural descriptors and clustering methods for use in diversity selection. *SAR QSAR Environ Res* 1998;**8**:23–39.
128. Martin YC, Kofron JL, Traphagen LM. Do structurally similar molecules have similar biological activities. *J Med Chem* 2002;**45**:4350–8.
129. LaBute P. Binary QSAR: a new method for the determination of QSARs. *Pac Sym Biocomp* 1999;**4**:444–5.
130. Clark RD, Sprous DG, Leonard JM. Progressive scrambling. In: Holtje H-D, Sippl W, editors. *Rational approaches to drug design*. S.A. Barcelona, Spain: Prous Science; 2001. p. 475–86.
131. Golbraik A, Tropsha A. Beware of q2. *J Mol Graph Model* 2002;**20**:269–76.
132. Golbraik A, Tropsha A. Predicted QSAR modeling. *J Comp Aided Mol Des* 2002;**16**:357–69.
133. Sheridan RP, Feuston BP, Maiorov VN, Kearsley SK. Similarity to molecules in training set is a good descriminator for prediction accuracy in QSAR. *J Chem Inf Comput Sci* 2004;**44**:1912–28.
134. Sprous DG. Fingerprint-based clustering applied to define a QSAR model use radius. *J Mol Graph Mod* 2008;**27**:225–32.
135. Van Drie J. Pharmacophore discovery: a critical review. In: Bultinck P, Winter H, Langenaeker J, Tollenare J, editors. *Computational medicinal chemistry for drug discovery.* New York: Marcel Dekker; 2004. p. 437–61.
136. Sprous DG, Palmer RK, Swanson JT, Lawless M. QSAR in the pharmaceutical research setting: QSAR models for broad large problems. *Curr Top Med Chem* 2010; (accepted).
137. Guner O. *Pharmacophore perception, development, and use in drug design.* La Jolla, CA, USA: International University Line; 2000.
138. Langer Y, Hoffman RD. *Pharmacophore and pharmacophore searches.* Weinheim, Germany: Wiley VCH Velag GmbH & Co. KgaA; 2006.
139. Discovery-Studio. Accelrys Inc.: San Diego, CA, USA.
140. Phase. Shrodinger Inc.: New York, NY, USA.
141. MDDR. MDL Information Systems: San Leandro, CA.
142. Aureus Biopharma. Paris, France.
143. GVK BIO. Hyperabad, India.
144. ZINC. University of California, San Francisco: San Francisco, CA, USA.
145. ACD. Symyx Corporation: Sunnyvale, CA, USA.
146. CAP. Accelrys Inc.: San Diego, CA, USA.
147. Asinex Inc.: Moscow, Russia.
148. ChemBridge Inc.: Moscow, Russia.
149. ChemDiv Inc.: Moscow, Russia.
150. Enamine Inc.: Kiev, The Ukraine.
151. BS Inc.: Moscow, Russia.
152. Maybridge Tintagel: Cornwall, UK.
153. LifeChem.Inc.: Kiev, The Ukraine.
154. Otava Inc.: Kiev, The Ukraine.
155. Pharmeks Inc.: Moscow, Russia.
156. Vitas Inc.: Moscow, Russia.

157. Baurin N, Baker R, Richardson C, Chen I, Foloppe N, Potter A, et al. Drug like annotation and duplicate analysis of a 23 supplier chemical database totaling 2.7 M compounds. *J Chem Inf Comp Sci* 2004;**44**:643–51.
158. Tropsha A, Reynolds CH. Designing focused libraries for drug discovery. *J Mol Graph Mod* 2002;**20**:427–8.
159. Olah MM, Bologna CG, Oprea TI. Strategies for compound selection. *Curr Drug Disc Tech* 2004;**1**:211–20.
160. Lipinski CA, Lombardo F, Dominy BW, Feeney PJ. *Adv Drug Deliv Rev* 1997;**23**:3–26.
161. Oprea TI, Davis AM, Teague SJ, Keeson PD. Is there a difference between leads and drugs. *J Chem Inf Comput Sci* 2001;**41**:1308.
162. Hahn MM, Oprea TI. Pursuing the leadlikeness concept in pharmaceutical research. *Curr Opin Chem Biol* 2004;**8**:255–63.
163. Seidler J, McGovern SL, Doman TN, Schoichet BK. Identification and prediction of promiscuous aggregating inhibitors among known drugs. *J Med Chem* 2003;**2003**(46):4477–86.
164. Ajay , Walter WP, Murcko MA. Can we learn to distinguish between "drug-like" and "nondrug-like" molecules? *J Med Chem* 1998;**41**:3314–24.
165. Sadowski J, Kubinyi H. A scoring scheme for discriminating between drugs and nondrugs. *J Med Chem* 1998;**41**:3325–9.
166. Burdock GA. The GRAS process. *Food Tech* 2003;**57**:17.
167. Woods LA, Doull J. GRAS evaluation of flavoring substances by the expert panel of FEMA. *Regul Toxicol Pharmacol* 1991;**14**:48–58.
168. Flavor-Base. Leffingwell & Associates: Canton Georgia (USA); 2004.
169. Sprous DG, Salemme FR. A comparison of the chemical properties of drugs and GRAS compounds. *Food Chem Toxicol* 2007;**45**:1419–27.
170. Prestwick Screening Library. Prestwick Chemical Company: Illkirch France.
171. Sprous DG, Stein P, Salemme FR, Palmer RK. *Development of a GRAS-like screening library from commercial suppliers.* 2008; [unpublished results].
172. Congreve M, Carr R, Murray C, Jhoti H. A rule-of-three for fragment based lead discovery. *Drug Discov Today* 2003;**8**:876–7.
173. Jimonet P, Jager R. Strategies for designing GPCR-focus libraries and screening sets. *Curr Opin Drug Discov Dev* 2004;**7**:325–33.
174. Horn F, Weare J, Beukers MW, Horsch S, Bairoch A, Chen W, et al. GPCRDB: an information system for G protein-coupled receptors. *Nucleic Acids Res* 1998;**26**:275–9.
175. Manning G, Whyte DB, Martinez R, Hunter T. The protein kinase complement of the human genome. *Science* 2002;**298**:1912–34.
176. Manallack DT, Pitt WR, Gancia W, Montana JG, Livingstone DT, Ford MG, et al. Selecting screening candidates for kinase and GPCR targets using neural networks. *J Chem Inf Comput Sci* 2002;**42**:1256–62.
177. Burden FR. Molecular identification number for substructure searches. *J Chem Inf Comput Sci* 1989;**29**:225–7.
178. Pearlman RS, Smith KM. Metric evaluation and the receptor-relecant subspace concept. *J Chem Inf Comput Sci* 1999;**39**:28–35.
179. Pearlman RS, Smith KM. Novel software tools for molecular diversity. *Perspect Drug Discov Des* 1998;**9**:339–53.
180. Balakin KV, Tkachenko SE, Lang SA, Okun I, Ivanshchenko AA, Savchuk NP. Property based design of GPCR targeted library. *J Chem Inf Comput Sci* 2002;**42**:1332–42.
181. Rolland C, Gozalbes R, Nicolai E, Paugam MF, Coussy L, Barbosa F, et al. GPCR affinity predicted based on the use of a profiling dataset. *J Med Chem* 2005;**48**:6563–74.
182. Givehchi A, Schneider G. Multispace classification for predicting GPCR ligands. *Mol Div* 2005;**9**:371–83.

183. Guo T, Hobbs DW. Privileged structure-based combinatorial libraries targeting G protein-coupled receptors. *Assay Drug Dev Tech* 2003;**1**:579–92.
184. Vasilyev DV, Shan QJ, Soloveva V, Nawoschik SP, Kaftan EJ, Mayer SC, et al. A novel HTS assay for HCN channel blockers using membrane potential sensitive dye and FLIPR. *J Biomol Screen* 2009;**14**:1119–28.
185. Haverkamp W, Breithardt G, Camm AJ, Janse MJ, Rosen MR, Antzelevitch C, et al. The potential for QT prolongation and pro-arrhythmia by non-anti-arrhythmic drugs. *Cardiovasc Res* 2000;**47**:219–33.
186. De Ponti F, Poluzzi E, Montanaro N. QT-interval prolongation by non-cardiac drugs. *Eur J Clin Pharmacol* 2000;**56**:1–18.
187. Redfern WS, Carlsson L, Davis AS, Lynch WG, MacKenzie I, Palethorpe S, et al. Relationships between preclinical cardiac electrophysiology, clinical QT interval prolongation and torsade de pointes for a broad range of drugs. *Cardiovasc Res* 2003;**58**:32–45.
188. Neher S, Sakmann B. The patch clamp technique. *Sci Am* 1992;**266**:44–51.
189. Wood C, Williams C, Waldron GJ. Patch clamping by numbers. *Drug Discov Today* 2004;**9**:434–41.
190. Hansen K, Rathke F, Schroeter T, Rast G, Fox T, Kriegl JM, et al. Bias-correction of regression models: a case study of hERG inhibition. *J Chem Inf Mod* 2009;**49**:1486–96.
191. Johnson SR, Yue H, Conder ML, Shi H, Doweyko AM, Lloyd J, et al. Estimation of hERG inhibition of drug candidates using multivariant property and pharmacophore SAR. *Bioorg Med Chem* 2007;**15**:6182–92.
192. Polak S, Wisniowska B, Brandys J. Collation, assessment and analysis of literature in vitro data on hERG receptor blocking potency for subsequent modeling of drugs cardiotoxic properties. *J Appl Toxicol* 2009;**29**:183–206.
193. Diller DJ, Hobbs DW. Understanding hERG inhibition with QSAR models based on a one-dimensional molecular representation. *J Comput Aided Mol Des* 2007;**21**:379–93.
194. Ekins S, Crumb WJ, Dustan-Sarazan R, Wikel JH, Wrighton SA. Three dimensional quantitative structure activity relationship for inhibition of human ether-a-go-go-related gene potassium channel. *J Pharm Exp Therap* 2002;**301**:427–34.
195. Song M, Clark M. Development and evaluation of an in-silico model for hERG binding. *J Chem Inf Model* 2006;**46**:392–400.
196. Thai KM, Eckert GF. A binary QSAR model for classification of hERG potassium channel blockers. *Bioorg Med Chem* 2008;**16**:4107–19.
197. Schiffman SS, Booth BJ, Carr BT, Losee ML, Sattely-Miller EA, Graham BG. Investigation of synergism of binary mixtures of sweeteners. *Brain Res Bull* 1994;**38**:105–20.
198. Biofocus: Saffron, Walden, UK.
199. Eskitis Institute: Brisbane, Queensland, Australia.
200. Phytomyco Research Corporation: Greenville, NC, USA.
201. Magellan Biosciences Group, Inc.: Tampa, FL, USA.
202. Albany Molecular: Albany, NY, USA.
203. Sprous DG, Stein P, Salemme FR, Palmer RK. *Internal RedPoint Bio review of natural product suppliers.* 2007; [unpublished data].
204. Gomtsyan A, Bayburt EK, Schmidt RG, Surowy CS, Honore P, Marsh KC, et al. Identification of (R)-1-(5-tert-butyl-2, 3-dihydro-1H-inden-1-yl)-3-(1H-indazol-4-yl)urea (ABT-102) as a potent TRPV1 antagonist for pain management. *J Med Chem* 2008;**51**:392–5.
205. Broad LM, Keding SJ, Blanco MJ. Recent progress in the development of selective TRPV1 antagonists for pain. *Curr Top Med Chem* 2008;**8**:1431–41.
206. Blum CA, Zheng X, Brielmann H, Hodgetts KJ, Bakthavatchalam R, Chandrasekhar J, et al. Aminoquinazolines as TRPV1 antagonists: modulation of druglike properties through the exploration of 2-position substitution. *Bioorg Med Chem Lett* 2008;**18**:4573–7.

207. Doherty EM, Fotsch C, Bannon AW, Bo Y, Chen N, Dominguez C, et al. Novel vanilloid receptor-1 antagonists. 2. Structure–activity relationships of 4-oxopyrimidines leading to the selection of a clinical candidate. *J Med Chem* 2007;**50**:3515–27.

208. Pomonis JD, Harrison JE, Mark L, Bristol DR, Valenzano KJ, Walker K. N-(4-tertiarybutylphenyl)-4-(3-cholorphyridin-2-yl)tetrahydropyrazine-1(2H)-carbox-amide (BCTC), a novel, orally effective vanilloid receptor 1 antagonist with analgesic properties. II. In vivo characterization in rat models of inflammatory and neuropathic pain. *J Pharmacol Exp Ther* 2003;**306**:387–93.

209. Kym PR, Kort ME, Hutchins CW. Analgesic potential of TRPV1 antagonists. *Biochem Pharmacol* 2009;**78**:211–6.

210. Cheung WS, Calvo RR, Tounge BA, Zhang SP, Stone DR, Brandt MR, et al. Discovery of piperidine carboxamide TRPV1 antagonists. *Bioorg Med Chem Lett* 2008;**18**:4569–72.

Challenges and Approaches for Assay Development of Membrane and Membrane-Associated Proteins in Drug Discovery

SCOTT GRIDLEY, ANTHONY L. SHROUT, AND EDWARD A. ESPOSITO

Blue Sky Biotech, Inc., Worcester, Massachusetts, USA

In addition to its role as a barrier between the cytoplasm and the extracellular milieu, the cell membrane is a scaffold for a diverse collection of receptors and enzymes. The organization afforded by this scaffold serves to ensure an efficient interaction between the components of the membrane. The desire to maintain this organization in solution is a challenge for the appropriate interrogation of these biochemical components. This chapter will discuss strategies that allow biochemical analysis of membrane-associated enzymes within standard biochemical reactions. The advantages of these screening strategies in identifying valuable compounds from compound libraries and in understanding the intricacies of complex multiprotein complexes (i.e., chemotaxis) will be discussed.

Progress in Molecular Biology
and Translational Science, Vol. 91
DOI: 10.1016/S1877-1173(10)91007-7

1877-1173/10 $35.00

I. Introduction

A diverse group of transmembrane receptors drive cellular responses to external stimulants via a multitude of signal transduction pathways. Transmembrane signaling is driven by the binding of an extracellular ligand to a transmembrane receptor which then typically activates the receptor, resulting in the recruitment of multiple proteins to the plasma membrane. Subsequent activation of the recruited protein signaling agents leads to the propagation of a multitude of specific signals throughout the cell in order to elicit a specific response to the initial stimulus. This dynamic, yet intricate, network relies primarily on the 2D nature of the plasma membrane that promotes the arrangement of these functional protein assemblies. Complex protein assemblies are not unique to the plasma membrane and have been observed on the ER, nuclear, and the Golgi membranes as well. While this discussion will largely focus on membrane-associated assemblies, we will also address their similarities with signaling complexes formed on protein scaffolds later in the discussion. Other proteins, which may or may not have membrane localization domains (MLDs), can also be involved in the localization of critical signaling proteins. Such adaptor proteins often have multiple protein recognition domains which enable temporal regulation of the recruitment of cytoplasmic proteins via bridging interactions between the activated membrane protein and their cytoplasmic signaling partners. These cytoplasmic partners may be recruited for activation or simply for colocalization with a low-affinity substrate. All-in-all, all organisms use their membranes as conduits to drive all their cell functions—it is the interrogation of these networks using a novel templating technique that will be the focus of this chapter. As will be discussed, while the membrane is an integral part of the signaling networks found in any cell, it is also the major hurdle that needs to be overcome in order to study these networks.

II. Strategies Used by Proteins to Target the Membrane

Proteins have adopted many strategies to establish associations with the cell membrane. The most stable interactions occur by actual transit of the peptide chain through the membrane, and range from conventional single-pass proteins to seven transmembrane protein receptors (7TM). Of more interest to this discussion is the fact that a number of proteins interact with the membrane scaffold via a targeting domain—domains that may serve as interesting drug targets.[1] Such membrane targeting may involve direct interaction with the cell membrane (myristoylation or palmitoylation; Ref. 2) or via other types of targeting domains (see Table I). Examples of these proteins include the receptor tyrosine kinases (RTKs; single transit proteins; e.g., insulin receptor and Epidermal Growth Factor (EGF) receptor), G-protein coupled receptors (GPCRs; 7TM, cytochrome p450), posttranslationally modified targets which are myristoylated or palmitoylated

TABLE I
SELECTED DOMAINS MODULATING INTERACTION BETWEEN PROTEINS AND THE MEMBRANE SCAFFOLD

	Definition	References
SH2	Bind specific to 1–6 residues c-term to pY	3,4
PTB	3–5 residues c-term to pY regulate specificity, w/ some ptb domains binding nonphospho Ys and still others recog and binding both pY and non-pY equally well (e.g., shc)	5
PH	Large family > 100 domains—small subset bind PIP2 or PIP3—Weak and nonspecific binding of PH domains compensated by oligomeric nature	5
SH3	Bind pro-rich PXXP	6
WW	Bind pro-rich PXPX	7,8
FYVE	PIP	5
PDZ	Bind hydrophob residues on C-term of targets	9,10
GPI-linked	Proteins targeted to cholesterol-rich membrane rafts	11

(Src-family kinases), those associated by lipid-binding domains such as the pleckstrin homology domain (BTK), and many others. Many proteins have more than one MLD to ensure colocalization with their respective binding partners.

For many of these proteins (e.g., the 7TM receptors), an association with the cell membrane is required for biochemical activity. Thus, assay methods using membrane fragments or whole cells have been developed to study these targets. For other proteins, molecular biological techniques can be employed to interrogate at least part of the biology of these targets. Thus, many membrane-associated protein kinases are evaluated as protein fragments in which the soluble catalytic domain is retained but the targeting domain is engineered away. This strategy has proved successful in the identification of orthosteric inhibitors that block the interaction with the enzyme's active site. However, this strategy assumes that the conformation of the catalytic domain is not affected by the targeting domain, a situation that may or may not be true. In our search for strategies to develop better therapeutic compounds, strategies that permit interrogation of the entire kinase protein are desired.

III. NanoDisc Technology

NanoDisc technology, developed by Sligar and coworkers, allows the researcher to use intact receptors rather than engineered fragments.[12–14] This method has shown much promise in the study of small numbers of transmembrane proteins and GPCRs, which are a large target class for drug discovery.[15]

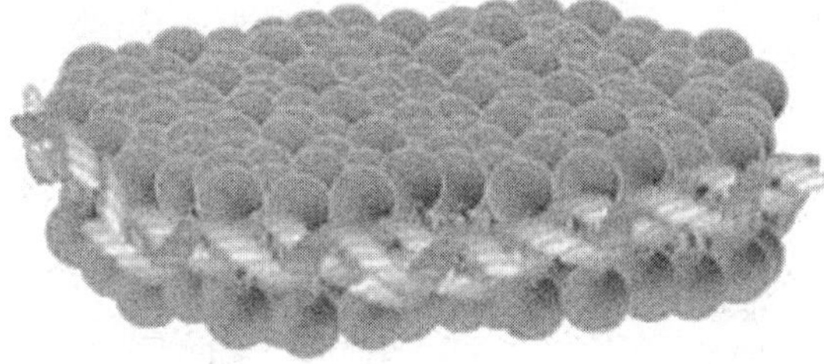

FIG. 1. Structure of protein NanoDisc.

The use of full proteins is the principle advantage in the use of NanoDisc technology; however, the methodology is significantly more laborious than the use of template-directed assembly (TDA) technology (see below) and requires special skills and equipment.[16] Additionally, the surface area on TDA liposomes allows for dynamic exchange between a large population of receptors, which occurs at the cell surface (and on TDA 2.0), while NanoDiscs only allow for a few complexes to form due to a smaller surface area.[12] NanoDisc technology (described in Fig. 1) is an outstanding method for the study of GPCR receptors as evidenced by a recent review by Ritchie et al. The authors describe the reconstitution of bacteriorhodopsin, rhodopsin, and other systems by this technique.[17] Lastly, Morrissey and coworkers employed both TDA 2.0 and NanoDiscs to study tissue factor (TF) complex assemblies and found extremely similar activity and kinetic results with both approaches.[18]

IV. TDA Technology

TDA technology was developed and patented based on work by Weis and coworkers; it was initially brought to market by P.A. Technologies, LLC, and is now under the exclusive license of Blue Sky BioProducts. TDA technology provides a simple method that facilitates the interrogation of membrane-associated proteins and more complex protein assemblies that form at the membrane surface—a common feature among all organisms. TDA makes use of the ubiquitous histidine tag, which has gained popularity in protein purification and structural studies over the last two decades.[19–22] Briefly, the TDA platform comprises stable liposomes that are decorated with Ni-NTA head groups. When these liposomes are combined with the His-tagged receptor fragments, the proteins reversibly anchor to the surface and the two-dimensional (2D) organization that normally occurs in the cell is imparted to these protein fragments. Proper enzymatic function and the ability of these proteins to form signaling complexes, attributes of proteins that are normally found at the membrane surface of the cell, can now be observed. Specifically, TDA technology regenerates the special features of the membrane environment, including (i) restriction of motion to two dimensions, (ii) the increased local concentration that accompanies such reduction in dimensionality, and (iii) facilitation of scaffolding interactions that lead to the formation of protein signaling complexes.

While the method is straightforward in concept, there are procedural and engineering concerns which must be addressed in order to obtain the maximum benefit. Chemical considerations include carefully developed buffer systems free from agents such as EDTA and large concentrations of detergents, as these disrupt the Ni-NTA-Ni^{2+}–protein interactions and destroy protein–protein interactions and membrane structure, respectively. Additionally, the engineering of membrane-associated proteins must be cautiously executed as these proteins are frequently sensitive to changes in their juxtamembrane domains and may need to be engineered specifically to function on the TDA platform. Lastly, protein cytoplasmic domains should be N-terminally His-tagged for proper orientation (C-terminal tag for recombinant ectodomain fragments). Thus, care must be taken when designing proteins and reagents for use in the TDA platform. The end result, however, is a more biologically relevant system where mobile arrays of cytoplasmic domains can interact as they do in the cell. Additional experimental benefits include lower concentrations of metal ions and the fact that researchers do not need to employ aggregating agents to stimulate enzymatic activity. Essentially, the TDA technology combines the ease of using engineered fragments with the ability to effectively interrogate cell-like arrays of receptors and protein complexes that exhibit increased function, activity, and cell-like behavior. See Fig. 2 for a scheme that depicts how this lipid-based nanosphere enables assembly of proteins.

A. TDA—Simple Solution for RTKs

RTKs have become important targets for drug discovery because of their roles in cancer and other diseases.[23,24] One significant challenge that researchers and drug developers face is the involvement of RTKs in multiple, overlapping, and branched signaling pathways, which makes it difficult to identify individual functions.[25–27] Furthermore, these proteins are intimately associated with the cell membrane which makes their isolation and study difficult. While practiced throughout the field, the removal of these proteins from the organizing environment of the cell membrane often results in loss of activity and an inability to form complexes with other proteins that are important for proper function. Recent data provides evidence for the formation of large clusters of proteins and suggests that these interactions contribute to the mechanism of activation for many RTKs, such as signal propagation and amplification.[28–31] For example, Ichinose and coworkers have demonstrated

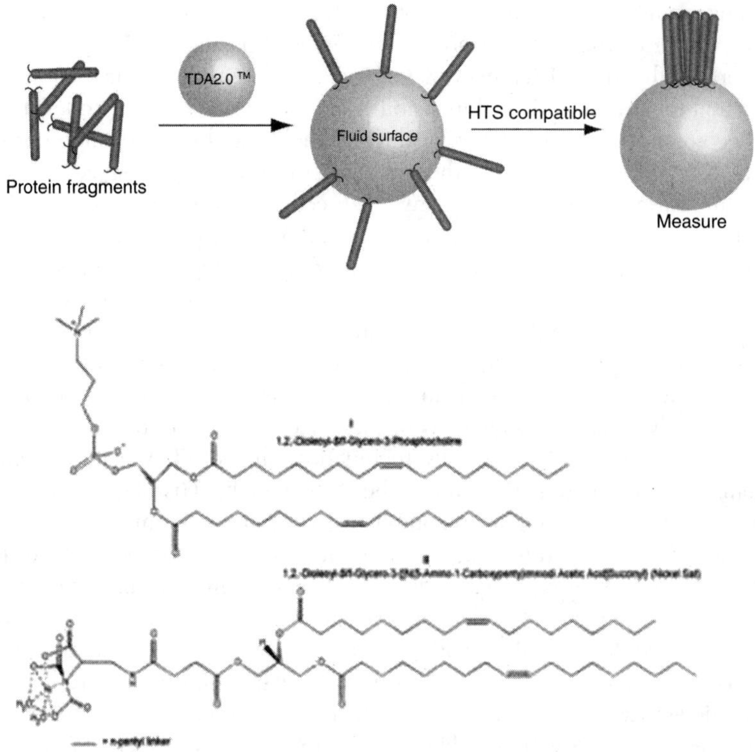

FIG. 2. Commercially available histidine tagged proteins bind to the template and are free to interact as they do in the cell, because of the fluid nature of the template. Below—the lipids used in early generations of template development.

that EGF receptors respond to ligand binding by dynamic clustering, which functions to amplify the signal initiated by ligand binding.[30] Assessing cooperative interactions between soluble receptor fragments of this class of proteins has not been possible using conventional enzyme *in vitro* assays. As RTKs and other membrane-associated proteins utilize the membrane as a scaffold to promote protein–protein interactions; these interactions are too weak to occur in solution and this was realized when engineered fragments demonstrated low or no activity in solution without the use of agents that promote close associations, similar to those that can be exhibited on the cell membrane surface. When the motion of molecules is limited to two dimensions, as it is in the TDA environment, or in the presence of the cell membrane, the increased local concentration and orientation promote clustering interactions to a degree that is not observed in solution, even at a comparable bulk concentration.

The TDA platform acts like the cell membrane in reducing the entropy of the system substantially; this reduction in entropy allows weak protein–protein interactions to occur.[32]

Drug discovery efforts to find specific small molecule inhibitors has focused almost exclusively on inhibitors that target the ATP-binding site, the substrate-binding site, or both; these sites reside in the cytoplasmic tyrosine kinase domain. As mentioned earlier, recombinant kinase domains are easily handled and are amenable to high throughput screening for specific inhibitors of enzyme activity; however, the soluble fragments lack the full function of intact receptors for a large part because they are no longer associated with the membrane and, therefore, do not form the interactions necessary for complete assembly and function. This issue is known industry-wide. Also, it has been noted that the use of soluble fragments frequently produces misleading screening results.[33] Our own work has demonstrated a difference in substrate specificity with and without the TDA platform. In spite of these concerns, soluble catalytic domains derived from RTKs are still widely used in research and drug discovery. Therefore, methods to improve the functional properties of these protein fragments for research and drug discovery are required.

As a means to reconstitute large portions of signaling pathways, TDA was first brought to practice by Shrout et al. for signaling proteins purified from the bacterial chemotaxis system.[34–39] TDA has proven effective for assembling the "receptosome" from the bacterial chemotaxis system which consists of four proteins: the cytoplasmic domain from the aspartate receptor (Tar-CD), the histidine kinase, CheA, the adaptor protein, CheW, and the response regulator protein, CheY.[34,35] Electron microscopic images of the surface-assembled complexes of Tar-CD, CheW, and CheA resemble the receptor arrays that occur in cells.[35] With the completion of this work, TDA has been successfully used to restore the biochemical activity in a variety of systems discussed in this chapter.

B. Template-Directed Assembly Regenerates Functional Interactions

In the cell, RTKs are confined to two dimensions, where the large local concentration of receptors promotes clustering. The interaction strengths among receptors are "tuned" to this environment such that intact receptors can engage each other and form higher order oligomers that are relevant for signaling.[30,40] Also, receptor clusters in both prokaryotes and eukaryotes act as scaffolds for adaptor proteins and downstream signaling proteins; these membrane-associated complexes that so assemble can exhibit cooperative and allosteric phenomena. Figure 3 illustrates how TDA facilitates functional interactions among the RTK domains that are not present in the conventional assay format. In the cell, the orientation and spatial arrangement of the kinase domains are

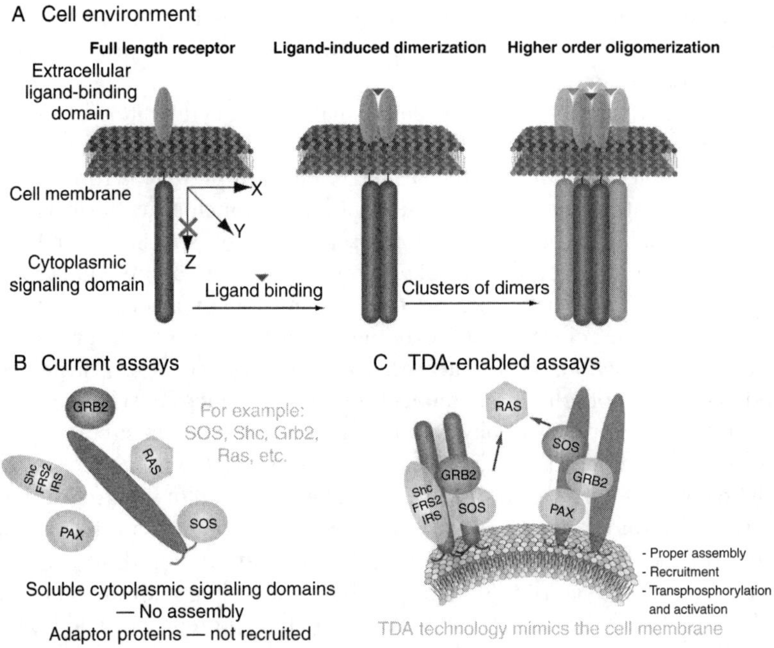

Fig. 3. Arrangements of signaling proteins in membranes, in solution, and on templates. (A) The membrane environment and receptor structure dictate receptor orientation and spacing. Activity is regulated by clustering and/or conformational changes induced by ligand binding. (Adaptor and downstream signaling elements are not shown for simplicity.) (B) Today's assays in solution do not form significant interactions among signaling proteins, for example engineered TK domains. These assays do not employ adaptor proteins or downstream enzymes, because they cannot assemble in solution. (C) The template-assembled environment imitates the membrane surface, where the natural interactions among proteins occur resulting in more biologically relevant function.

regulated by the membrane (Fig. 3A). For example, regulated interactions among dimers of cytoplasmic domain are necessary for their transphosphorylation and stimulated substrate activity. In some cases, signal enhancement may occur via higher order oligomerization.[30] These interactions, among TK domains, do not occur without the template (Fig. 3B) and interactions with adaptor proteins and downstream signaling elements are similarly ineffective. Figure 3C depicts the patented approach used by Blue Sky Bioproducts to assemble TK domains in an environment that imitates the membrane surface. The TDA technology can also be used to assemble downstream signaling proteins.

We now understand that all membrane-based signaling systems have the same organizing principles regardless of the origin of the cell, so it is plausible to expect that TDA will improve the activity and function of eukaryotic signaling proteins.[41]

V. Lipoparticle Approach

Another approach of interest for studying membrane-associated proteins is the use of retroviral pseudotypes. These virus-like particles (VLPs) consist of a protein core surrounded by cell membrane with intact integral membrane proteins. They are produced by co-expressing a target membrane-associated protein with a retroviral core protein (Gag) in mammalian cells.[42–45] The viral Gag protein self assembles and buds from the infected cell surrounded by the host lipid-membrane and its integral proteins, within which the transmembrane proteins retain proper orientation.[46] The resulting noninfectious VLPs are $\sim$ 150 nm in size, are relatively easy to isolate, and the enriched membrane proteins retain their function. VLPs have been utilized in biosensor assays as either attached to the surface or as a soluble component.

A. Retention of Function

VLPs expressing the cellular receptors for the Rous sarcoma virus (Tva) or the murine leukemia virus (MCAT-1) were used to determine whether the expressed viral receptors could target infected cells by presenting the appropriate viral glycoproteins.[45] Tva and MCAT-1 were chosen because they do not require additional cofactors for recognition or subsequent infection. Additionally, Tva and MCAT-1 represent Type I and multipass transmembrane receptors, respectively. Following isolation of the Tva-VLPs and confirmation of Tva expression, the Tva-VLPs were added to chronically infected cells, which expressed the RSV-A envelope glycoprotein (EnvA), and to uninfected cells. In uninfected cells, Tva-VLPs were not incorporated, however, infected cells that expressed EnvA were readily infected by the Tva-VLPs. Similarly, cells transfected with an EnvA vector encoding a functional EnvA were also efficiently infected with the Tva-VLPs. EnvA mutants with known entry defects were not infected or poorly infected in control experiments.

The multipass TM viral receptor MCAT-1 was also isolated as a VLP and assayed for function in cells expressing either the full-length MLV envelope protein (MLV env) or MLV env with a C-terminal deletion that has been reported to increase fusogenicity. In these experiments, cells expressing the C-term deletion were 100 to 1000 times more susceptible to MCAT-VLP infection than the cells expressing the wild-type protein, demonstrating, not only receptor specificity but also biologically relevant function.

B. Attachment to Optical Sensor

Functionalization of surfaces with intact, properly oriented TM proteins has long been desirable for the purpose of identifying binding partners and/or inhibitors of binding partners, as well as for the kinetic study of such interactions. Utilizing VLPs, Hoffman et al. have defined a general method by which intact TM proteins, in their native lipid environment, can be attached to optical biosensors.[42] Utilizing standard Biacore chips (C1 or F1), the researchers could successfully attach VLPs containing a number of overexpressed chemokine receptors for both the Type I and 7TM families. Importantly, the VLP-coated chips could be regenerated several times and the individual receptors retained their ability to bind antibodies and the native ligand. To demonstrate the utility of this technique, the authors examined interactions between the HIV-1 envelope protein (EnvH) and its 7TM coreceptors and between collapsin-1 and a single-pass receptor, neuropilin-1 (NP1).

In addition to binding CD4, different HIV viral strains bind to unique coreceptors on the host cell surface to gain entry in to the cell.[47,48] EnvH-coreceptor affinity, for some viral strains, has been difficult to measure using the conventional cell-surface binding assays.[42,49,50] The improved ability to measure these binding events and inhibit this binding offer tremendous therapeutic potential.[48] To achieve this aim, an EnvH gp120 subunit responsible for CD4-independent coreceptor binding (8xgp120) was used as a capture ligand for surface-attached VLPs expressing the CXCR4 coreceptor. After determining appropriate conditions for specific binding and demonstrating antibody-specific inhibition, a dose–response experiment was performed to calculate K_d. The calculated K_d of 506 ± 101 nM justifies the authors' difficulty in determining this value using standard cell-surface binding assays, because most of the ligand is released before completion of the wash steps. In a subsequent experiment, CD4-dependant gp120 was utilized and, appropriately, binding of gp120 to CXCR4 was not observed without addition of CD4 to the running buffer.

NP1 is a Type I TM protein receptor involved in axonal guidance, and the binding of collapsin to NP1 triggers repulsion and redirection.[51] VLPs generated from cells overexpressing NP1 were attached to biosensor chips, and collapsin-1 was shown to bind specifically to the surface-anchored VLPs. These results suggest that a diverse group of receptors can not only be incorporated into VLPs, but can also retain their cellular structure and functional capacity.

In an alternate approach, Willis et al. have attached an antibody to the Biacore chips and demonstrated the ability to quantitatively define antibody–receptor interactions. VLPs expressing seven different 7TM receptors, a type I receptor, and a Type II receptor were assayed against surface-anchored antibodies to their respective receptors. In each experiment, only the VLPs with the cognate receptors were specifically captured by the anchored antibodies. Then, using VLPs generated from cells over expressing different CCR5 receptor constructs, this group screened a dozen commercially available CCR5 antibodies and quantified the relative binding efficiency of each antibody.[43] One of the commercial antibodies, CTC5, is known to interact with an epitope present on the intact N-terminus of CCR5.[52] When an N-terminal HA-tagged CCR5 was incorporated into VLPs, the VLPs could not be captured by the immobilized CTC5 antibody, thus demonstrating retention of epitope specificity.

Taken together, these results demonstrate that over expressed receptors in VLPs represent a more facile method of studying binding interactions with TM receptors. It will be interesting to see whether the protein core in the VLPs can limit protein mobility and the ability to form functional oligomers such as those formed by the TF complexes. Additionally, it will be of tremendous value if these VLPs can be used to generate functional assays which are amenable to the more common, high throughput screening methods.

Many systems have now been studied using the TDA approach and we describe the breadth of utility that has been recently demonstrated by several different research groups. A future use of the technology may be to assemble biologically relevant signaling complexes as targets for compound screening. Such complexes may better reflect the biology of a particular target of therapeutic interest, and this approach may be better suited to find the most relevant candidate compounds for medicinal chemistry. The potential of this approach is seen in the work that investigated a complex chemomechanistic system that drives a bacterial response to environmental cues that are critical for the organism's survival—the bacterial chemotaxis system.

VI. Brief Review of the Chemotaxis System—Pioneering Work

The two main requirements for an organism's survival are, at least on the surface, quite simple: successfully seek nutrients and move away from harmful agents. Members of the Enterobacteriaceae family of bacteria, including *Escherichia coli*, *Salmonella*, *Shigella*, and many others, possess a two-component sensory system that allows the cell to constantly monitor a wide range of chemical cues in their immediate environment and adjust their swimming behavior accordingly.[53] In fact, nearly all motile prokaryotes possess a two-component system that allows strict control of motor activity.[54] This behavior, termed chemotaxis, was first observed by Englemann and Pfeffer in the 1880s. The modern study of chemotaxis, via genetics, was initiated by Julius Adler in the mid 1960s.[55–58] These pioneers observed that bacteria could move away or toward chemicals by a mechanism that is now known to involve a biased random walk. Today, through a series of biochemical, genetic, and experimental advances, the two-component chemotaxis signaling systems of *E. coli* and *Salmonella typhimurium* are the most thoroughly studied of all biological signaling systems and is, in part, due to the relatively few proteins involved. Most importantly, these systems exhibit many of the same fundamental processes for signal transduction found in more complex systems.[54,59–66] More to the point, the kinase responsible for signal generation in *E. coli* shares some of the same protein architecture of proteins found in human signaling, like the SH3 domain.[67,68] While this system seems simple, it possesses and displays many remarkable properties such as extremely high sensitivity to stimuli, signal amplification, and also the ability to assess and process many different environmental stimuli simultaneously[53]; the same attributes and functions as human signaling systems. The cell's ability to sense and respond to extremely low concentrations of chemical stimuli ($\sim 10^{-7}$ M) and also adapt to the presence of high concentrations requires further investigation into how the signaling proteins in this system interact and how these complex interactions affect the cell's ability to behave and respond in such a manner.[69–72]

While an exhaustive explanation of the chemosensory system in bacteria is inappropriate for this publication, the authors do wish to provide a basic overview of the system in order to establish the driving-force for development of the TDA technology, which is the central focus of this article.

A. The Two-Component System—Brief Overview

Attractants, for example amino acids and sugars, and repellents, for example metal ions and acids, are sensed by one of four methylatable chemotaxis proteins (MCPs) which are ~ 60 kDa monomers when initially expressed. In *E. coli* these are: serine (Tsr), aspartate, ribose and galactose (Trg), and the dipeptide receptor (Tap).[73,74] These transmembrane receptors are expressed as monomeric units, however many studies show that they are functional only in a dimeric state. Tsr, Tar, Trg, and Tap are also ~ 85% homologous (pair-wise) in the cytoplasmic region of the protein and have essentially the same topology.[59,74] Thus, it is not surprising that they all interact with the same collection of cytosolic signaling proteins. Evidences for the dimeric state of the receptor include crystallographic determinations of the periplasmic domain (ligand-binding region of the receptor) both in the presence and absence of the ligand. This domain was organized as a dimer with the binding pocket at the interface of the two monomers.[75–77] Furthermore, biochemical evidence, including cysteine cross-linking analysis showing that the receptors are primarily dimers, has demonstrated the dimeric state of the receptors, does not form heterodimers, and is stably associated even in the presence of detergents.[78] Perhaps the most important evidence over the last decade was the demonstration that these receptors and associated cytosolic proteins are cooperatively linked—acting in concert while at the cell membrane.[36,38,39,63,64,79–87]

B. The Two-Component System—Set in Motion

The two components of the chemotaxis system in *E. coli* and related bacteria are the transmitters of the signal CheA, a histidine kinase, and the response regulator/receiver CheY, an aspartate kinase.[60,67,82,83] The chemosensory system is regulated by a ternary complex consisting of CheA (142 kDa, dimer), CheW (18 kDa, scaffold protein), and a dimeric MCP, as depicted in Fig. 4.[84,85] All of these proteins are complexed to the receptors at the inner leaflet of the cell membrane. When neither an attractant nor a repellant are present, CheA is

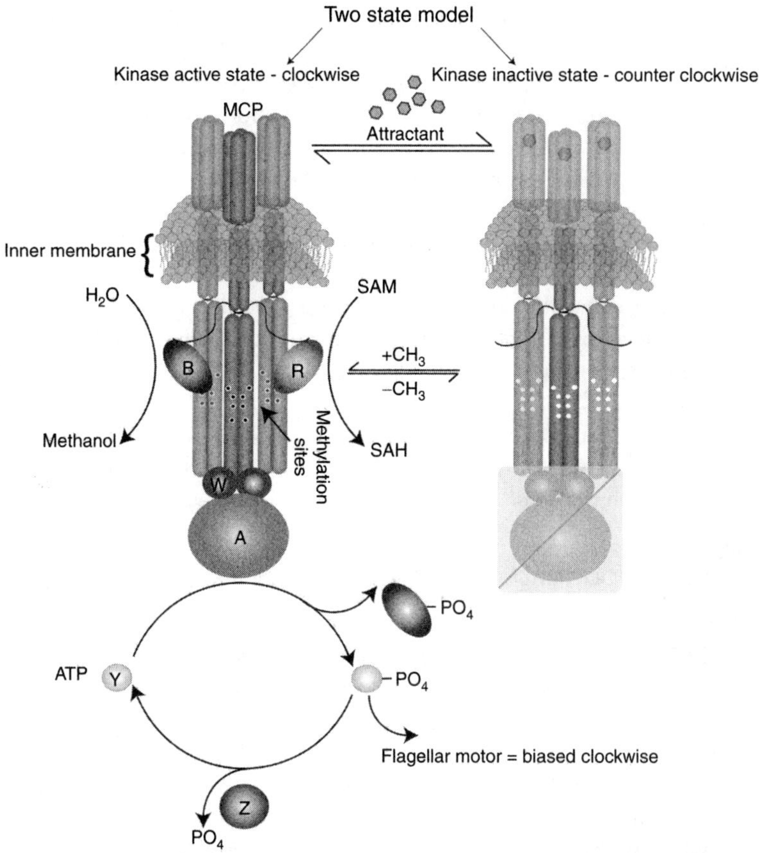

FIG. 4. Chemosensory circuit diagram, with single letter designations for all cytosolic proteins involved in the pathway presented.

constantly involved in the phosphorylation of the regulator protein, CheY. CheY-phosphate then diffuses through the cytoplasm to interact with the 37 kDa flagellar motor switch, fliM.[86,87] The flagella are long, left handed protein structures that, when rotated CCW, bundle and propel the cell forward. Consequently, when the direction of the motor switches, these unwind briefly and cause a tumble, leading to a random change in the direction of movement.[88]

To adapt to the ever-changing stimuli in the environment, the chemotaxis system has developed a mechanism of adaptation which is analogous to how our own eyes adapt to a darkened room.[53,66,89,90] The adaptation branch is mainly controlled by CheR, a 32 kDa methyltransferase, and CheB, a 35 kDa methylesterase.[91–93] Prior to the addition of an attractant the tumble frequency, and therefore the CheY-phosphate concentration, is determined by the kinase activity of the *adapted* state. When a stimulus is added, an immediate decrease in kinase activity occurs, and this results in a decrease in the concentration of CheY-phosphate and CheB-phosphate (the active form of the esterase). Additionally, the receptors become better substrates for methylation via CheR.[94,95] Increased methylation, over time, restores the kinase activity and increases tumble frequency to prestimulus levels, thereby allowing the cell to adapt to the current conditions, continue its progress up an attractant gradient, *and* detect further changes in the environment. Repellents, or removal of attractants, have the opposite effect on the system, leading to increased kinase activity and therefore an increase in tumble frequency and CheB-Pi concentrations, which serves to remove the methyl groups from the MCPs. Ultimately, the relative rates of CheR and CheB determine the steady-state methylation level of the system and allow the cell to move toward attractants or away from repellents.[66,90,96] Figure 4, below, illustrates the proteins involved in the chemosensory system of *E. coli*.

C. Signal Propagation—Conformational Changes and Clusters

Subtle conformational changes in the ligand-binding subunits have been thought to be most responsible for the propagation of a signal. Careful study of the crystal structures of the ligand-binding domain, with and without aspartate, has revealed a subtle rotation between subunits and a piston-type motion.[77,97] The use of spin labels to measure distance changes upon ligand binding using EPR showed that ligand binding generated a ~ 1 Å intrasubunit piston motion or shift within the receptor dimmer; this is consistent with other experimental evidences of small structural perturbations that were earlier observed.[59,98–101] How these subtle changes are relayed from the periplasmic portion of the receptor down through the membrane to the kinase, which is approximately 300 Å away, is not yet understood.

Over the last two decades, mounting evidence shows that receptor clustering plays a significant role in signal propagation, adaptation, and amplification of signaling in the chemosensory pathway.[81,84,102–105] Specific evidence for the existence of receptor clusters was first provided by a number of electron microcopy studies that showed receptors localized at the cell poles.[84,106–109] Allosteric interactions among the dimers are now thought to significantly influence how information is sensed and processed, by propagating conformational changes to neighboring dimers within the patch. Evidence for such communication between neighboring receptors has been collected through the use of mutations in the serine receptor that were suppressed by mutations in the aspartate receptor and also by employing synthetic multivalent ligands that specifically bound to the Trg receptor but enhanced the response to *serine*.[110–113] Additionally, it has been demonstrated that fragments of the Tar receptor, engineered with leucine zipper motifs, spontaneously associate, in solution, with a well-defined number of subunits involved to form large kinase-activating complexes.[104,114–116]

Throughout the decades of research on this seemingly simple system, there has been a need for simple reconstitution of the main protein players. While the molecular puzzle pieces of the system have all been identified there had not yet existed a simple experimental procedure to assemble the system in a way that enabled easy characterization. Recent studies from the Weis lab and from others have indicated that clustering gives rise to kinase activation via a highly cooperative process suggesting that teams of receptors act together to bind and activate the kinase.[63,64,81,105]

Empirical observations and a number of theoretical endeavors have been completed on the composition of receptor clusters and how these receptors communicate and respond within the context of a cluster or a patch.[109,110,117–124] Collectively, these data, along with *in vivo* observations demonstrating that receptors localize at the cell poles, have provided a driving-force for a method that can reconstitute well-defined samples of clustered receptors. Subtle changes in structure within a dimer upon ligand binding are most likely propagated via clustering, thus providing lateral communication between dimers and clusters of dimers. Therefore, reconstituting arrays of receptors rapidly and in a defined manner is a worthy effort.

D. Surface Templating Approach

In vitro systems have been used to study the behavior of heterogeneous protein complexes of bacteria for many years. Purified *E. coli* inner membranes containing overexpressed MCPs have been shown to activate the kinase in the presence of CheW.[81,125–129] However, purified inner membranes cannot be easily used for experiments that require the manipulation of receptor surface density, that is, degree of clustering, or experiments that require stoichiometric control over protein reagents. Furthermore, in traditional membranous samples, it is often difficult to characterize the accessibility to CheA and CheW, and perhaps most importantly, the vectoral orientation of the receptor.[130,131] These challenges are common to all transmembrane signaling systems because of the 2D organization of receptors and the complexes formed between them and a number of scaffolding and regulatory proteins.

TDA was developed by Weis and coworkers in 2003 as a means of providing an experimental system that can recreate the 2D organization of transmembrane signaling systems while overcoming many of the difficulties normally associated with membrane receptor samples.[34] NTA-DOGS-Ni^{+2} and DOPC lipids were used to form vesicles which can anchor a His-tagged cytoplasmic domain fragment, CF, which can also be identified by its level of modification at the major sites of methylation: CF_{EEEE}, CF_{QEQE}, and CF_{QQQQ}, where Q = glutamine and E = glutamate.[34] This Ni-NTA lipid was commercially available and was originally developed for applications in the 2D protein crystallization and biosensor development.[19–21,132–136]

Addition of CheA and CheW to the surface-assembled fragments allowed the formation of active ternary complexes (see Fig. 5). It was demonstrated that the kinase, CheA, was bound to the surface-assembled fragments along with CheW; and ultimately the level of covalent modification led to stark differences in complex stability. Additionally, it was observed that the templated CFs also served as a suitable substrate for the methyltransferase, CheR, thereby generating methylated receptor fragments that demonstrated a close association with fragments on the surface as it is already known that methylation activity is an interdimer event.[103]

E. Continued TDA Success

After the first publication in 2003, Weis and coworkers have furthered the utility of the TDA system significantly.[36–39] For example, known signaling mutations located in the receptor were employed to determine how competitive and cooperative interactions *between* receptors contributed to the control of the kinase, CheA.[39] These experiments highlight a major advantage of TDA—its ability to prepare controlled samples of heterogeneous receptor fragments on the surface. Fragments of the CF that were known to not stimulate kinase activity exhibited the same behavior with the TDA platform. When mixtures of nonactivating fragments were titrated with kinase stimulating fragments (wt) the TDA platform allowed the determination of *how* kinase activity was diminished. The mutants were sorted into three classes: (i and ii) CheW and CheA bound more weakly, or more tightly, to mutants than to wt-CF, and (iii) mutations inhibited kinase activity in a cooperative manner. These data support a model of CF/A/W interaction that is thought to be both cooperative and competitive. These data also are in agreement with studies carried out with intact receptor systems.

Weis and coworkers have also compared and contrasted SUVs (small unilamellar vesicles) and LUVs (large unilamellar vesicles) and how different receptor fragments are formed on these different templates.[37] This work also provided a first look at these complexes via electron microscopic analysis. Perhaps the most intricate work using the TDA system has been provided by Besschetnova et al. Both the excitation and adaptation branches of the chemotaxis system were studied and shown to be interlinked, thus exerting a counterbalancing effect on one another.[36] It was demonstrated that receptor density affected kinase and methyltransferase activity significantly, but in opposite directions. The TDA system allowed the preparation of samples wherein the 2D concentration of the anchored receptor fragments could be varied significantly. It was demonstrated that large concentrations of receptors favored an

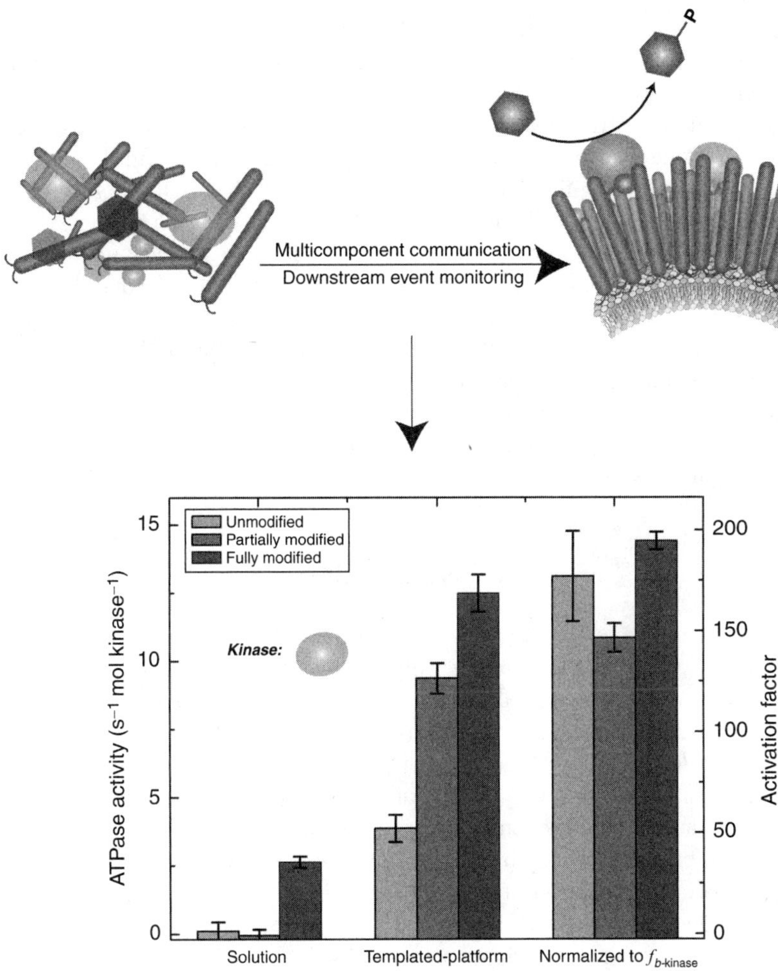

FIG. 5. *Top*: Assembly scheme based on published data employing proteins from the chemosensory system in *E coli*. Proteins were mixed together with the template. This results in arrays of signaling complexes at the membrane surface, receptor fragment (blue rods), kinase (green ovals), scaffolding protein (magenta ovals), and response regulator (purple hexagon). *Bottom*: Activities of surface-assembled complexes, error bars represent SEM or three experiments. The three receptors used, EEEE, QEQE, and QQQQ, represent three different levels of covalent modification. These experiments revealed a difference in complex stability.[34]

increase in kinase activity, while low concentration favored higher CheR activity. This observation can be placed in the larger context of the chemosensory stimulation of ligand binding. It has been shown that ligand binding increases the distance between receptors, thereby lowering the effective concentration of receptors and lowering kinase activity, while this same event causes an increase in methylation activity.

Furthermore, FRET experiments were carried out using fluorescently labeled C-fragments. These data demonstrated that an increase in the surface concentration of unmodified CF led to an increase in kinase activity from a low 2D concentration (5% mole Ni lipid) to a high 2D concentration (60% mole Ni lipid). Employing the same strategy with fully modified CF showed no increase in activity from the initial low 2D samples and the activity remained at the level over all surface concentrations. These results support the observation that kinase activity depends on both surface concentration and the degree to which the receptor is modified (Ref. 36, see also Ref. 137). These types of experiments can lead to an in-depth understanding of how the various units communicate within the context of the 2D membrane environment while monitoring distance and activity simultaneously.

VII. Tissue Factor

This exemplary work demonstrated that the TDA approach can be utilized for studying extracellular domains as well as cytoplasmic domains, and that the approach can be used for assembling complex signaling components in a simple format that is also amenable to high throughput assays. Most importantly, this work clearly demonstrated that TDA of purified components yields cell-like results that cannot be generated in the absence of complex assembly.[138]

TF is a noncatalytic transmembrane protein which is responsible for regulating the blood clotting cascade. TF binding to the plasma serine protease, factor VIIa (fVIIa), imparts allosteric activation and the resulting TF-VIIa complex activates downstream factors. Factors IX and X (fIX and fX, respectively) are membrane-bound protease zymogens which are activated by the TF-VIIa complex. Previous studies using a recombinant membrane-localized TF ectodomain (mTF, ectodomain and transmembrane domain) had shown that membrane anchoring is essential for pro-coagulant activity, but the reconstitution processes were quite laborious, time consuming, and difficult to control. In this work, a soluble His-tagged ectodomain fragment (shtTF) that enabled facile expression and purification was used. This soluble fragment lacks the transmembrane segment and can be docked onto liposomes which have the Ni-NTA headgroups (Fig. 6; Ref. 138).

Importantly, the authors note that they had expressed two soluble His-tagged ectodomains with different C-termini. In these experiments, a C-terminal His-tag is appropriate as it positions the ectodomain juxtamembrane residues adjacent to the liposome, thereby, most likely, mimicking their orientation in native assembly. One construct had residues 1–217 with six histidines encoded immediately

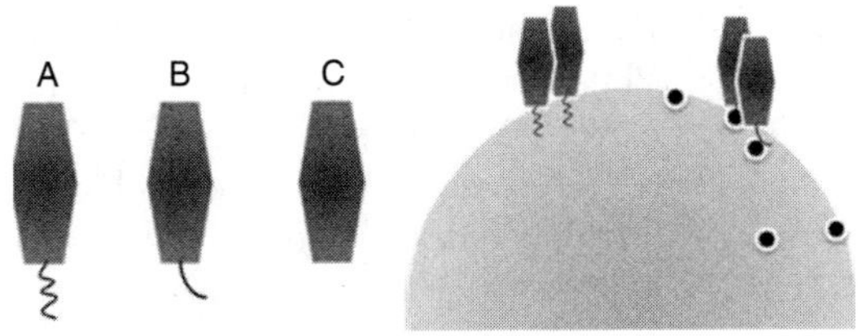

FIG. 6. Simple description of tissue factor constructs described in text where the hexagonal shape represents the extracellular domain. (A) Cloned fragment retaining the transmembrane domain which can be targeted to prepared liposomes retaining its membrane-associated properties (mTF), (B) soluble cloned fragment with a C-terminal hexahistidine tag which can be colocalized on liposomes with Ni-NTA headgroups (shtTF), and (C) a soluble cloned extracellular fragment of tissue factor (sTF). On right, a diagram representing the assembly of either mTF or shtTF on liposomes. Black circles represent Ni-NTA headgroups which bind the his-tagged fragments.

adjacent to residue 217, and the other construct (1–219) had five amino acids (a GGAAG flexible spacer) inserted between residue 219 and the hexahistidine affinity tag. Both constructs were active but the latter exhibited more than ten times higher specific activity in clotting assays. The authors propose that the increase in activity exhibited by the latter construct might be due to the linker which allowed better alignment of the ectodomains on the liposome surface. It has been our experience when designing RTK cytoplasmic fragments that we must occasionally employ a shotgun approach, adding flexible linker regions or shortening the juxtamembrane domains to generate active constructs. It is difficult to predict a priori the most functional construct; therefore, it is best to begin with a couple or several recombinant fragments for the initial assessment. Additionally, the liposomes in these studies have phosphoserine headgroups because such negatively charged phospholipids are required for optimal TF-VIIa activity; pointing to a need for assay development.[139]

Purified plasma-derived fVII, fX, and fXa were used to compare the activity of mTF incorporated into liposomes with that of shtTF bound to the Ni-NTA liposomes. Initial assays examined the binding affinity of TF with fVIIa. Previous experiments had shown that wild-type TF, when incorporated into a phospholipid bilayer with the appropriate charge profile, binds to the fVIIa ligand with extremely high affinity ($K_d < 50$ pM), but a recombinant soluble TF fragment (sTF) bound to fVIIa with about 100 times less affinity ($K_d \approx 5$ nM).[140] Upon comparing mTF incorporated into PCPS liposomes with shtTF attached to metal-chelating NiPCPS liposomes, the authors found each bound fVIIa with nearly identical affinities ($K_d \approx 10$ and 10.8 pM, respectively); when the isolated shtTF ectodomain was bound to the liposome surface via an Ni-His interaction, its fVIIa-binding ability was indistinguishable from the membrane spanning mTF construct.

A more relevant test of full function is the ability of the TF-fVIIa complex to activate factor X (fX). When compared with the membrane-associated TF, the soluble ectodomain does not support the same level of fX activation, with or without the addition of PCPS liposomes because the soluble ectodomain is not incorporated or bound by the liposomes.[140] However, when comparing shtTF assembled on NiPCPS liposomes with the membrane-anchored TF, the activation rates of fX were essentially identical, with the catalytic efficiency (k_{cat}/K_m) of the liposome-assembled shtTF being slightly higher.

In the assays described above, autoactivated fVIIa was used. A further test of functionality is to compare the effectiveness of the different TF fragments to enable fVII autoactivation. Earlier work has shown that a membrane-associated, purified, wild-type TF could support fVII autoactivation while a soluble fragment failed to support detectable levels of fVII autoactivation in the presence or absence of liposomes, even after 2 h. By contrast, in this assay, shtTF and mTF both supported fVII autoactivation to a similar extent when tested at identical surface densities, indicating that the TF ectodomain need not be integrated into the membrane, as had been previously suggested. Taken together, these results clearly demonstrate that the soluble TF fragment, when colocalized on a fluid membranous matrix with an appropriate charge profile, can be utilized in biologically relevant assays that use multiple interacting cascade components.

In a final test of relevant TF function, the authors measured the procoagulant activities of sTF, shtTF, and mTF by measuring clotting times in human plasma using a coagulometer. Both the shtTF and mTF showed similar clotting rates after a modification to the nickel-liposomes; this was reasoned to be necessary because of the Ni-NTA interference. Interestingly, the authors also demonstrated that immobilized liposomes could additionally be used for purification of the soluble His-tagged ectodomain from crude mixtures. The captured ectodomain was subsequently assayed *in situ* for its ability to promote clotting, with excellent results.

In a study to understand the effects of local changes in the phospholipid environment, the Morrissey group utilized the NanoDisc approach discussed earlier.[18] The NanoDisc technology was useful in clearly demonstrating a preference for phosphoserine headgroups in the local lipid during fX activation by the NanoDisc-integrated TF-fVIIa complexes. An identical phosphoserine preference had been observed when generating mTF liposome complexes and shtTF complexes with Ni-NTA liposomes.[139,140] In this study the authors directly compared the proteolytic activity of TF-fVII with mTF-liposomes and wt TF in NanoDiscs. Interestingly, the NanoDiscs required a higher local concentration of phosphoserine to generate similar activity; this may be attributed to selectivity differences between membrane-associated fX and fX in solution.[141] On the liposomes, fX can accumulate at higher fX to TF ratios, effectively shuttling the fX to membrane-bound TF-fVIIa complexes. In the case of the NanoDiscs, the number of fX molecules that can bind to TF-NanoDiscs are smaller (ca. 5–6 fX/Nanodisc leaflet) compared to mTF-liposomes. This work also suggests that signaling components requiring larger oligomeric arrays or dynamic exchange of cofactors for full function might be better suited to the TDA approach where a larger number of protein units can be effectively assembled on the lipid scaffold.

VIII. EGFR

The epidermal growth factor family of transmembrane RTKs is an important group of enzymes, largely because of their link to cancers and other diseases. The four members of this family include EGFR/ErbB1/Her1, ErbB2/Her2, ErbB3/Her3, and ErbB4/Her4. EGFR and ErbB4 are both activated in a fairly conventional manner, by ligand-induced dimerization,[142] though the precise mechanism of this activation has eluded researchers for some time. Her2 has no known ligand, yet can be activated by other members of the family, and is the preferred heterodimerization partner for all family members. Her3 is catalytically inactive, but can activate other members as heterodimers or higher order oligomers.[143–145] Any of these receptors can activate themselves or other family members via heterodimerization, when over expressed.

In a recent structural study by Zhang et al., an allosteric mechanism of activation was proposed for the EGFR kinase domain.[146] Though not the focus of their study, they utilized template-assembled EGFR kinase domains to correlate their structural findings with biochemical activity. Using soluble recombinant fragments of the wild-type or mutant cytoplasmic kinase domains, they challenged their model of allosteric activation via the formation of an asymmetric dimer.

The isolated kinase domain of EGFR is autoinhibited and can be activated by mutating the leucine residue at position 834 to an arginine; an activating mutation observed in lung cancer patients. Alternatively, the isolated kinase domain can be activated by increasing its local concentration in a controlled manner or by confining the kinase domain to the 2D surface of liposomes, thus mimicking the membranous matrix in which they normally reside. These membrane-bound kinase domains are free from regulatory control of the extracellular domains. Consequently, this local high concentration ensures transphosphorylation of the activation loop. One might interpret this activation as being the result of an increase in trans/autophosphorylation. In this interpretation, EGFR is its own substrate, and colocalization of the activating enzyme with its enzyme substrate, another EGFR, increases the rate at which the substrate can be phosphorylated. Phosphorylation of Y845 in the activation loop of EGFR, however, is not required for activity, and a nonphosphorylatable

Y845F mutant is activated to a similar degree when colocalized. This result is consistent with the proposal that the assembled constructs are activated by allosteric interactions that result from productive dimer formation rather than a decrease in the kinetic barrier to transphosphorylation. This is analogous to the natural activation that occurs upon release of the ectodomain via a sheddase activity or alternate translation which leaves the cytoplasmic domain anchored to the membrane, and yet unregulated (no longer dependant on ligand-induced dimerization).

In their structural analysis, the authors discuss the relationship of the asymmetric dimer observed in the EGFR structure to the CDK/cyclin complex, where the C-terminal lobe of one EGFR molecule plays a role analogous to cyclin in the activated CDK/cyclin complexes. This C-terminal to N-terminal interaction offers a better explanation for the observation that many activating heterodimer pairs can form between the EGFR family members. ErbB3, despite having an inactive kinase domain, is capable of activating all other EGFR family members, and the sequence of the C-terminal (activating) lobe, in the region described to interact with the N-terminal lobe, is very similar to that of the other family members. In subsequent studies, the authors continue to correlate this template-assembled biochemical analysis to structural analysis and have probed the possibility of identifying inhibitors of the asymmetric dimer interface.

MIG6 (mitogen-induced gene 6; gene symbol ERRFI1) is a 51 kDa cytoplasmic protein which had previously been identified as an inhibitor of EGFR activity.[147,148] Structural analysis of EGFR-MIG6 complexes revealed that a MIG6 fragment of approximately 25 residues binds to the C-terminal (activating) lobe of the EGFR kinase domain[149]. In subsequent cell-based assays, the authors showed that while wild-type MIG6 did indeed inhibit EGFR activity, MIG6 constructs with introduced point mutations were unable to inhibit EGFR activity. These point mutations were introduced in the region corresponding to the segment which had been identified as structurally interacting with the C-terminal lobe of EGFR. In biochemical assays, template-assembled EGFR kinase domains can also be inhibited by the addition of peptides which block the formation of the asymmetric dimer. Interestingly, we think that this suggests that template-assembled enzymes are well-suited for identifying inhibitors of protein–protein interactions.

Using a similar approach, Monsey et al. have recently characterized the remaining ErbB family members, demonstrating that the catalytically inactive Her3 readily activates Her2, and that ErbB4 can be activated by TDA in a manner similar to the EGF receptor.[150] This work paves the way for future analysis of these interactions and demonstrates the plausibility of targeting this family of receptors in a simple, yet meaningful, format that is also readily amenable to the currently used high throughput methods.

IX. Polio Virus Fusion *In Vitro* Using TDA

The creation of a chemically defined system where a biological membrane acts not only as a surface for organization of proteins but also as an active component in a molecular machine has been technically difficult. The picornaviridae, such as the polio virus, are uncoated viruses which, due to lack of a lipid coat must deliver their charged RNA genome directly through the plasma membrane of the target cell.

Only a single protein from the host cell, specifically the ectodomain of the polio virus receptor (Pvr), is required for viral genome entry.[151] Thus, the major components of the machine which delivers the viral genome are virally encoded, and enable the viral genome to cross the membrane in the absence of any host transmembrane proteins. Upon binding to the membrane viral proteins, VP1 and VP4 undergo an irreversible, two-step conformational change which results in genome delivery.[152,153] However, neither the structures derived from VP1 and VP4 conformers in solution in the absence of a membrane nor an electron microscopic analysis of virus entry in cells have been sufficient to elucidate the physical mechanism of this process.

To address these limitations, Dr. Hogel's group at Harvard employed TDA to create a chemically defined molecular machine which reproduces the process of viral genome entry into cells. NTA-liposomes were decorated with recombinant Pvr with a carboxy-terminal poly-histidine tag. Intact polio virus was shown to bind to the NTA-liposomes in a Pvr-dependent manner.[154] Cryo-EM and tomographic studies of the virus-Pvr-liposome complex provided compelling results that supported the proposed model that the fivefold symmetry of the virus particle is responsible for contacting five receptor molecules on the surface of the cell membrane. This study also demonstrated that the association of the virus with the ectodomain of the receptor, even in the absence of the transmembrane domains, produces a deformation of the lipid bilayer, an intermediate step required for genome delivery.[155,156]

This body of work represents a major extension of the utility of TDA; specifically, the creation of a "model membrane system" which facilitates the functional, structural, and mechanistic study of interactions among membrane receptors with their ligands, in a fully defined system.[155,156]

X. Future Directions

The pharmaceutical industry is in the midst of important new challenges. With a premium on effective, safe, and inexpensive therapeutics, research organizations must focus on finding the best compound for each therapeutic indication, early in the discovery process. This will hopefully eliminate the costs associated with extensive medicinal chemistry efforts on compounds that fail to achieve the desired efficacy in clinical settings. With no guarantee that a particular therapeutic target will reach that goal, all that development teams can do is ensure that each screening campaign evaluates the target in as natural an environment as possible, using native assay conditions that mirror the in vivo state as accurately as possible. The strategies discussed in the preceding chapter describe a set of tools that can be employed to study membrane-associated targets using high throughput biochemical tools that have significantly advanced in the last 10 years. We hope that this combination of an optimized enzymatic assay system with the high throughput biochemistry and automation will help improve the effectiveness of the industry.

References

1. Sudhahar CG, Haney RM, Xue Y, Stahelin RV. Cellular membranes and lipid-binding domains as attractive targets for drug development. *Curr Drug Targets* 2008;**9**(8):603–13.
2. Hayashi N, Titani K. N-myristoylated proteins, key components in intracellular signal transduction systems enabling rapid and flexible cell responses. *Proc Jpn Acad B Phys Biol Sci* 2010;**86**(5):494–508.
3. Filippakopoulos P, Müller S, Knapp S. SH2 domains: modulators of nonreceptor tyrosine kinase activity. *Curr Opin Struct Biol* 2009;**19**(6):643–9.
4. Songyang Z, Cantley LC. ZIP codes for delivering SH2 domains. *Cell* 2004;**116**(2 Suppl.): S41–3, 2 p following S48.
5. DiNitto JP, Lambright DG. Membrane and juxtamembrane targeting by PH and PTB domains. *Biochim Biophys Acta* 2006;**1761**(8):850–67.
6. Kaneko T, Li L, Li SS. The SH3 domain—a family of versatile peptide- and protein-recognition module. *Front Biosci* 2008;**13**:4938–52.
7. Del Mare S, Salah Z, Aqeilan RI. WWOX: its genomics, partners, and functions. *J Cell Biochem* 2009;**108**(4):737–45.
8. Kuriyan J, Cowburn D. Modular peptide recognition domains in eukaryotic signaling. *Annu Rev Biophys Biomol Struct* 1997;**26**:259–88.
9. Wang NX, Lee HJ, Zheng JJ. Therapeutic use of PDZ protein-protein interaction antagonism. *Drug News Perspect* 2008;**21**(3):137–41.
10. Gomperts SN. Clustering membrane proteins: It's all coming together with the PSD-95/SAP90 protein family. *Cell* 1996;**84**(5):659–62.
11. Nichols B. Endocytosis of lipid-anchored proteins: excluding GEECs from the crowd. *J Cell Biol* 2009;**186**(4):457–9.
12. Boldog T, Grimme S, Li M, Sligar SG, Hazelbauer GL. Nanodiscs separate chemoreceptor oligomeric states and reveal their signaling properties. *Proc Natl Acad Sci USA* 2006;**103**:11509–14.
13. Denisov IG, Grinkova YV, Lazarides AA, Sligar SG. Directed self-assembly of monodisperse phospholipid bilayer Nanodiscs with controlled size. *J Am Chem Soc* 2004;**126**:3477–87.
14. Civjan NR, Bayburt TH, Schuler MA, Sligar SG. Direct solubilization of heterologously expressed membrane proteins by incorporation into nanoscale lipid bilayers. *Biotechniques* 2003;**35**:556–60, 553–562.

15. Lim WK. GPCR drug discovery: novel ligands for CNS receptors. *Recent Pat CNS Drug Discov* 2007;**2**:107–12.
16. Bayburt TH, Leitz AJ, Xie G, Oprian DD, Sligar SG. Transducin activation by nanoscale lipid bilayers containing one and two rhodopsins. *J Biol Chem* 2007;**282**:14875–81.
17. Ritchie TK, Grinkova YV, Bayburt TH, Denisov IG, Zolnerciks JK, Atkins WM, et al. Chapter 11—Reconstitution of membrane proteins in phospholipid bilayer nanodiscs. *Methods Enzymol* 2009;**464**:211–31.
18. Shaw AW, Pureza VS, Sligar SG, Morrissey JH. The local phospholipid environment modulates the activation of blood clotting. *J Biol Chem* 2007;**282**:6556–63.
19. Celia H, Wilson-Kubalek E, Milligan RA, Teyton L. Structure and function of a membrane-bound murine MHC class I molecule. *Proc Natl Acad Sci USA* 1999;**96**:5634–9.
20. Kubalek EW, Le Grice SF, Brown PO. Two-dimensional crystallization of histidine-tagged, HIV-1 reverse transcriptase promoted by a novel nickel-chelating lipid. *J Struct Biol* 1994;**113**:117–23.
21. Thess A, Hutschenreiter S, Hofmann M, Tampe R, Baumeister W, Guckenberger R. Specific orientation and two-dimensional crystallization of the proteasome at metal-chelating lipid interfaces. *J Biol Chem* 2002;**277**:36321–8.
22. Terpe K. Overview of tag protein fusions: from molecular and biochemical fundamentals to commercial systems. *Appl Microbiol Biotechnol* 2003;**60**:523–33.
23. Pawson T, Warner N. Oncogenic re-wiring of cellular signaling pathways. *Oncogene* 2007;**26**:1268–75.
24. Sharma PS, Sharma R, Tyagi T. Receptor tyrosine kinase inhibitors as potent weapons in war against cancers. *Curr Pharm Des* 2009;**15**:758–76.
25. Hunter T. Tyrosine phosphorylation in cell signaling and disease. *Keio J Med* 2002;**51**:61–71.
26. Scott JD, Pawson T. Cell signaling in space and time: where proteins come together and when they're apart. *Science* 2009;**326**:1220–4.
27. Jin J, Xie X, Chen C, Park JG, Stark C, James DA, et al. Eukaryotic protein domains as functional units of cellular evolution. *Sci Signal* 2009;**2**:ra76.
28. Carraway 3rd KL, Koland JG, Cerione RA. Visualization of epidermal growth factor (EGF) receptor aggregation in plasma membranes by fluorescence resonance energy transfer. Correlation of receptor activation with aggregation. *J Biol Chem* 1989;**264**:8699–707.
29. Gadella Jr. TW, Jovin TM. Oligomerization of epidermal growth factor receptors on A431 cells studied by time-resolved fluorescence imaging microscopy. A stereochemical model for tyrosine kinase receptor activation. *J Cell Biol* 1995;**129**:1543–58.
30. Ichinose J, Murata M, Yanagida T, Sako Y. EGF signalling amplification induced by dynamic clustering of EGFR. *Biochem Biophys Res Commun* 2004;**324**:1143–9.
31. Pellicena P, Kuriyan J. Protein-protein interactions in the allosteric regulation of protein kinases. *Curr Opin Struct Biol* 2006;**16**:702–9.
32. Grasberger B, Minton AP, DeLisi C, Metzger H. Interaction between proteins localized in membranes. *Proc Natl Acad Sci USA* 1986;**83**:6258–62.
33. Ji QS, Mulvihill MJ, Rosenfeld-Franklin M, Cooke A, Feng L, Mak G, et al. A novel, potent, and selective insulin-like growth factor-I receptor kinase inhibitor blocks insulin-like growth factor-I receptor signaling in vitro and inhibits insulin-like growth factor-I receptor dependent tumor growth in vivo. *Mol Cancer Ther* 2007;**6**(8):2158–67.
34. Shrout AL, Montefusco DJ, Weis RM. Template-directed assembly of receptor signaling complexes. *Biochemistry* 2003;**42**:13379–85.
35. Montefusco DJ, Asinas AE, Weis RM. Liposome-mediated assembly of receptor signaling complexes. *Methods Enzymol* 2007;**423**:265–98.
36. Besschetnova TY, Montefusco DJ, Asinas AE, Shrout AL, Antommattei FM, Weis RM. Receptor density balances signal stimulation and attenuation in membrane-assembled complexes of bacterial chemotaxis signaling proteins. *Proc Natl Acad Sci USA* 2008;**105**:12289–94.

37. Montefusco DJ, Shrout AL, Besschetnova TY, Weis RM. Formation and activity of template-assembled receptor signaling complexes. *Langmuir* 2007;**23**:3280–9.
38. Montefusco DJ, Asinas AE, Weis RM. Liposome-mediated assembly of receptor signaling complexes. *Methods Enzymol* 2007;**423**:267–98.
39. Asinas AE, Weis RM. Competitive and cooperative interactions in receptor signaling complexes. *J Biol Chem* 2006;**281**:30512–23.
40. Schlessinger J. Cell signaling by receptor tyrosine kinases. *Cell* 2000;**103**:211–25.
41. Pawson T, Scott JD. Signaling through scaffold, anchoring, and adaptor proteins. *Science* 1997;**278**:2075–80.
42. Hoffman TL, Canziani G, Jia L, Rucker J, Doms RW. A biosensor assay for studying ligand-membrane receptor interactions: binding of antibodies and HIV-1 Env to chemokine receptors. *Proc Natl Acad Sci USA* 2000;**97**:11215–20.
43. Willis S, Davidoff C, Schilling J, Wanless A, Doranz BJ, Rucker J. Virus-like particles as quantitative probes of membrane protein interactions. *Biochemistry* 2008;**47**:6988–90.
44. Endres MJ, Jaffer S, Haggarty B, Turner JD, Doranz BJ, O'Brien PJ, et al. Targeting of HIV- and SIV-infected cells by CD4-chemokine receptor pseudotypes. *Science* 1997;**278**:1462–4.
45. Balliet JW, Bates P. Efficient infection mediated by viral receptors incorporated into retroviral particles. *J Virol* 1998;**72**:671–6.
46. Young JA, Bates P, Willert K, Varmus HE. Efficient incorporation of human CD4 protein into avian leukosis virus particles. *Science* 1990;**250**:1421–3.
47. Wu L, Gerard NP, Wyatt R, Choe H, Parolin C, Ruffing N, et al. CD4-induced interaction of primary HIV-1 gp120 glycoproteins with the chemokine receptor CCR-5. *Nature* 1996;**384**:179–83.
48. Berger EA, Murphy PM, Farber JM. Chemokine receptors as HIV-1 coreceptors: roles in viral entry, tropism, and disease. *Annu Rev Immunol* 1999;**17**:657–700.
49. Etemad-Moghadam B, Sun Y, Nicholson EK, Fernandes M, Liou K, Gomila R, et al. Envelope glycoprotein determinants of increased fusogenicity in a pathogenic simian-human immunodeficiency virus (SHIV-KB9) passaged in vivo. *J Virol* 2000;**74**:4433–40.
50. Doranz BJ, Baik SS, Doms RW. Use of a gp120 binding assay to dissect the requirements and kinetics of human immunodeficiency virus fusion events. *J Virol* 1999;**73**:10346–58.
51. He Z, Tessier-Lavigne M. Neuropilin is a receptor for the axonal chemorepellent Semaphorin III. *Cell* 1997;**90**:739–51.
52. Lee B, Sharron M, Blanpain C, Doranz BJ, Vakili J, Setoh P, et al. Epitope mapping of CCR5 reveals multiple conformational states and distinct but overlapping structures involved in chemokine and coreceptor function. *J Biol Chem* 1999;**274**:9617–26.
53. Baker MD, Wolanin PM, Stock JB. Signal transduction in bacterial chemotaxis. *Bioessays* 2006;**28**:9–22.
54. Stock AM, Robinson VL, Goudreau PN. Two-component signal transduction. *Annu Rev Biochem* 2000;**69**:183–215.
55. Engelmann T. *Pfluegers Arch Gesam* 1881;**25**:285.
56. Adler J. Chemotaxis in *Escherichia coli*. *Cold Spring Harb Symp Quant Biol* 1965;**30**:289–92.
57. Pfeffer W. Untersuch. *Botn Inst Tübingen* 1888;**2**:582–9.
58. Adler J. Chemotaxis in bacteria. *Science* 1966;**153**:708–16.
59. Falke JJ, Bass RB, Butler SL, Chervitz SA, Danielson MA. The two-component signaling pathway of bacterial chemotaxis: a molecular view of signal transduction by receptors, kinases, and adaptation enzymes. *Annu Rev Cell Dev Biol* 1997;**13**:457–512.
60. Falke JJ, Hazelbauer GL. Transmembrane signaling in bacterial chemoreceptors. *Trends Biochem Sci* 2001;**26**:257–65.
61. Stock JB, Levit MN, Wolanin PM. Information processing in bacterial chemotaxis. *Sci STKE* 2002;PE25.

62. Wolanin PM, Stock JB. Bacterial chemosensing: cooperative molecular logic. *Curr Biol* 2004;**14**:R486–7.
63. Parkinson JS, Ames P, Studdert CA. Collaborative signaling by bacterial chemoreceptors. *Curr Opin Microbiol* 2005;**8**:116–21.
64. Sourjik V, Berg HC. Functional interactions between receptors in bacterial chemotaxis. *Nature* 2004;**428**:437–41.
65. Webre DJ, Wolanin PM, Stock JB. Modulated receptor interactions in bacterial transmembrane signaling. *Trends Cell Biol* 2004;**14**:478–82.
66. Bren A, Eisenbach M. How signals are heard during bacterial chemotaxis: protein-protein interactions in sensory signal propagation. *J Bacteriol* 2000;**182**:6865–73.
67. Bilwes AM, Alex LA, Crane BR, Simon MI. Structure of CheA, a signal-transducing histidine kinase. *Cell* 1999;**96**:131–41.
68. Pawson T. SH2 and SH3 domains in signal transduction. *Adv Cancer Res* 1994;**64**:87–110.
69. Segall JE, Block SM, Berg HC. Temporal comparisons in bacterial chemotaxis. *Proc Natl Acad Sci USA* 1986;**83**:8987–91.
70. Berg HC, Tedesco PM. Transient response to chemotactic stimuli in *Escherichia coli*. *Proc Natl Acad Sci USA* 1975;**72**:3235–9.
71. Mesibov R, Ordal GW, Adler J. The range of attractant concentrations for bacterial chemotaxis and the threshold and size of response over this range. Weber law and related phenomena. *J Gen Physiol* 1973;**62**:203–23.
72. Mao H, Cremer PS, Manson MD. A sensitive, versatile microfluidic assay for bacterial chemotaxis. *Proc Natl Acad Sci USA* 2003;**100**:5449–54.
73. Stock J, Levit M. Signal transduction: hair brains in bacterial chemotaxis. *Curr Biol* 2000;**10**: R11–4.
74. Le Moual H, Koshland Jr. DE. Molecular evolution of the C-terminal cytoplasmic domain of a superfamily of bacterial receptors involved in taxis. *J Mol Biol* 1996;**261**:568–85.
75. Aksamit R, Koshland Jr. DE. A ribose binding protein of *Salmonella typhimurium*. *Biochem Biophys Res Commun* 1972;**48**:1348–53.
76. Yeh JI, Biemann HP, Pandit J, Koshland DE, Kim SH. The three-dimensional structure of the ligand-binding domain of a wild-type bacterial chemotaxis receptor. Structural comparison to the cross-linked mutant forms and conformational changes upon ligand binding. *J Biol Chem* 1993;**268**:9787–92.
77. Yeh JI, Biemann HP, Prive GG, Pandit J, Koshland Jr. DE, Kim SH. High-resolution structures of the ligand binding domain of the wild-type bacterial aspartate receptor. *J Mol Biol* 1996;**262**:186–201.
78. Milligan DL, Koshland Jr. DE. Site-directed cross-linking. Establishing the dimeric structure of the aspartate receptor of bacterial chemotaxis. *J Biol Chem* 1988;**263**:6268–75.
79. Lai RZ, Manson JM, Bormans AF, Draheim RR, Nguyen NT, Manson MD. Cooperative signaling among bacterial chemoreceptors. *Biochemistry* 2005;**44**:14298–307.
80. Sourjik V. Receptor clustering and signal processing in *E. coli* chemotaxis. *Trends Microbiol* 2004;**12**:569–76.
81. Li G, Weis RM. Covalent modification regulates ligand binding to receptor complexes in the chemosensory system of *Escherichia coli*. *Cell* 2000;**100**:357–65.
82. Stock AM, Martinez-Hackert E, Rasmussen BF, West AH, Stock JB, Ringe D, et al. Structure of the Mg(2+)-bound form of CheY and mechanism of phosphoryl transfer in bacterial chemotaxis. *Biochemistry* 1993;**32**:13375–80.
83. Smith RA, Parkinson JS. Overlapping genes at the cheA locus of *Escherichia coli*. *Proc Natl Acad Sci USA* 1980;**77**:5370–4.
84. Maddock JR, Shapiro L. Polar location of the chemoreceptor complex in the *Escherichia coli* cell. *Science* 1993;**259**:1717–23.

85. Gegner JA, Graham DR, Roth AF, Dahlquist FW. Assembly of an MCP receptor, CheW, and kinase CheA complex in the bacterial chemotaxis signal transduction pathway. *Cell* 1992;**70**:975–82.
86. Parkinson JS, Parker SR, Talbert PB, Houts SE. Interactions between chemotaxis genes and flagellar genes in *Escherichia coli*. *J Bacteriol* 1983;**155**:265–74.
87. Kuo SC, Koshland Jr. DE. Roles of cheY and cheZ gene products in controlling flagellar rotation in bacterial chemotaxis of *Escherichia coli*. *J Bacteriol* 1987;**169**:1307–14.
88. Khan S, Macnab RM, DeFranco AL, Koshland Jr. DE. Inversion of a behavioral response in bacterial chemotaxis: explanation at the molecular level. *Proc Natl Acad Sci USA* 1978;**75**:4150–4.
89. Martin AC, Wadhams GH, Shah DS, Porter SL, Mantotta JC, Craig TJ, et al. CheR- and CheB-dependent chemosensory adaptation system of Rhodobacter sphaeroides. *J Bacteriol* 2001;**183**:7135–44.
90. Springer MS, Goy MF, Adler J. Protein methylation in behavioural control mechanisms and in signal transduction. *Nature* 1979;**280**:279–84.
91. Stock JB, Koshland Jr. DE. A protein methylesterase involved in bacterial sensing. *Proc Natl Acad Sci USA* 1978;**75**:3659–63.
92. Springer WR, Koshland Jr. DE. Identification of a protein methyltransferase as the cheR gene product in the bacterial sensing system. *Proc Natl Acad Sci USA* 1977;**74**:533–7.
93. Yonekawa H, Hayashi H, Parkinson JS. Requirement of the cheB function for sensory adaptation in *Escherichia coli*. *J Bacteriol* 1983;**156**:1228–35.
94. Chalah A, Weis RM. Site-specific and synergistic stimulation of methylation on the bacterial chemotaxis receptor Tsr by serine and CheW. *BMC Microbiol* 2005;**5**:12.
95. Antommattei FM, Munzner JB, Weis RM. Ligand-specific activation of *Escherichia coli* chemoreceptor transmethylation. *J Bacteriol* 2004;**186**:7556–63.
96. Jasuja R, Keyoung J, Reid GP, Trentham DR, Khan S. Chemotactic responses of *Escherichia coli* to small jumps of photoreleased L-aspartate. *Biophys J* 1999;**76**:1706–19.
97. Chi YI, Yokota H, Kim SH. Apo structure of the ligand-binding domain of aspartate receptor from *Escherichia coli* and its comparison with ligand-bound or pseudoligand-bound structures. *FEBS Lett* 1997;**414**:327–32.
98. Ottemann KM, Xiao W, Shin YK, Koshland Jr. DE. A piston model for transmembrane signaling of the aspartate receptor. *Science* 1999;**285**:1751–4.
99. Chervitz SA, Falke JJ. Lock on/off disulfides identify the transmembrane signaling helix of the aspartate receptor. *J Biol Chem* 1995;**270**:24043–53.
100. Chervitz SA, Lin CM, Falke JJ. Transmembrane signaling by the aspartate receptor: engineered disulfides reveal static regions of the subunit interface. *Biochemistry* 1995;**34**:9722–33.
101. Chervitz SA, Falke JJ. Molecular mechanism of transmembrane signaling by the aspartate receptor: a model. *Proc Natl Acad Sci USA* 1996;**93**:2545–50.
102. Long DG, Weis RM. Oligomerization of the cytoplasmic fragment from the aspartate receptor of *Escherichia coli*. *Biochemistry* 1992;**31**:9904–11.
103. Li J, Li G, Weis RM. The serine chemoreceptor from *Escherichia coli* is methylated through an inter-dimer process. *Biochemistry* 1997;**36**:11851–7.
104. Francis NR, Levit MN, Shaikh TR, Melanson LA, Stock JB, DeRosier DJ. Subunit organization in a soluble complex of tar, CheW, and CheA by electron microscopy. *J Biol Chem* 2002;**277**:36755–9.
105. Studdert CA, Parkinson JS. Crosslinking snapshots of bacterial chemoreceptor squads. *Proc Natl Acad Sci USA* 2004;**101**:2117–22.
106. Briegel A, Ortega DR, Tocheva EI, Wuichet K, Li Z, Chen S, et al. Universal architecture of bacterial chemoreceptor arrays. *Proc Natl Acad Sci USA* 2009;**106**:17181–6.
107. Wang H, Wingreen NS, Mukhopadhyay R. Self-organized periodicity of protein clusters in growing bacteria. *Phys Rev Lett* 2008;**101**:218101.

108. Kentner D, Sourjik V. Spatial organization of the bacterial chemotaxis system. *Curr Opin Microbiol* 2006;**9**:619–24.
109. Bray D, Levin MD, Morton-Firth CJ. Receptor clustering as a cellular mechanism to control sensitivity. *Nature* 1998;**393**:85–8.
110. Bray D, Duke T. Conformational spread: the propagation of allosteric states in large multiprotein complexes. *Annu Rev Biophys Biomol Struct* 2004;**33**:53–73.
111. Ames P, Studdert CA, Reiser RH, Parkinson JS. Collaborative signaling by mixed chemoreceptor teams in *Escherichia coli*. *Proc Natl Acad Sci USA* 2002;**99**:7060–5.
112. Gestwicki JE, Kiessling LL. Inter-receptor communication through arrays of bacterial chemoreceptors. *Nature* 2002;**415**:81–4.
113. Gestwicki JE, Strong LE, Borchardt SL, Cairo CW, Schnoes AM, Kiessling LL. Designed potent multivalent chemoattractants for *Escherichia coli*. *Bioorg Med Chem* 2001;**9**:2387–93.
114. Levit MN, Liu Y, Stock JB. Mechanism of CheA protein kinase activation in receptor signaling complexes. *Biochemistry* 1999;**38**:6651–8.
115. Surette MG, Stock JB. Role of alpha-helical coiled-coil interactions in receptor dimerization, signaling, and adaptation during bacterial chemotaxis. *J Biol Chem* 1996;**271**:17966–73.
116. Cochran AG, Kim PS. Imitation of *Escherichia coli* aspartate receptor signaling in engineered dimers of the cytoplasmic domain. *Science* 1996;**271**:1113–6.
117. Goldman J, Andrews S, Bray D. Size and composition of membrane protein clusters predicted by Monte Carlo analysis. *Eur Biophys J* 2004;**33**:506–12.
118. Shimizu TS, Aksenov SV, Bray D. A spatially extended stochastic model of the bacterial chemotaxis signalling pathway. *J Mol Biol* 2003;**329**:291–309.
119. Bray D. Bacterial chemotaxis and the question of gain. *Proc Natl Acad Sci USA* 2002;**99**:7–9.
120. Morton-Firth CJ, Shimizu TS, Bray D. A free-energy-based stochastic simulation of the Tar receptor complex. *J Mol Biol* 1999;**286**:1059–74.
121. Duke TA, Bray D. Heightened sensitivity of a lattice of membrane receptors. *Proc Natl Acad Sci USA* 1999;**96**:10104–8.
122. Abouhamad WN, Bray D, Schuster M, Boesch KC, Silversmith RE, Bourret RB. Computer-aided resolution of an experimental paradox in bacterial chemotaxis. *J Bacteriol* 1998;**180**:3757–64.
123. Mello BA, Tu Y. Quantitative modeling of sensitivity in bacterial chemotaxis: the role of coupling among different chemoreceptor species. *Proc Natl Acad Sci USA* 2003;**100**:8223–8.
124. Rao CV, Frenklach M, Arkin AP. An allosteric model for transmembrane signaling in bacterial chemotaxis. *J Mol Biol* 2004;**343**:291–303.
125. Levit MN, Grebe TW, Stock JB. Organization of the receptor-kinase signaling array that regulates *Escherichia coli* chemotaxis. *J Biol Chem* 2002;**277**:36748–54.
126. Borkovich KA, Alex LA, Simon MI. Attenuation of sensory receptor signaling by covalent modification. *Proc Natl Acad Sci USA* 1992;**89**:6756–60.
127. Borkovich KA, Kaplan N, Hess JF, Simon MI. Transmembrane signal transduction in bacterial chemotaxis involves ligand-dependent activation of phosphate group transfer. *Proc Natl Acad Sci USA* 1989;**86**:1208–12.
128. Bornhorst JA, Falke JJ. Attractant regulation of the aspartate receptor-kinase complex: limited cooperative interactions between receptors and effects of the receptor modification state. *Biochemistry* 2000;**39**:9486–93.
129. Barnakov AN, Barnakova LA, Hazelbauer GL. Comparison in vitro of a high- and a low-abundance chemoreceptor of *Escherichia coli*: similar kinase activation but different methyl-accepting activities. *J Bacteriol* 1998;**180**:6713–8.
130. Ninfa EG, Stock A, Mowbray S, Stock J. Reconstitution of the bacterial chemotaxis signal transduction system from purified components. *J Biol Chem* 1991;**266**:9764–70.
131. Bogonez E, Koshland Jr. DE. Solubilization of a vectorial transmembrane receptor in functional form: aspartate receptor of chemotaxis. *Proc Natl Acad Sci USA* 1985;**82**:4891–5.

132. Levy D, Mosser G, Lambert O, Moeck GS, Bald D, Rigaud JL. Two-dimensional crystallization on lipid layer: A successful approach for membrane proteins. *J Struct Biol* 1999;**127**:44–52.
133. Barklis E, McDermott J, Wilkens S, Fuller S, Thompson D. Organization of HIV-1 capsid proteins on a lipid monolayer. *J Biol Chem* 1998;**273**:7177–80.
134. Nieba L, Nieba-Axmann SE, Persson A, Hamalainen M, Edebratt F, Hansson A, et al. BIACORE analysis of histidine-tagged proteins using a chelating NTA sensor chip. *Anal Biochem* 1997;**252**:217–28.
135. Crowe J, Dobeli H, Gentz R, Hochuli E, Stuber D, Henco K. 6xHis-Ni-NTA chromatography as a superior technique in recombinant protein expression/purification. *Methods Mol Biol* 1994;**31**:371–87.
136. Hochuli E, Dobeli H, Schacher A. New metal chelate adsorbent selective for proteins and peptides containing neighbouring histidine residues. *J Chromatogr* 1987;**411**:177–84.
137. Yu P, Pettigrew DW. Linkage between fructose 1, 6-bisphosphate binding and the dimer-tetramer equilibrium of *Escherichia coli* glycerol kinase: critical behavior arising from change of ligand stoichiometry. *Biochemistry* 2003;**42**(14):4243–52.
138. Waters EK, Morrissey JH. Restoring full biological activity to the isolated ectodomain of an integral membrane protein. *Biochemistry* 2006;**45**:3769–74.
139. Morrissey JH, Neuenschwander PF, Huang Q, McCallum CD, Su B, Johnson AE. Factor VIIa-tissue factor: functional importance of protein-membrane interactions. *Thromb Haemost* 1997;**78**:112–6.
140. Neuenschwander PF, Morrissey JH. Roles of the membrane-interactive regions of factor VIIa and tissue factor. The factor VIIa Gla domain is dispensable for binding to tissue factor but important for activation of factor X. *J Biol Chem* 1994;**269**:8007–13.
141. Krishnaswamy S, Field KA, Edgington TS, Morrissey JH, Mann KG. Role of the membrane surface in the activation of human coagulation factor X. *J Biol Chem* 1992;**267**(36):26110–20.
142. Yarden Y, Schlessinger J. Self-phosphorylation of epidermal growth factor receptor: evidence for a model of intermolecular allosteric activation. *Biochemistry* 1987;**26**:1434–42.
143. Tzahar E, Waterman H, Chen X, Levkowitz G, Karunagaran D, Lavi S, et al. A hierarchical network of interreceptor interactions determines signal transduction by Neu differentiation factor/neuregulin and epidermal growth factor. *Mol Cell Biol* 1996;**16**(10):5276–87.
144. Graus-Porta D, Beerli RR, Daly JM, Hynes NE. ErbB-2, the preferred heterodimerization partner of all ErbB receptors, is a mediator of lateral signaling. *EMBO J* 1997;**16**(7):1647–55.
145. Olayioye MA, Graus-Porta D, Beerli RR, Rohrer J, Gay B, Hynes NE. ErbB-1 and ErbB-2 acquire distinct signaling properties dependent upon their dimerization partner. *Mol Cell Biol* 1998;**18**(9):5042–51.
146. Zhang X, Gureasko J, Shen K, Cole PA, Kuriyan J. An allosteric mechanism for activation of the kinase domain of epidermal growth factor receptor. *Cell* 2006;**125**:1137–49.
147. Xu D, Makkinje A, Kyriakis JM. Gene 33 is an endogenous inhibitor of epidermal growth factor (EGF) receptor signaling and mediates dexamethasone-induced suppression of EGF function. *J Biol Chem* 2005;**280**:2924–33.
148. Hackel PO, Gishizky M, Ullrich A. Mig-6 is a negative regulator of the epidermal growth factor receptor signal. *Biol Chem* 2001;**382**:1649–62.
149. Zhang X, Pickin KA, Bose R, Jura N, Cole PA, Kuriyan J. Inhibition of the EGF receptor by binding of MIG6 to an activating kinase domain interface. *Nature* 2007;**450**:741–4.
150. Monsey J, Shen W, Schlesinger P, Bose R. Her4 and Her2/neu tyrosine kinase domains dimerize and activate in a reconstituted in vitro system. *J Biol Chem* 2010;**285**(10):7035–44.
151. Tosteson MT, Wang H, Naumov A, Chow M. Poliovirus binding to its receptor in lipid bilayers results in particle-specific, temperature-sensitive channels. *J Gen Virol* 2004;**85**:1581–9.
152. De Sena J, Mandel B. Studies on the in vitro uncoating of poliovirus. II. Characteristics of the membrane-modified particle. *Virology* 1977;**78**:554–66.

153. De Sena J, Mandel B. Studies on the in vitro uncoating of poliovirus. I. Characterization of the modifying factor and the modifying reaction. *Virology* 1976;**70**:470–83.
154. Tuthill TJ, Bubeck D, Rowlands DJ, Hogle JM. Characterization of early steps in the poliovirus infection process: receptor-decorated liposomes induce conversion of the virus to membrane-anchored entry-intermediate particles. *J Virol* 2006;**80**:172–80.
155. Bubeck D, Filman DJ, Hogle JM. Cryo-electron microscopy reconstruction of a poliovirus-receptor-membrane complex. *Nat Struct Mol Biol* 2005;**12**:615–8.
156. Bostina M, Bubeck D, Schwartz C, Nicastro D, Filman DJ, Hogle JM. Single particle cryoelectron tomography characterization of the structure and structural variability of poliovirus-receptor-membrane complex at 30 A resolution. *J Struct Biol* 2007;**160**:200–10.

Index

D

E

F

G

H

I

K

L

M

V

X

Y

Z